CUMBRIA: AN HISTORICAL GAZETTEER

Edited by
Angus J. L. Winchester

Lancaster University

Regional Heritage Centre

in association with Cumbria County History Trust

2016

Cumbria: An Historical Gazetteer

Published by the Regional Heritage Centre, Lancaster University,
in association with the Cumbria County History Trust

Designed, typeset, printed and bound by Badger Press Ltd, Bowness-on-Windermere, Cumbria

British Library Cataloguing-in-Publication Data

A CIP catalogue record for this book is available from the British Library

ISBN 978-1-86220-330-3

Cover: extract from the 'Quartermaster's Map' of northern England, 1644, one of a series of maps engraved by Wenceslaus Hollar intended to be 'Portable for euery Mans Pocket, Vsefull for all Comanders for Quareteinge of Souldiers'. Courtesy of Dr W. D. Shannon.

The publishers gratefully acknowledge the financial support of Dr Philip Welch.

An electronic version of this book is available as a searchable PDF file and may be purchased from Regional Heritage Centre, Department of History, Lancaster University LA1 4YT, see http://www.lancaster.ac.uk/users/rhc/books/index.htm

Lancaster University Regional Heritage Centre

in association with Cumbria County History Trust

CONTENTS

ACKNOWLEDGEMENTS

The initial historical digests on which the Gazetteer is based were compiled in 2012-13 by the following volunteers contributing to the Victoria County History of Cumbria project: Adrian Allan, Elizabeth Amos, Eric Apperley, Roger Asquith, Margaret Ball, Robert Baxter, Susan Beachell, John Berry, Mark Blackett-Ord, David Bradbury, Richard Brockington, Ian Campbell, Joe Cann, Ian Caruana, †Paul Cheesbrough, Maggie Clowes, Marilyn Cole, Andrew Connell, Kathleen Cook, Barbara Copeland, Frank Cosgrove, Rachel Cosgrove, Tony Cousins, Stanley Darke, Rob David, Edward and Stella Davis, Anne Dick, Christopher Donaldson, Helen Dunham, John Dunning, Maureen Fisher, Sheila Fletcher, Jennifer Forsyth, Chester Forster, Emmeline Garnett, Patricia Garside, Margaret and John Gowling, Lydia Gray, Edwin Haddon, Valerie Halbert, June Hall, Stuart Heatherington, Jan Hicks, Wendy Higgins, Tiffany Hunt, Michael Jackson, Bernard Kirkbride, Fiona and Charles Lambrick, Felicity Lawler, Marion McClintock, Betty Marshall, David Metcalfe, Nigel Mills, Stuart Nicholson, Alan Ormrod, Linda Peel, Jane Penman, Verena Power, Leslie Price, Helen Pugh, Stephen Read, Ted Relph, Colin Richards, Peter Riddle, Joseph Ritson, Alastair Robertson, Pat Rowland, Charles Rowntree, Ann Sandell, Merrilyn Sawrey-Cookson, Bill Shannon, Sandra Shaw, Christine Shearin, Alan Smith, John Thorley, Michael Turnbull, James Underwood, Henry Vane, Mary Wane, Jean Williams, Val Winchester, Fay Winkworth, Janice Wrench, and members of the following societies: Keswick History Group, Lorton and Derwentfells Local History Society and Sedbergh and District History Society. It is a pleasure to record my thanks to all for undertaking the groundwork on which this book is based. The entries have been substantially re-written and I take full responsibility for them as they now appear. I owe a special debt of gratitude to James Perry, who prepared the reference maps with great care, and to Bill Shannon who provided the cover image.

Publication of the Gazetteer has been made possible by a grant from the Department of History at Lancaster University towards the preparation of the maps and a generous donation towards production costs from Dr Philip Welch, a longstanding supporter of the Regional Heritage Centre.

I should also like to record my thanks to the officers of the Cumbria County History Trust, core funders of the Victoria County History of Cumbria project, who have enthusiastically supported the publication of this Gazetteer and offered valuable advice during its long period of gestation. Finally, it is a pleasure to be able to record my thanks to my colleague Dr Sarah Rose, who not only oversaw the preparation of the original digests but drafted several of them herself and has been an unfailingly supportive and committed colleague on the project.

AJLW

ABBREVIATIONS AND GLOSSARY

ancient parish	territory paying tithes and other dues to maintain a priest who had cure of souls across the parish. The boundaries of ancient parishes had generally become fixed by 1200. Some ancient parishes coincided with a single township but in Cumbria many parishes embraced several townships. Kendal ancient parish covered 25 townships, for example.
barony	in medieval Cumbria one of the great units of overlordship in Cumberland and Westmorland, covering an extensive block of land. Estates which were manors in their own right were often held under such an overarching lordship.
c.	*circa*, i.e. 'around'.
caput	literally 'head'; used of administrative headquarters of an estate.
CE	Church of England.
chapel of ease	a church, subsidiary to a parish church, built to serve the needs of an outlying part of an ancient parish.
chapelry	a section of an ancient parish over which a chapel of ease had acquired quasi-parochial rights, particularly the right to bury in its own churchyard.
constablewick	the territory of a constable, often (but not necessarily) the same as a township.
CP	civil parish, the basic unit of civil administration in England since 1866. Some covered the same area as an ancient parish but in northern England many CPs were coterminous with townships.
CWAAS	Cumberland & Westmorland Antiquarian & Archaeological Society.

detached portion	part of one township or ancient parish surrounded by territory belonging to another. Under the Divided Parishes Act 1883, detached portions became part of the administrative unit by which they were surrounded.
extra-parochial	a place which did not fall within the bounds of an ancient parish. From 1858 extra-parochial places became CPs for secular administrative purposes.
ha	hectares.
hamlet	the core meaning is a place subsidiary to larger administrative unit. A place might be a hamlet within a larger manor, or a township might be divided into hamlets for local administrative purposes, such as highway maintenance.
honour	a group of manors held under a lord paramount.
manor	a landed estate over which the owner ('lord of the manor') held certain privileges, notably the right to hold a manor court, through which he exercised jurisdiction over his tenants holding land in the manor.
MB	Municipal Borough; an incorporated borough under the terms of the Municipal Corporations Act 1835.
moiety	a half share.
OS	Ordnance Survey.
quarter	a section of a parish or township (not necessarily one of four). Quarters of parishes often appointed their own churchwardens; quarters of townships were often sub-divisions for highway maintenance.
q.v.	*quod vide* ('which see'), i.e. see the entry for the place named.
RC	Roman Catholic.
township	the basic unit of civil administration for the purposes of taxation, justice etc. from the medieval period.
UD	Urban District, established under the Local Government Act 1894.
ward	administrative division of the historic counties of Cumberland and Westmorland, comparable to the hundred or wapentake in other English counties.

PREFACE

This book is a spin-off from the early phases of the Victoria County History of Cumbria (hereafter 'VCH Cumbria') project, based in the Department of History at Lancaster University and supported and overseen by the Cumbria County History Trust (CCHT) since 2010. The VCH is a national work of reference, definitive and highly respected, which aims eventually to produce historical 'biographies' of every community in England. It is run from the Institute of Historical Research at the University of London. The VCH Cumbria project was set up to revive work in Cumbria, with a view to preparing the detailed local histories, written to a standard template, parish by parish, which characterise the VCH.

One of the first steps taken by the VCH Cumbria project was to build a website (**www.cumbriacountyhistory.org.uk/**) to house a databank of local historical material by creating a web page for every community in the county. To give each web page a 'front door', a brief historical digest for the place in question was prepared, providing basic historical data drawn from a checklist of key sources. This was a large undertaking, since Cumbria is divided for VCH purposes into 348 places, normally defined as the Civil Parishes (hereafter 'CPs') as they existed in the 1890s. The work was carried out in 2012-13, largely by volunteer local historians from across the county: a total of 87 individuals, as well as members of three local history groups, contributed to the exercise. The digests they prepared formed the starting point for this Gazetteer but further research has been undertaken and they have been substantially re-written for publication, in an attempt to ensure accuracy and consistency.

The entries below contain brief factual statements arranged under a standard set of headings to provide information on the administrative status of the place in question; its acreage and a note of any boundary changes; the extent of common land, with the date of enclosure; population numbers; landownership; economic activity; places of worship; schools; and other community institutions, such as libraries, almshouses and village halls. In order to present a standardised body of information, the entries are based on a standard set of sources (listed at the end of the book in the section entitled 'How the Gazetteer was compiled').

It must be stressed that the entries below ought not to be considered definitive or comprehensive – they rely on a comparatively small range of standard sources and most places await the thorough research in primary sources which will be undertaken for the VCH Cumbria project. As a result, the following pages will almost certainly contain errors and omissions. VCH Cumbria is a long-term venture, stretching over many decades. A start has been made: several detailed parish articles have been completed in draft and are posted on the project website. In these cases, the entry in the Gazetteer is marked by an asterisk and draws on the detailed research carried out for the full article. Although this Gazetteer is *not* VCH Cumbria, we hope that it will act as a useful work of ready reference for all those interested in the history of Cumbria and prove to be a stepping stone to the full work.

Angus J L Winchester
Department of History, Lancaster University
August 2016

REFERENCE MAPS

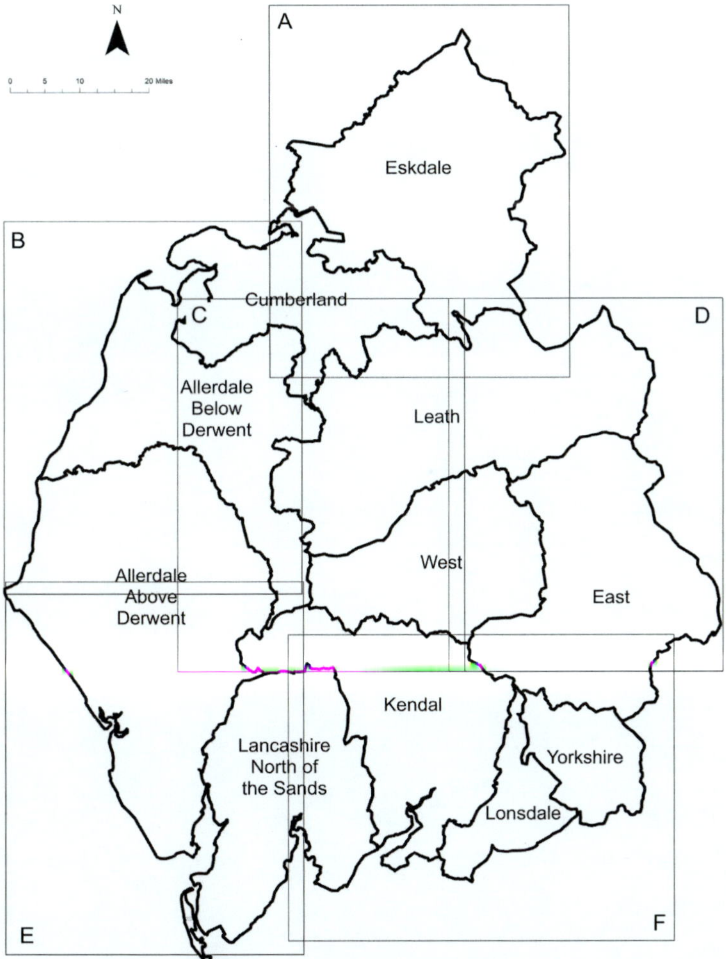

Cumbria: wards, with key to Maps A-F

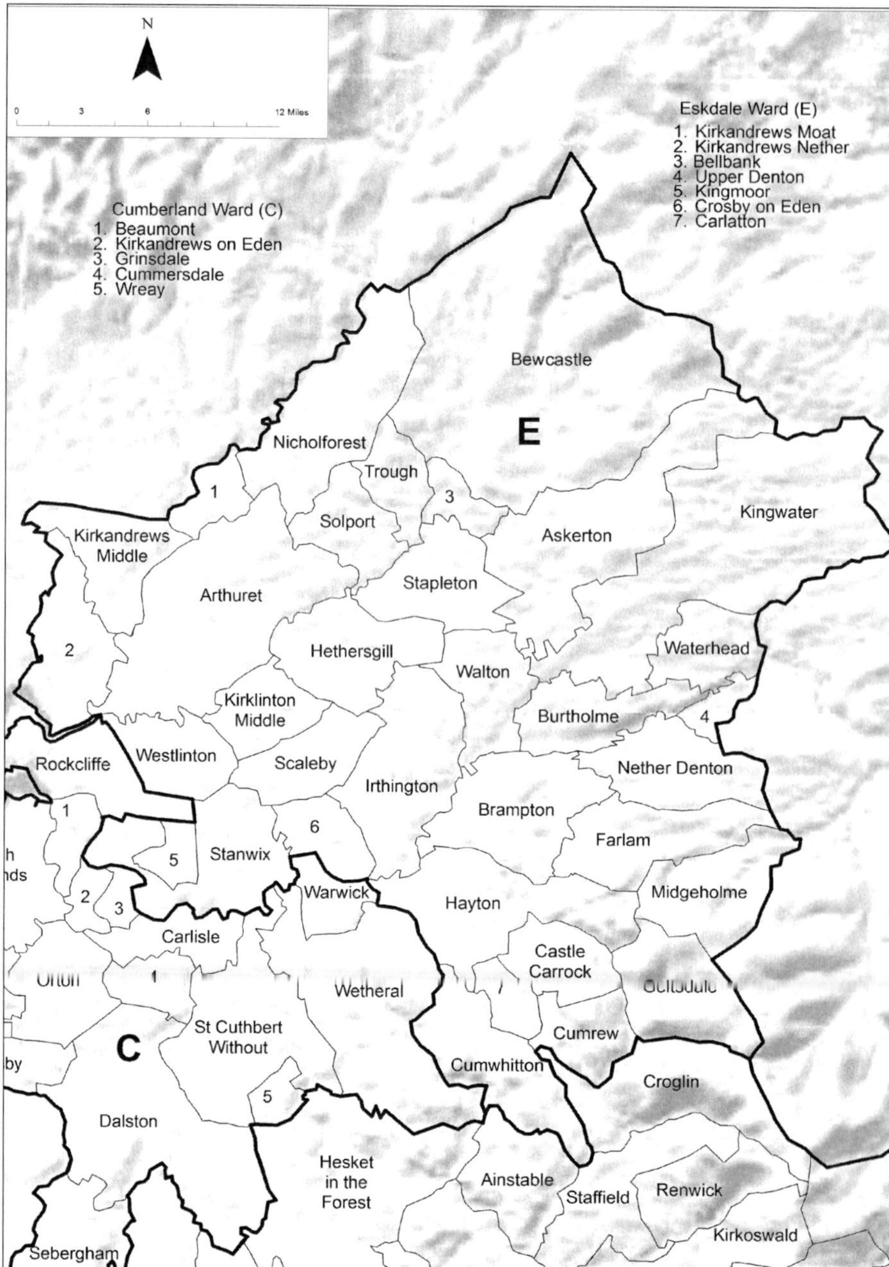

Map A: North Cumbria: Carlisle and Eskdale Ward

Map B: North-west Cumbria: Allerdale and the Solway

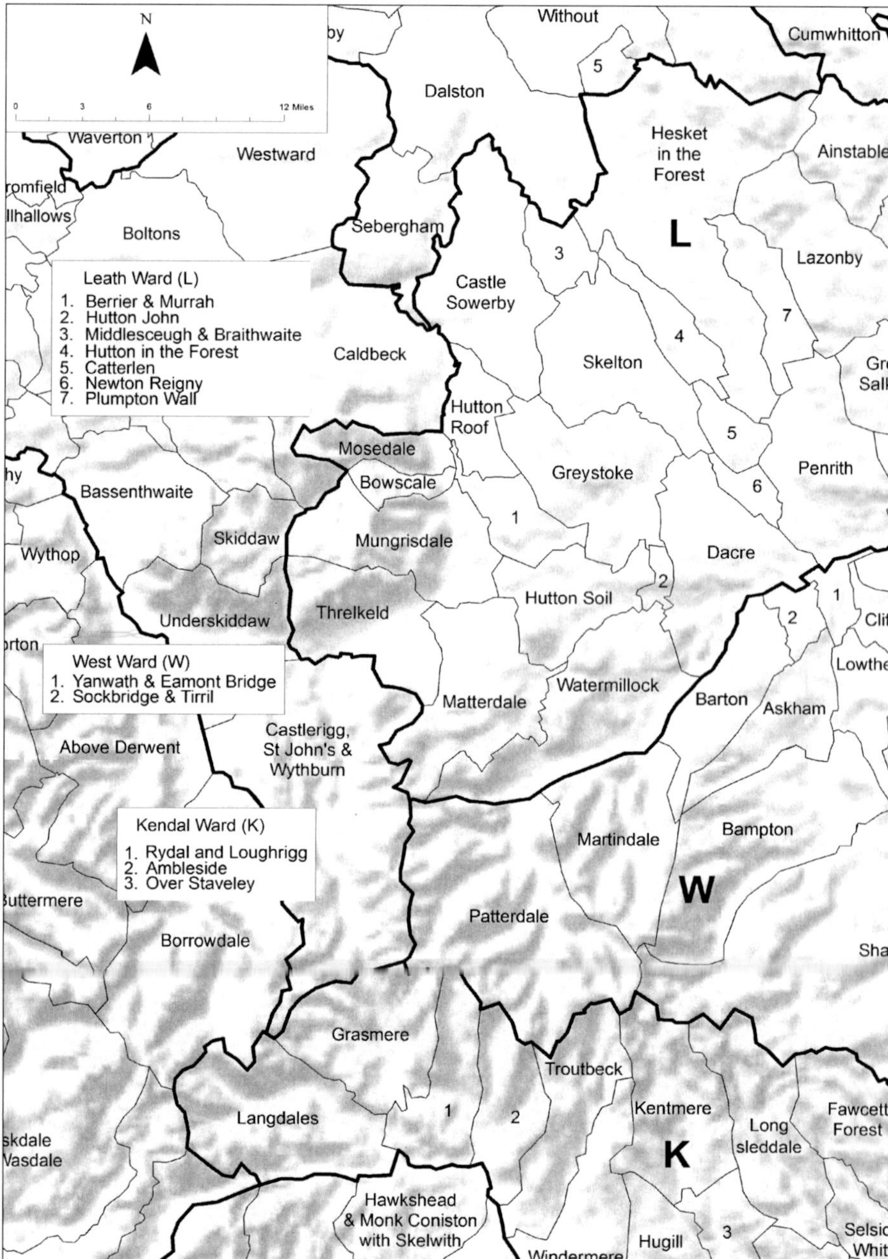

Map C: Central Cumbria: Northern Lake District

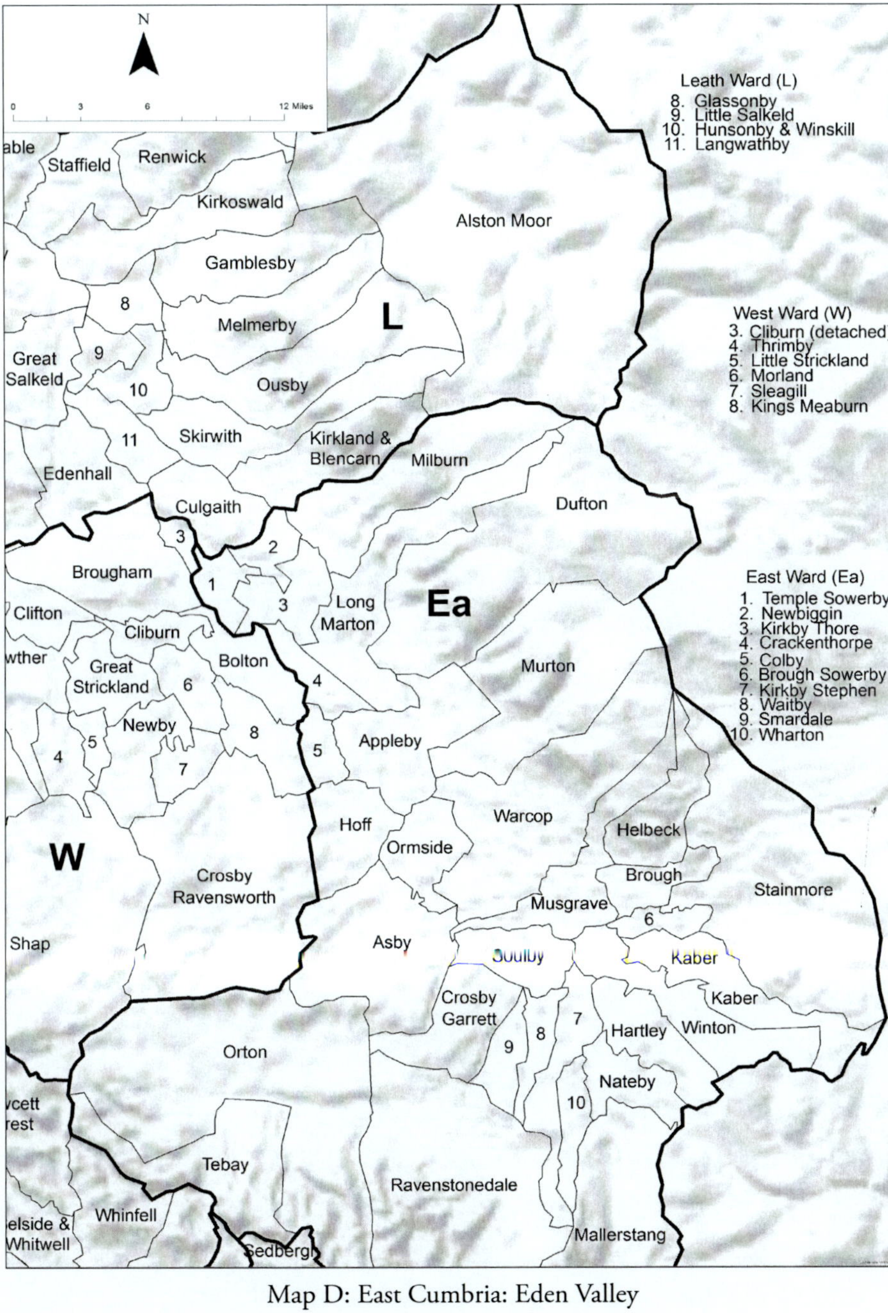

Map D: East Cumbria: Eden Valley

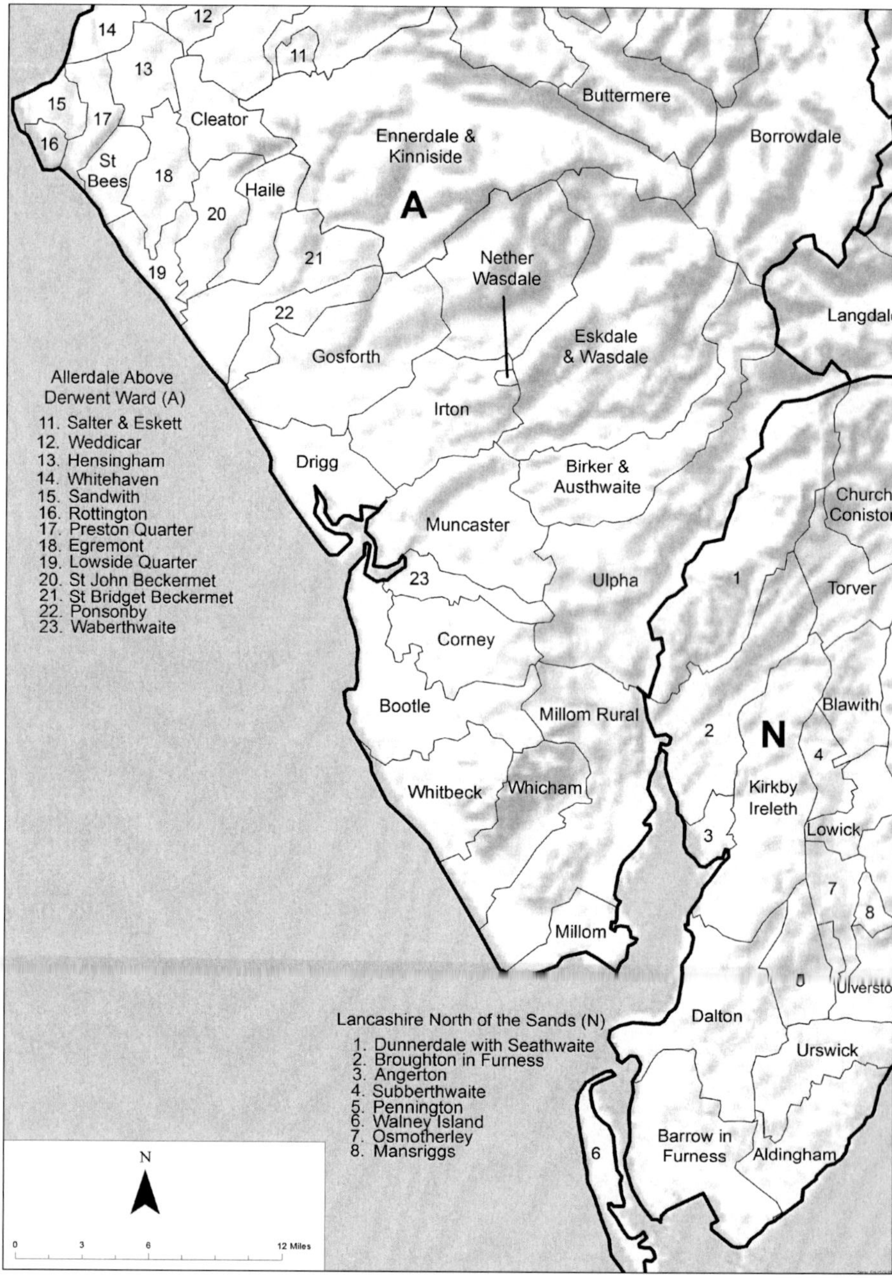

Map E: South-west Cumbria: Copeland and Furness

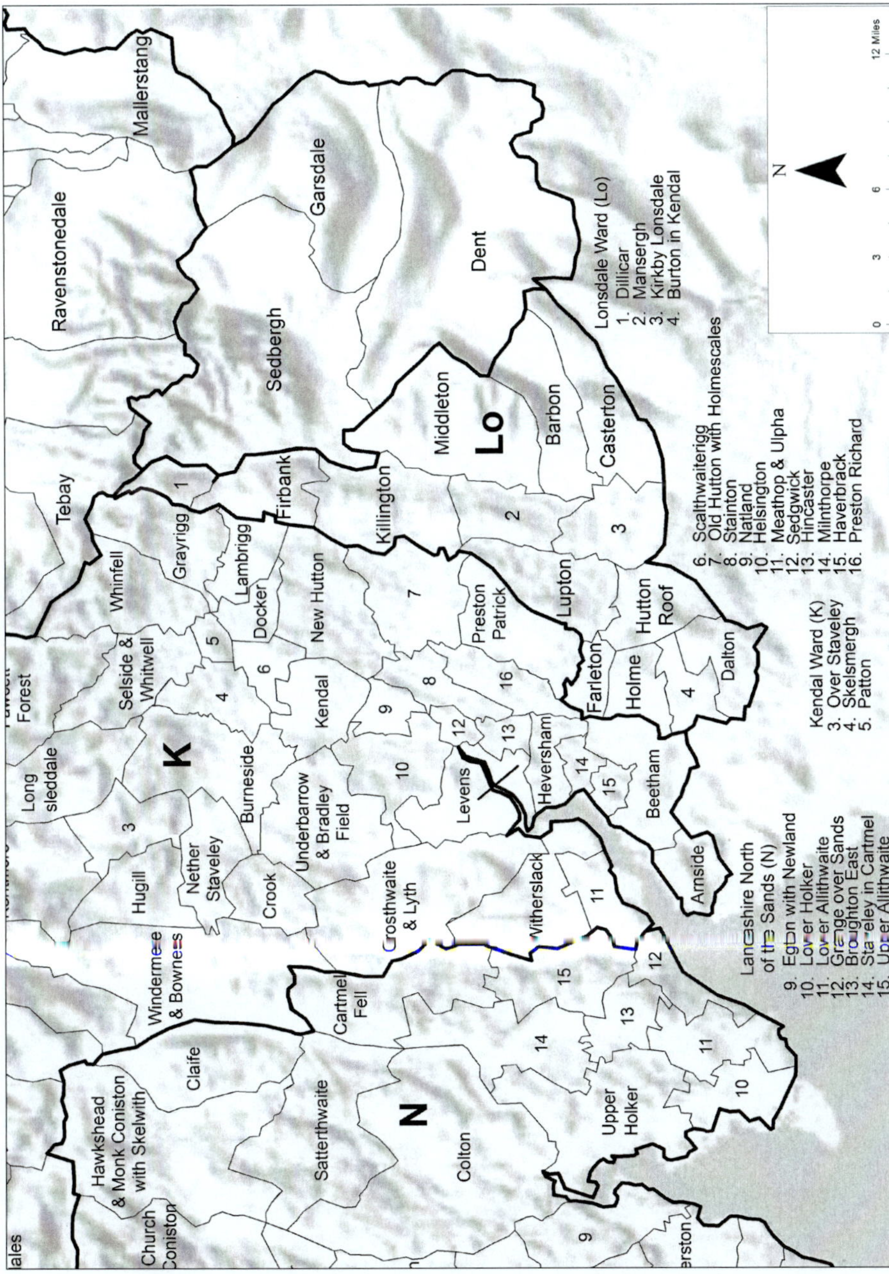

Map F: South Cumbria: Kendal, South Lakes and Sedbergh

CUMBRIA: A TO Z

ABBEY TOWN *see* Holme Abbey

ABOVE DERWENT

Township in Crosthwaite parish, Allerdale above Derwent ward, Cumberland. Above Derwent was a division of the parish for poor law purposes, embracing medieval townships of Braithwaite, Coledale (or Portinscale), Thornthwaite and Newlands (or Rogersett).

Acreage: 14,750 acres [5,969 ha], including large acreages of unenclosed fell commons. Swinside pasture (203 acres [82 ha]) enclosed 1814. Commons in Thornthwaite manor (1,273 acres [515 ha]) enclosed 1814; Ladstocks common field in Thornthwaite (76 acres [31 ha]) divided by private agreement 1843. ***Population:*** estimated at 615 in 1688; rising across early 19th century from 668 in 1811 to 1,115 by 1851 and then stable in range 900-1,100 until 1911; resident population stood at c.1,200 in late 20th century. ***Landownership:*** part of forest of Derwentfells. *Manor of Thornthwaite* held by Curwens of Workington from 13th century, until bought by Ratcliffes in 1628, passing with their estates to Greenwich Hospital and Marshall family (see Keswick). Braithwaite, Coledale and Newlands lay within *manor of Braithwaite and Coledale*, part of honour of Cockermouth (q.v.). ***Economy:*** hill farming; forestry at Thornthwaite and Whinlatter (Hospital Plantation from 1815; expansion after acquisition by Forestry Commission from 1920). Lead and copper mining, recorded from 13th century: Goldscope (in Newlands) was principal mine of Elizabethan Mines Royal. Lead mining continued at Thornthwaite, Brandelhow and Force Crag into 20th century; barytes mined until 1991. Coppice woodland supported mining and tanning to 18th century; bobbin mill at Thornthwaite in 19th-century. Woollen mill at Coledale, built 1824; wool carding and spinning mill at Stair by 1847. Pencil making at Braithwaite by 1847; woollen mill at Coledale became Cumberland Pencil Co. from 1867 to 1898 and mill at Stair converted to pencil making 1889. Tourism developed from 1770s as circuit of Derwentwater became popular and grew in importance across 20th century. ***Places of worship:*** chapels of ease at Thornthwaite (now St Mary's church) from c.1240 (rebuilt 1832 and enlarged 1852) and at Newlands by 1571 (rebuilt 1843; restored 1885). St Herbert's church (Anglican), Braithwaite, built 1900, replacing mission room. Methodist chapel at Braithwaite, built 1850; rebuilt 1876; now used by Keswick Orthodox church. ***Schools and other institutions:*** schools taught in chapels at Thornthwaite and Newlands in early 19th century. School at Braithwaite built 1842; now Braithwaite CE Primary School. National infant school at Thornthwaite by 1898 (closed). School at west end of Newlands church built 1877; closed 1967. Village halls at Portinscale (built 1925) and Braithwaite (Institute built as Memorial Hall 1927); recreation room at Stair in 1920s. Isolation hospital (former Traveller's Rest inn on Whinlatter Pass) from later 19th century to 1940.

AGLIONBY *see* Warwick

AIKTON

Ancient parish in Cumberland ward, Cumberland, encompassing townships of Aikton, Biglands & Gamelsby, Wampool and Wiggonby.

Acreage: 6,172 acres [2,497 ha], divided between constituent townships thus: Aikton: 1,829 acres [740 ha]; Biglands & Gamelsby: 917 acres [371 ha]; Wampool: 1,022 acres [414 ha]; Wiggonby: 2,404 acres [973 ha]. Common land totalling 992 acres [401 ha] survived until 19th century, namely Aikton Common & Green (500 acres [202 ha]), enclosed 1816; Whitrigglees common (196 acres [79 ha]), enclosed 1811; Leathes common (130 acres [53 ha]), enclosed 1825; and Gamelsby & Biglands common (166 acres [67 ha]), enclosed 1856. Common land at Laverockstone enclosed in 17th century. ***Population:*** estimated at 530 in 1688. Rose from 582 in 1801 to peak of 856 in 1851, then declined from 808 in 1871 to 449 in 1971; slight recovery to stand at 480 by 2001. ***Landownership:*** *manor of Aikton* (which included Wiggonby) held by Colville family in 13th century, conveyed to Daniel family 1348, and sold to Dacres in 15th century, thus uniting it with barony of Burgh (q.v.). *Manor of Biglands and Gamelsby* held by Crookdake family from 1290s; divided between two daughters on death of John Crookdake 1322-3, one moiety passing through several families to Dentons of Cardew in 1522; other moiety descended through Boyvill family, passing by marriage to Highmores of Harby Brow in 15th century, who sold it to Dacres. ***Economy:*** predominantly agricultural; market gardening by c.1900. Wiggonby airfield (RAF Great Orton) opened 1942; closed 1952; wind farm on site (airfield used for burial of animal carcases during foot and mouth disease outbreak 2001). ***Places of worship:*** medieval parish church of St Andrew; restored 1732 and again 1869. ***Schools and other institutions:*** school at Aikton endowed 1764 (closed). Wiggonby Grammar School endowed 1802; now Wiggonby CE Primary School. School at Biglands by 1920s; closed 1959. Aikton Parish Room, built as mission room 1891; now village hall.

AINSTABLE

Ancient parish in Leath ward, Cumberland. CP enlarged by addition of most of Croglin and part of Cumwhitton CPs 1934.

Acreage: 4,433 acres [1,826 ha] before boundary changes. 285 acres [115 ha] of common land enclosed 1698; and a further 2,000 acres [809 ha] 1821. ***Population:*** 441 in 1801; peaking at 569 in 1831; had dropped to 361 by 1931 (last census year before boundary changes). ***Landownership:*** *manor of Ainstable* initially held by Vallibus (Vaux) family; passed by marriage to Burdon and Denton of Cardew families; sold to Dacres 1447; forfeited 1570 and in crown hands to 1649; awarded to Howards of Naworth and Arundel 1651, and assigned to Howards of Naworth (later earls of Carlisle) 1654. *Manor of Nunnery* (otherwise Armathwaite or Nunnery and Ruckcroft) originated in endowment of nunnery of Armathwaite, probably in 12th century; after Dissolution granted to William Graham 1552; sold 1690 to Sir John Lowther; exchanged 1694, passing to John Aglionby, whose family held it to 1901. ***Economy:*** predominantly agricultural; freestone quarrying; fisheries in river Eden. ***Places of worship:*** medieval parish church of St Michael; rebuilt 1870-1; tower taken down 1983. Wesleyan Methodist chapel built 1861. United Methodist (Free Church) chapel near Armathwaite, built 1876. ***Schools and other institutions:*** endowed school 1743, rebuilt as Board school 1874; closed late 20th century; building used as library and converted to dwelling. Parochial library founded under will of Barnabas Olney, 1685; few books survived by 1747. Church Institute built 1906. Swimming pool and leisure centre between Ainstable and Armathwaite, early 21st century.

ALDINGHAM

Ancient parish in Lonsdale hundred, Lancashire North of the Sands, embracing townships of Aldingham, Gleaston and Leece with Dendron. Parish includes settlements of Scales, Roosebeck, Baycliff, Newbiggin and Sunbrick.

Acreage: 4,707 acres [1,904 ha], divided between constituent townships thus: Aldingham: 2,463 acres [997 ha]; Gleaston: 1,055 acres [427 ha]; Leece with Dendron: 1,157 acres [468 ha]. Skelding Moor (90 acres [36 ha]), partly in Urswick parish, enclosed 1822. Deep Meadows (113 acres [46 ha]) in Gleaston enclosed 1859. ***Population:*** rising from 633 in 1801 to peak of 1,152 by 1881; then dropping to 891 by 1931 before rising again in later 20th century to stand at 1,187 in 2001. ***Landownership:*** part of the *manor of Muchland* held by Michael le Fleming in 1127, which descended to Harrington family by 1306. They made their seat at Gleaston, where castle built before 1389. Manor then descended by marriage to Grey family, marquesses of Dorset, in 15th century, reverting to Crown after execution of Lady Jane Grey 1554. Remained Crown manor into 20th century. *Gleaston castle* and lands, sometimes treated as separate manor, came to Prestons of Furness Abbey in 17th century. ***Economy:*** predominantly agricultural; lime-burning in 19th century. ***Places of worship:*** medieval parish church of St Cuthbert; extended 1845. Chapel of ease at Dendron (now St Mathew's church) built 1644; rebuilt 1776; enlarged and restored 1891. Wesleyan Methodist chapel at Roosebeck opened 1879; closed 1962 (later demolished). United Methodist chapel at Leece opened 1881; closed 1912 (later demolished). Congregational chapel and school at Gleaston built 1887; still open 1924, but closed shortly after (now private house). ***Schools and other institutions:*** chapel at Dendron used as school in early 18th century; school room built opposite church 1833, later used as Sunday school and parish hall. Replaced by new school at North Hill, built 1864, later becoming Dendron CE School; closed 1994 (converted to dwelling). Aldingham parochial school at Scales (later Scales Primary School), converted from barn 1875; closed 1994 (likewise converted to housing). Reading room (Aldingham Working Men's Club) at Scales built 1883; functioned until c.1930. Aldingham parish hall, Scales, converted from former malt kiln. Village halls at Leece (later 20th century) and Gleaston (built 1970s).

ALLERDALE (barony) *see* Cockermouth

ALLHALLOWS

Ancient parish in Allerdale below Derwent ward, Cumberland.

Acreage: 1,692 acres [685 ha]. Allhallows Common enclosed under Aspatria and Brayton enclosure award 1825. ***Population:*** 173 in 1801 rising slowly to 222 in 1871; then substantial increase to 968 in 1911 as result of coal mining and creation of Fletchertown colliery village. Had fallen to around 500 in late 20th century. ***Landownership:*** *Upmanby* descended with Blennerhassett (q.v.). *Manor of Whitehall* came into hands of Lancelot Salkeld (d. 1610) and remained with his descendants until passing via marriage of Henry Salkeld (d. 1749) and Margaret Charlton of Hesleyside (Northumb.) to her nephew William Charlton. By 1861 manor house at Whitehall was in hands of George Moore. *Harby Brow* was another small estate, held by Highmore family by 15th century. Part was sold in 1595 by Francis Highmore; later owners included Blencow and Steel families; subsequently passed to Charltons, owners of manor of Whitehall. ***Economy:*** farming; coal mining by 1860s. Priestcroft colliery closed 1863 but was succeeded by Allhallows Colliery, sunk 1874 by I. and W. Fletcher, whose name was perpetuated in colliery village of Fletchertown; initially struggled to make a profit; finally closed 1928. ***Places of worship:*** medieval parish church of All Saints, recorded from 12th century but described 1425 as a 'chapel'. Replaced by new church of All Hallows at Fletchertown 1898. Wesleyan chapel at Fletchertown built 1894; closed, then re-opened 1949; converted to dwelling 2010. ***Schools and other institutions:*** school recorded 1818 but had closed by c.1830; parish school built by subscription 1855; closed 1984. George Moore Memorial Hall erected 1879 as memorial to the merchant and philanthropist, containing 'reading and recreation rooms, library and a room for religious purposes'. Former school building converted to community centre 1993.

ALLITHWAITE, LOWER

Township in Cartmel parish, Lonsdale hundred, Lancashire North of the Sands, including village of Allithwaite and eastern part of Cartmel village (formerly 'Churchtown'). Part of Lower Allithwaite transferred to help create Grange CP (q.v.) 1894 and further land transferred to Grange 1914. CP enlarged by absorbing Upper Holker CP 1949.

Acreage: 3,211 acres [1,299 ha] before boundary changes; reduced to 2,962 acres [1,198 ha] after loss of territory to Grange CP. Common land in Lower

Allithwaite enclosed under Cartmel enclosure award 1809. ***Population:*** increasing from 589 in 1801 to peak at 1,009 in 1871; stood at 974 in 1891 (last census year before boundary changes). Population of enlarged CP stood at 1,758 in 2001. ***Landownership:*** *manor of Cartmel* (which extended across whole of ancient parish) granted to Cartmel Priory on its foundation 1189; passing to Crown on Dissolution and in 1667 to George Preston of Holker; subsequently by marriage to Lowther family and then Cavendish family also by marriage. *Manor of Allithwaite* was in hands of Harrington family from c.1200 to at least 1375. Part of estate, known as Wraysholme (centred on 15th-century tower house), granted to George Lord Strange 1489, passing through earls of Derby to 1594, when sold to Dicconson family. ***Economy:*** farming and fishing; limestone quarrying and lime-burning. Corn mill at Allithwaite converted to brewery in 19th century. Tourism increasingly important, particularly at Cartmel, in 20th century. Nursing homes (including Boarbank Hall since 1921). ***Places of worship:*** medieval parish church of St Mary and St Michael, the former Augustinian priory at Cartmel Churchtown, founded c.1189. St Mary's Church, Allithwaite, built at Templand 1865. Wesleyan Methodist chapel at Cartmel built 1871-2. Quaker meeting house there built 1859. United Reformed Church at Allithwaite from 1919; closed 2002. Kingdom hall of Jehovah's Witnesses, Allithwaite, 1985-2009. Roman Catholic house of Augustinian nuns at Boarbank Hall from 1921. ***Schools and other institutions:*** Cartmel Grammar School originated when parish purchased priory gatehouse 1624 for use as school; further endowment 1689; moved to new site in Upper Holker (q.v.) 1790. CE Primary School at Cartmel, established 1861. Cartmel Priory Secondary School opened 1958 on new site to south of village. Allithwaite Institute built as a 'chapel school' 1853; replaced by new school by St Mary's church, built 1865; now Allithwaite CE Primary School. Allithwaite Institute continued to be used as community hall; another village hall (Mary Lambert Hall) built near St Mary's church 1908 and another beside Quarry Lane playing fields, built 1982. Cartmel Institute, in substantial 18th-century house in village, founded 1865.

ALLITHWAITE, UPPER

Township in Cartmel parish, Lonsdale hundred, Lancashire North of the Sands, including villages of Lindale and Newton (High and Low). Part of CP transferred to help create Grange CP 1894.

Acreage: 2,675 acres [1,083 ha] before boundary changes; 3,436 acres [1,391 ha] after absorbing numerous formerly detached portions of other townships in Cartmel parish (but also loss of land to Grange CP). Common land enclosed under Cartmel enclosure award 1809. ***Population:*** 541 in 1801, rising to 771 by 1821; then remaining stable across remainder of 19th century. After boundary changes, rose from 675 in 1951 to 824 by 2001. ***Landownership:*** lordship probably descended with Lower Allithwaite (q.v.) in medieval period. Cartmel Priory held lands in Newton. Lindale was separate estate, granted to Knights of St John of Jerusalem in 12th century; sold by Crown 1611 and purchased 1636 by Robert Rawlinson, descending with Cark Hall (in Lower Holker) thereafter until 1860. ***Economy:*** farming; quarrying and lime-burning in 19th century. Seed mill at Lindale in 1912; four car showrooms there from late 20th century. ***Places of worship:*** chapel of ease in Lindale recorded 1627; rebuilt 1828 as St Paul's Church; chancel added 1864. Quaker meeting house at Height built 1677; meetings discontinued c.1772; short revival c.1910; sold 1922. Plymouth Brethren had meeting place at Lindale in early 20th century. ***Schools and other institutions:*** Lindale church school built 1838; enlarged 1894; now Lindale CE Primary School. High Newton school built 1875; closed (building then used as village hall). Lindale Institute built as reading room 1862; sold and converted to housing 2001. Apostolic school, established by Fathers of the Holy Ghost, at Castle Head, during 20th century. Field study centre established at Castle Head in 1980s.

ALLONBY

Township in Bromfield parish, Allerdale below Derwent ward, Cumberland. United with Westnewton to become West Newton and Allonby CP in 1866 but union was dissolved and Allonby became CP in its own right 1894.

Acreage: 1,340 acres [542 ha] including 490 acres [198 ha] of commons enclosed 1820. Sea banks and foreshore (44 acres [18ha]) remain common land. ***Population:*** over 700 in early 19th century, declining steadily to less than 400 in 1981, but rising to 462 by 2001. ***Landownership:*** *manor of Allonby* belonged to Allonby family, passing by marriage to Flimbys and from them to Blennerhassetts of Flimby, who held it in 17th century; sold 1700 to William Thomlinson, who enfranchised the tenants. ***Economy:*** saltpans and sea fishing by later 17th century. Herring fishery and fish curing grew in importance in later 18th century; c.50 boats involved in herring fishery by mid-19th century but it had declined by 1900. Allonby had become resort for sea bathing by mid-18th century: suite of baths built 1835 (defunct by 1901). ***Places of worship:*** chapel of ease (now Christ Church) built 1744; enlarged 1845. Quaker meeting house converted from cottage 1703; closed 1990. Congregational chapel built 1844; closed. ***Schools and other institutions:*** school endowed 1755; rebuilt as National school beside chapel of ease 1837; closed and later used as Sunday school. British and Foreign school built 1840; now Allonby Primary School. Almshouses for poor Quakers at North Lodge (built by Thomas Richardson early 19th century), established under scheme of 1854. Reading room and library built 1862; closed 1970s. Leisure centre, built 1992.

ALSTON MOOR

Ancient parish in Leath ward, Cumberland. Includes market town of Alston, chapelry of Garrigill and mining village of Nenthead.

Acreage: 36,967 acres [14,960 ha], of which 20,224 acres [8,185 ha] was Pennine moorland, enclosed 1820. Smaller areas of common at Crookburn Moor (129 acres [52 ha]) and Wanwoodside (211 acres [85 ha]), enclosed 1843 and 1856 respectively. ***Landownership:*** *manor of Alston* held by de Veteripont (Vipont) family from mid-12th century until early 15th century. Settled by trial in Chancery on Stapleton family 1426 and passed by marriage to Hiltons of Sunderland later in 15th century. Leases for 1,000 years granted to tenants by Henry Hilton 1611; he sold manor to Edward Radcliffe of Dilston (Northumb.) 1629. Forfeited to Crown 1716 after Jacobite rising and passed to Greenwich Hospital 1734; sold to Catholic Trust for Charitable Purposes 1964.

Economy and growth of settlement: pastoral farming and lead mining and smelting from medieval period until 20th century. Lead mines, famous by 17th century, expanded across 18th century, particularly after London Lead Co. leased mineral rights between 1750 and 1765. Population, estimated at 555 in 1688, had reached 4,746 by 1801 and rose to peak of 6,858 in 1831. Town of Alston grew up haphazardly around parish church; tightly-clustered knot of lanes had evolved by 1770s. Small weekly market there by later 17th century. Nenthead, by contrast, originated as model village, laid out by London Lead Co. from 1825. Other extractive industries included quarrying of both sandstone and limestone and coal mining. Falling price of lead from 1870s (resulting in withdrawal of London Lead Co. 1882) led to long period of stagnation and decline. Population had fallen to 3,134 by 1901 and continued to decline across 20th century to low of 1,916 in 1971, before rising slightly to stand at 2,156 by 2001. Although some mining continued into 20th century – particularly for zinc, barytes and flourspar – mining had almost ceased by 1950. Other industrial activity included textiles (new mill at Alston on River Nent, built c.1802 for 'spinning cotton and flax'; used for woollens from 1820s and for carpet-making from c.1870; closed c.1889); brewing from 1780s to 1880s; tile-making in mid-19th century; hosiery manufacture in former brewery by 1906; and commercial lime-burning 1938 to c.1960. Steel foundry established on site of woollen mill, to manufacture munitions during Second World War; closed 1989. Precision metal castings from 1947 to present.

Places of worship: medieval parish church of St Augustine; rebuilt 1770 and again 1869-70. Chapel of ease (St John) at Garrigill, first recorded 1215: building of 1790 reconstructed 1888-90. Church of St John, Nenthead, built 1845. Numerous nonconformist places of worship, earliest being Independent chapel at Loaning Head, near Garrigill (erected c.1690; replaced 1756 by new chapel at Redwing, near Garrigill 1756; closed 1977) and Quaker meeting houses in Alston (built 1732; discontinued 1902; reopened 1981) and Wellgill, near Nenthead (1724, out of use before 1800). Wesleyan Methodist chapels in Alston (first built 1760; rebuilt on new site Back o' the Burn 1797 and enlarged 1825; rebuilt as St Paul's Church on prominent site at Townhead 1864; closed by 1990s); Nenthead (met in school which was rebuilt as chapel and school 1816; rebuilt on new site 1827; rebuilt again 1873; closed c.2002); Garrigill (met in former Independent chapel at Loaning Head from c.1763; chapel at Low Houses built 1795; rebuilt 1859; closed and converted to dwelling); Nentsberry (built 1825); Low Brownside near Leadgate (built 1849; closed c.1968); Nest (built 1844; closed 1935); and Tynehead (built c.1823). Primitive Methodist chapels at Whitehall, Nenthead (1823; rebuilt 1847; renovated 1893, closed 1938); Alston (1825; rebuilt 1843; closed 1932); Gatehead, Garrigill (1825; rebuilt 1856 and again 1885; closed 2004); Nentsberry (1830; rebuilt 1868); and Blagill. Congregationalist chapel in The Butts, Alston (built 1804; closed and converted to art gallery by 1980s). Mormon church from 1837 to 1850s but no place of worship recorded. St Wulfstan's RC church founded 1953 in former town gaol; church later shared with Methodists.

Schools: Alston Grammar School, recorded 1775; rebuilt 1828, closed 1909. Free school built 1811; became elementary school 1909; closed c.1957. National school for girls built 1844, closed c.1957. Infant school built 1845; closed c.1957. Samuel King's Grammar School built 1909; rebuilt on new site

1957 (now Samuel King's High School); former building became Alston Primary School. At Garrigill endowed school in village recorded early 19th century, and girls' school at Gatefoot, built 1850; combined into newly-built mixed school at Gatefoot c.1875, which closed in 1960s. At Nenthead non-denominational school built by London Lead Co. by 1818; replaced by imposing new building 1864 (now used as village hall). County Council school opened 1899; now Nenthead Primary School. Outlying schools at Nenthall (built 1843 and supported by Hudgill Burn Mining Co.; demolished in 1950s); Leadgate (from c.1770; but described 1847 as having been built (with a library) 'a few years since'; closed 1932); High Plains (closed 1873); and Tynehead (held in Wesleyan chapel; closed c.1933).

Other institutions: Alston subscription library, founded 1821. Mechanics' Institute established 1847 and had library, reading room and museum by late 1850s. It was housed in Alston Town Hall, opened 1858, which also contained public room, board room and 'gentleman's reading room'. At Nenthead London Lead Co. built reading room (1833; rebuilt 1855), shambles, public washhouse and covered market. Reading room at Gatefoot, Garrigill, mid-19th century. Garrigill village hall converted from church institute 1989. Workhouse at Fairhill, Alston, recorded 1777; enlarged after parish became Poor Law Union, 1837. Ruth Lancaster James Cottage Hospital, named after principal benefactress, built 1908 and extended several times since.

AMBLESIDE

Chapelry straddling boundary between parishes of Grasmere and Windermere, in Kendal Ward, Westmorland, consisting of two townships: Ambleside Above Stock, in Grasmere parish, and Ambleside Below Stock, in Windermere parish. Ambleside became UD 1894 and was subsumed into Lakes UD 1935.

Acreage. 4,371 acres [1,769 ha] including extensive fell commons in Scandale and Stock Ghyll valleys, 961 acres [389 ha] of which were enclosed 1865, though 270 acres [109 ha] on Wansfell survives as common land. Bird House common (12 acres [5 ha]), mainly open field, enclosed 1853.

Landownership. Part of manor of Windermere (q.v.). Prominent landowners included Braithwaite family in 16th and 17th centuries (seat at Ambleside Hall).

Origins and growth of the town. Flat ground at head of Windermere chosen as site of Roman fort of GALAVA. Ambleside grew as focal point for local woollen industry in later middle ages. In 1650 it gained charter for weekly market on Wednesdays and cattle fairs at Whitsuntide and in October and was described in 1671 by Sir Daniel Fleming as 'a market cheifly for wool and yarne'. Several fulling mills by 16th century and paper mill on Scandale Beck by 1681. Large mill for manufacture of woollen goods opened c.1797 but woollen trade had declined by 1807, when Ambleside was described as 'a paltry little market town, with little trade'; wool market closed 1825. Fulling mills closed as textile industry declined, to be replaced by bobbin turning from mid-19th to mid-20th centuries. In 19th century Ambleside also had tannery, brewery and saw mill. Short-lived knitwear and toy factories in later 20th century after decline of traditional industries. Quarries outside town, both in Coniston limestone and green slate, the latter notably at Pets Quarry, Kirkstone, in 20th century.

Ambleside's position on main arterial route through Lake District linking Kendal and Cockermouth both fostered its market function and placed it on tourist route from later 18th century; tiny Bridge House (built 1723) became an iconic tourist site. Population, which stood at 538 in 1801, more than quadrupled across 19th century, rising to 1,592 in 1851 and to 2,360 in 1891. As tourism came to dominate economy, visitor accommodation expanded. Salutation Inn had been coaching inn since 18th century; Low Wood Hotel, on lake shore, served Lakes tourists by early 19th century. Hotels and guest houses proliferated from mid-19th century; restaurants and retailing in 20th. Several villas (including The Knoll, Wanlass Howe and Wansfell) were built in 1840s for wealthy incomers. Waterhead, on lake to south of town, developed in 1890s. Growth continued across 20th century as town became popular tourist resort: by 1991 Ambleside's population was 4,414 (of whom nearly one quarter were recorded as visitors) and by 2001 it stood at 5,127.

Places of worship. Chapel (later known as St Anne's church) made parochial 1675 and rebuilt 1812 (converted to flats c.1984). Replaced by parish church of St Mary, on new site, built 1853-4. Wesleyan Methodist chapel, Rydal Road, by 1860 (closed; converted to flats); replaced by Methodist church, Millans Park, built 1898; closed 2005 and converted to housing. Roman Catholic chapel ('Shap Cell') opened 1893, as first wing of planned priory of Premonstratensian canons. Replaced by new church of Mater Amabilis, built 1933. Former police station in Rydal Road converted to Quaker meeting house 1964; closed c.2008. Baptist church in converted house in Compston Road.

Schools. Grammar school, free for boys, endowed by will of John Kelsick 1721, on site at Hill Top next to chapel. Moved to new premises (former

Infectious Diseases Hospital) to east of town 1908; closed 1965. Also at Hill Top in 19th century a private academy for girls run by Miss Dowling (1814 to at least 1859). National school and a girls' school opened in 19th century.

Other institutions. Courthouse (built 1858) and market hall (built 1863). Lecture room built 1872; enlarged 1879; contained library, reading room, school-room and bank. Charlotte Mason (1842-1923), the educationist, opened private teacher training college 1892; later known as Charlotte Mason College, it became part of Lancaster University 1992, passing to St Martin's College (now University of Cumbria) 1996. Ambleside Book Club (founded 1828) and Ambleside Ruskin Library (founded 1882 by Canon Rawnsley) formed basis of Armitt Library and Museum (opened 1912; moved to purpose-built premises 1997). Infectious Diseases Hospital by 1897 (building later used for Kelsick Grammar School).

ANGERTON*

Extra-parochial place in Lonsdale hundred, Lancashire North of the Sands. Became CP following Extra-Parochial Places Act 1857, and merged with Seathwaite with Dunnerdale and Broughton West CPs to form Duddon CP 1976.

Acreage: 959 acres [388 ha] in 1921 Census, but fluctuated as result of coastal change and changes to Duddon channel. Around one-third to half had been enclosed, much of it recently, by 1587. Remainder mostly enclosed between 1796 and 1818. ***Population:*** 36 in 1841, falling to 21 by 1951, and to 14 in 2001. ***Landownership:*** granted to Furness abbey 1299. After Dissolution leased to Preston family, who subsequently purchased it, together with rest of Abbey's demesnes, 1608. ***Economy:*** peat digging from at least 13th century; gravel pit at Waltham Hill in mid-19th century. ***Places of worship:*** none. ***Institutions:*** none.

ANTHORN *see* Bowness-on-Solway

APPLEBY

Borough, market town and county town of Westmorland, in East ward, Westmorland. This entry covers medieval borough itself (Boroughgate; in Appleby St Lawrence parish) and townships of Bongate (in Appleby St Michael parish) and Scattergate (in Appleby St Lawrence parish). The ancient borough, Scattergate township and part of Bongate township were combined to form Appleby MB, created 1885; remainder of Bongate was subsumed into Appleby 1908.

Acreage. Boroughgate (the medieval borough) covered a mere 48 acres [19 ha]; Bongate township covered 3,448 acres [1,395 ha] and Scattergate township 947 acres [383 ha]. After 1908 area of Appleby MB was 1,876 acres [759 ha]. Bongate and Burrells Moor enclosed 1774. Gallows (Fair) Hill, which became part of Appleby MB 1885, remained common land until 1912, when commoners conveyed it to Appleby Corporation, which then enclosed it.

Landownership. Appleby was caput of *barony of Appleby or Westmorland* from 12th century. Appleby castle may have been built by Ranulph Meschin in early 12th century and was in hands of Sir Hugh de Morville in reign of Henry II. Robert de Veteripont (or Vipont) was granted barony 1203; it descended to his granddaughters Isabel (d. 1291), who married Roger de Clifford, and Idonea, who married Roger de Leyburne. Both heiresses' shares of barony were reunited in Robert de Clifford in 1308. Cliffords held barony until 17th century, when it passed to Lady Anne Clifford (1590-1676) and through her daughter Margaret Sackville by marriage to Tufton family, earls of Thanet, who rebuilt parts of castle 1686. In 1881 the then owner, Sir Henry James Tufton, was created Baron Hothfield; his descendants remained in possession until 1961 when castle and estates were sold by auction. By 18th century local influence of earls of Thanet was challenged by acquisition of land and property in Appleby by Lowther family, later earls of Lonsdale.

Origins and growth of the town. Originated as planted borough, laid out, probably in 12th century, at foot of baronial castle. New town consisted of wide, planned street of Boroughgate with its own parish church of St Lawrence; earlier settlement of Appleby lay across River Eden in St Michael's parish, in Bongate (literally 'the street of the peasants or unfree tenants'), referred to in 1276 as 'old Appleby where the villeins dwell' – it was not within area granted borough status by Henry II 1179. As caput of barony of Westmorland, borough became county town of Westmorland. Its growth was limited and it suffered badly from Scottish raids and economic depression in troubled late-medieval centuries; in 1528 Leland described it as 'the Shire town, but now a poor village'. In 1671 Sir Daniel Fleming described Appleby as being 'so mean' and devoid of manufacturing that it would be 'little better than a village' were it not for its county town functions. However, 17th-century Appleby owed much to widowed Lady Anne Clifford, who made a refurbished Appleby Castle her principal home and rebuilt both parish churches and much of Boroughgate.

Appleby's principal role from 12th century was as market centre for upper Eden valley, with important cattle fairs every fortnight from Whitsuntide to Michaelmas, held on the Sands. After 1775 they were gradually eclipsed by 'New Fair' held in June on Gallows Hill, outside borough boundary and not subject to corporation control. Following opening of auction mart next to Appleby West railway station in 1880, New Fair became largely a horse fair, increasingly dominated by travellers until by late 20th century it was recognised as largest gypsy/traveller gathering in Britain.

Appleby's second main function was as county town for Westmorland and, as such, it was location of shire hall and county gaol and sittings of both Quarter Sessions and Assize courts. Inns to serve these gatherings and lawyers' practices were features of the town. However, though nominally Westmorland's shire town, it was overshadowed by Kendal by 18th century. Some manufacturing industry developed in 19th century: in 1829 there were two breweries, a rope and twine manufacturer, three bellows makers and a bag maker; two linen manufacturers were located in Bongate. Despite arrival of railway (Appleby East station on North Eastern Railway opened 1862 (closed 1962); Appleby West station on Midland Railway opened 1876), Victorian Appleby remained very small country town. Ancient borough had 711 inhabitants in 1801, rising to peak of 960 in 1861 (coinciding with railway building) and falling back to 850 in 1901; population of whole of Appleby MB was a mere 1,764 in 1901. In 1910 town's main industries were recorded as brewery and mineral water manufactory. After 1931 Express Dairy, handling milk eggs and manufacturing cheese, became major employer until it closed 1997. Population had dropped to 1,618 in 1931 but numbers steadily rose during second half of 20th century, reaching 2,862 in 2001. Ceased to be borough and county town 1974, on creation of new county of Cumbria, its name was subsequently changed to Appleby-in-Westmorland. Economy has focused on tourism from later 20th century.

Places of worship. Two medieval parish churches: St Michael, Bongate, probably earlier of the two (evidence of pre-Conquest fabric), closed 1974 and converted to house and studio. Borough's medieval parish church of St Lawrence, Boroughgate, remains in use. Carmelite Friary (Whitefriars) founded at Battlebarrow 1281. Wesleyan Methodist chapel built 1823; rebuilt on The Sands 1888 and extended 1904. United Methodist Free Church (formerly Wesleyan Reform Tabernacle Chapel) built 1837; rebuilt 1872 and Sunday School added

1884; now youth centre. Chapel on the Rock Primitive Methodist church built 1872; later used as Oddfellows' Hall and, since 1995, as Kingdom Hall of Jehovah's Witnesses. Our Lady of Appleby RC church on Garth Heads Road built 1958. Municipal cemetery with mortuary chapel opened c.1908.

Schools. Appleby Grammar School, in existence by 1453, originated as chantry school; re-founded by Elizabeth I 1574; moved from town centre to present site on Battlebarrow 1887 (became co-educational 1956 and comprehensive 1962). National school in St Lawrence's parish recorded 1818 as having ceased due to want of school room. Bongate National School built 1844 and subsequently enlarged. British school in St Lawrence's parish built 1849 and enlarged 1875. St Lawrence's elementary school for girls built by subscription 1872; remained open until 1962. British and National schools closed 1910 after opening of county elementary school on site between the two railway stations; now Appleby Primary School. Appleby Training and Heritage Centre for further education established 1996, using former Midland Railway buildings.

Other institutions. St Nicholas' leper hospital founded on outskirts of town before 1235. Almshouses for single elderly ladies, known as St Anne's Hospital, founded 1653 by Lady Anne Clifford and endowed with estate at Brougham and lands formerly belonging to St Nicholas' hospital. Moot Hall overlooking market place built 1596 (upper storey still used as council chamber; ground floor for Tourist Information Centre and shops). Cloister at foot of Boroughgate built 1811; used as market house in 19th century; its function replaced by Market House, erected by corporation 1867 and enlarged 1880. Shire Hall built on The Sands 1770-1, as court house for Assizes, Quarter Sessions, Petty Sessions and County Court. Its judicial functions ceased in late 20th century and, after refurbishment, now used for business premises. County Gaol behind Shire Hall disused by 1885; part of building subsequently used as vagrant ward for East Ward Poor Law Union, and as police station since 1971.

Mechanics' Institute founded 1848, with library and reading room in Boroughgate. News room in King's Head Inn and book club recorded 1860. Church Institute, The Sands, built 1862. Appleby Golf Club founded 1894 and transferred to present site on Brackenber Moor, two miles from town, 1903. King George V playing field donated by Lowther family for public recreation. Open air swimming pool built as Second World War memorial; replaced by indoor leisure centre 1994. Sports hall on land next to Grammar School opened as a millennium project.

ARLECDON

Ancient parish in Allerdale above Derwent ward, Cumberland, comprising townships of Arlecdon, Frizington and Whillimoor. Parish became UD 1894, reverting to CP status 1934, when it was renamed Arlecdon and Frizington CP.

Acreage: 5,556 acres [2,248 ha], divided between constituent townships thus: Arlecdon: 1,570 acres [635 ha]; Frizington: 2,080 acres [842 ha] and Whillimoor: 1,907 acres [772 ha]. Frizington common (610 acres [247 ha]) enclosed 1820; Arlecdon, Winder and Whillimoor commons (totalling 10,000 acres [4,047 ha]) enclosed 1823. ***Population:*** estimated at 430 in 1688. Rose from 354 in 1801 to 643 in 1851; then expanded dramatically with development of iron-ore mining and quarrying in 1850s, more than doubling to 1,550 by 1861 and reaching peak of 6,651 in 1881. Bulk of population lived in mining villages of Frizington and Rowrah. Thereafter population declined, dropping to 4,247 by 1951 and to 3,678 by 2001. ***Landownership:*** *manor of Arlecdon* was one of manors held by Fleming family from medieval period. *Manor of Frizington* sold to John Leigh in early 15th century; later owned by Williamson and Patrickson families; sold to Sir James Lowther, thus becoming part of Lonsdale estates. Other prominent landowners included Dixon family of Rheda in 19th and 20th centuries. ***Economy:*** farming; cattle fairs held three times a year in Arlecdon until early 20th century. Coal mining by 1720s, when there was an ironworks at Frizington. Limestone quarrying, lime-burning and tile-making by mid 19th century. Rapid expansion in mining, especially for high-quality haematite iron ore, from 1850s. After demise of mining, some diversification: factory making Kangol hats in Frizington late 20th century. ***Places of worship:*** medieval parish church of St Michael, between Arlecdon and Asby; rebuilt 1829 and again 1904. Services held at schoolroom in Frizington by 1847; St Paul's church built next to school 1867; building divided 2000 to create community space (the Lingla Centre). Mission room at Asby opened 1875. Wesleyan Methodist chapel at Chapel Row, Rowrah, built 1829; replaced by new building in village centre 1895. Wesleyan chapel at Frizington built 1859; rebuilt 1884; replaced by modern brick building (now St John's Methodist Church). Primitive Methodist chapels at Asby (built 1875; open 1927) and Frizington (built 1876; replaced by adjacent new building 1894). Roman Catholics in Frizington worshipped in chapel at Catholic school from 1875; St Joseph's RC Church built as an iron church 1890; rebuilt in stone 1896. Gospel hall of Plymouth Brethren at Frizington, built 1892. ***Schools and other institutions:*** school had been built by parishioners by 1778; private school at New Asby, recorded 1847. St Paul's

National School, Frizington, originated in parish school built 1838; rebuilt 1887; closed 1971. Board schools built at Frizington 1877 and Arlecdon 1878; now Frizington Community Primary School and Arlecdon Primary School. St Joseph's RC School, Frizington, built 1871; replaced 1968. Reading rooms established at Frizington (Working Men's Reading and Recreation Room 1862; St Joseph's 1893), Arlecdon (Liberal Reading Room 1891) and Rowrah (1896). Council Chambers, erected 1898, included free library.

ARMATHWAITE *see* Hesket-in-the-Forest

ARNSIDE

Hamlet in Beetham parish, Kendal ward, Westmorland, which became separate CP 1897.

Acreage: CP contained 2,489 acres [1,007 ha]. ***Population:*** 'a mere handful' before construction of Ulverston & Lancaster Railway 1857. Small coastal resort developed close to railway station and population had risen to 844 by 1901. Continued to rise across 20th century, reaching 1,678 by 1921 and 2,301 by 2001. ***Landownership:*** part of manor of Beetham (q.v.) ***Economy:*** rural (farming; fishing; salt-making) until mid-19th century; limestone quarrying and lime-burning in mid-19th century. Pier built 1860 after railway viaduct barred navigation upstream (rebuilt 1984). Resort function grew quickly in later 19th century: refreshment rooms, boating, boarding houses and two hotels by 1880s. Holiday camp at Far Arnside for Leeds Poor Children opened 1905; rebuilt 1919; demolished 2000. ***Places of worship:*** St James' Church built 1864-6; enlarged 1884 and 1905 and again 1914. Wesleyan Methodist chapel built 1875; extended 1888; new nave built 1899-1900. Cemetery with mortuary chapel, 1904. Our Lady of Lourdes RC church converted from former National school 1977. ***Schools and other institutions:*** British school, Church Hill, built 1873; now Arnside Educational Institute. National school built 1880; moved to new site on Church Hill (now Arnside National CE School). Private schools, including Wesleyan boarding school at Earnseat by 1885; re-established 1900; closed 1979. Institute (containing reading room, billiard room, lecture room and small gymnasium) built 1890. Women's Institute and Village Hall, Orchard Road, built 1928.

ARTHURET

Ancient parish in Eskdale ward, Cumberland, containing townships of Brackenhill, Longtown, Lineside and Netherby.

Acreage: 12,955 acres [5,243 ha], divided between constituent townships thus: Brackenhill: 4,644 acres [1,879 ha]; Longtown: 2,429 acres [983 ha]; Lineside: 1,903 acres [770 ha]; and Netherby: 3,980 acres [1,611 ha]. Common moorlands enclosed by Grahams of Netherby in later 18th century. ***Population:*** parish population, estimated at 600 in 1688, fluctuated around 2,500-3,000 for much of 19th and 20th centuries. Bulk of inhabitants lived in Longtown, which contained 2,049 of parish's 2,903 inhabitants in 1831, remainder being divided between other townships as follows: Brackenhil: 391; Lineside: 137; Netherby: 326. Longtown's population reached peak of 2,863 in 1861, numbers swollen by recently-founded bobbin mill and temporary presence of railway workers. Parish population fell back from 3,714 in 1861 to 2,145 in 1931, before rising during second half of 20th century to stand at 2,434 in 2001. ***Landownership:*** part of *barony of Liddel or Esk* (see Kirkandrews Moat). ***Economy:*** rural parts largely agricultural; several brick and tile works in 19th century, notably Sandysike brick works. Second World War military airfield at Hallburn (RAF Longtown) from 1941 to 1946. Haulage industry in later 20th century.

LONGTOWN was market town, probably originating in grant of market and fair at Arthuret 1307 (though market and fairs held by William de Stutevill in Cumberland 1200 might have been at Longtown). Described by Thomas Denton in 1688 as containing only a dozen houses, mostly built of clay, except 'courthouse or townhall', which was stone-built and slated. Market was then 'of more benefit to Scotland then to this countrey'. Town laid out afresh from 1757 by Robert Graham of Netherby; by early 19th century principal occupation was weaving. Bobbin mill established 1831; by c.1900 it employed 70-100 people; closed 1936. Market had ceased by 1900. Livestock auction mart established 1926 by Cumberland & Dumfriesshire Auction Mart Ltd.

Places of worship: medieval parish church of St Michael; rebuilt 1609; restored 1868. Former medieval church at Easton (separate parish in 13th century). Scottish Presbyterian chapel, Netherby Street, Longtown, built 1799. United Presbyterian chapel, Bridgend, Longtown (now St Andrew, Church of Scotland), built 1834. United Methodist Free church, Longtown, built 1865. Roman Catholic church of Our Lady of Good Counsel, Longtown. ***Schools and other institutions:*** free school at Longtown, founded 1754; endowed school at Easton; replaced by new school 1872 (now community hall). British (later Board) school at Longtown, built 1857. Lady Hermione Graham's Girls' School, Longtown, converted into nurses' home, 1895. Secondary school (Lochinvar School, Longtown) built 1964; closed 2008. Modern primary school survives. Longtown Union Workhouse, built 1847 on site of farmhouse, east of the town; demolished 1950s. Isolation Hospital to east of town (now Virginia Lodge care home) by 1950s. Diamond Jubilee Nursing Home, Graham Street, recorded 1901. Mechanics Hall, Longtown, built 1851; library and reading room, moved to Moot Hall by 1900. Arthuret Parish Memorial Hall built at Longtown 1924; extended 1981; now Longtown Community Hall.

ASBY (Great Asby)

Ancient parish in East ward, Westmorland, embracing townships of Asby Coatsforth, Asby Winderwath and Little Asby.

Acreage: 8,497 acres [3,439 ha], divided between constituent townships thus: Asby Coatsforth: 2,259 acres [914 ha]; Asby Winderwath: 4,024 acres [1,628 ha]; Little Asby: 2,215 acres [896 ha]. Common land at Asby Mask (1,248 acres [505 ha]), enclosed 1855, and Asby Winderwath (1,850 acres [749 ha]), enclosed 1874. Asby High Intake and Asby Low Intake, totalling 265 acres [107 ha], enclosed 1849. ***Population:*** 357 in 1801, rising to peak of 496 in 1881; decline thereafter, falling to low of 238 in 1981; stood at 280 in 2001. ***Landownership:*** parish contained three manors. *Manor of Asby Winderwath*, held by Asby (Askeby) family before passing successively to Moresbys and Pickerings, who sold it to Fletchers, from whom it was later purchased by Bagots. *Manor of Little Asby* held by English family in 13th century and descended by marriage to Standishes in 14th. It then passed to Honeywoods, before being sold to earl of Thanet. *Manor of Asby Coatsforth* held by Cottesfords from 12th to 15th centuries. Later it descended to Musgraves, then passed through several hands by purchase, before becoming property of Earl Bective. Byland Abbey had grange here. ***Economy:*** predominantly agricultural. Limestone quarrying and lime-burning; marble mill for polishing limestone 19th century. Copper mine in 19th century; short-lived. Caravan park at Little Asby established later 20th century. ***Places of worship:*** medieval parish church of St Peter in Great Asby; rebuilt c.1865-6. Medieval chapel of St Leonard in Little Asby (foundations survive). Methodist chapel at Great Asby built 1859; still in use. Baptist chapel at Great Asby built 1862; still in use. Congregational chapel at Little Asby built 1889; now a dwelling. ***Schools and other institutions:*** grammar school at Great Asby, endowed 1682 and 1688; rebuilt 1853; enlarged and re-opened as public elementary school 1880; now community primary school (Asby Endowed School). Village hall at Great Asby built as 'Great Asby Club' 1908. St Helen's almshouses by St Peter's church founded 1811 by Joseph and William Fairer and built by 1814.

ASKERTON

Township in Lanercost parish, Eskdale ward, Cumberland. CP includes Kirkcambeck, formerly extra-parochial.

Acreage: Askerton township contained 11,021 acres [4,460 ha]; extra-parochial place of Kirkcambeck 281 acres [114 ha]. Askerton commons enclosed with others in Lanercost parish, totalling c.6,000 acres [c.2,400 ha], 1807. Side Fell (1,326 acres [537 ha]) remains common land. ***Population:*** in range 400-500 in first half of 19th century, standing at 479 in 1851; steady downward trend thereafter to low point of 127 in 1971; slight recovery in late 20th century, to stand at 162 in 2001. ***Landownership:*** part of barony of Gilsland (see Brampton). Askerton castle rebuilt by Thomas Lord Dacre in 1490s. *Kirkcambeck* was separate manor, held of Gilsland barony by succession of families in medieval period before coming back into hands of Dacres. ***Economy:*** largely agricultural; tile works near Kirkcambeck in 19th century; afforestation late 20th century. ***Places of worship:*** medieval church of St Kentigern, Kirkcambeck, rebuilt 1885 (arch of medieval church survives); now closed. ***Schools and other institutions:*** National school at Lees Hill built 1876; now Lees Hill CE Primary School. School at Kirkcambeck recorded 1862.

ASKAM-IN-FURNESS *see* Dalton-in-Furness

ASKHAM

Ancient parish in West ward, Westmorland, embracing townships of Askham and Helton.

Acreage: 4,484 acres [1,815 ha], divided between townships thus: Askham: 1,839 acres [744 ha]; Helton: 2,645 acres [1,070 ha]. Askham High Field (242 acres [98 ha]) enclosed 1853; Helton Cruise and Gatelands Pasture,

Helton, (36 acres [15 ha]) enclosed 1853. Askham Fell and Helton Fell (1,887 acres [764 ha]) remain unenclosed common land. ***Population:*** rising from 448 in 1801 to peak of 635 in 1841; then falling (explained 1861 as due to 'migration of labourers in search of employment'). Continued to fall through 20th century from 513 in 1901 to 370 in 2001. ***Landownership:*** *manor of Askham* belonged to Helbeck family in 13th century, passing to Swinburnes, and then in 1375 to Sandford family (builders of Askham Hall) who held it until 1724 when it passed by marriage to Tathams. Then sold to Lowther family who remain major landowners. *Manor of Helton* passed through Morvill, Wessington and English families to Sandfords, who sold it to Sir John Lowther 1680. ***Economy:*** predominantly agricultural; limestone quarrying and lime-burning in 19th century. Tourism increasingly important from later 20th century. ***Places of worship:*** medieval parish church of St Peter; rebuilt 1832-3. Wesleyan Methodist chapel in Helton built 1867; went out of use in late 20th century. ***Schools and other institutions:*** school in Askham founded by Jane Bowman 1779; endowment enlarged 1809 by public subscription; closed 1980s. Two day schools for girls c.1830. School in Helton in 19th century. Reading room in Askham during late 19th and early 20th centuries. Village hall built as Memorial Hall 1920s (extended and refurbished 2004). Open air swimming pool.

ASPATRIA AND BRAYTON

Township in Aspatria parish, Allerdale below Derwent ward, Cumberland, which became UD 1894. Reconfigured 1934 to become Aspatria CP, with loss of territory to Hayton & Mealo and Oughterside & Allerby CPs.

Acreage: 3,550 acres [1,436 ha] before boundary changes. Outfields enclosed 1758-9; East and West Moors, extensive common moorland across parish containing 4,300 acres [1,740 ha], enclosed 1825. ***Population:*** the 'long dry village', as Thomas Denton called Aspatria in 1688, grew into small town with arrival of coal mining by mid-19th century. Population rose from 327 in 1801 to 3,521 in 1921; then declined to c.2,700 by end of 20th century. New colliery village at Harriston built c.1870. ***Landownership:*** *manor of Aspatria* was retained in hands of lords of Allerdale barony, descending with honour of Cockermouth (q.v.). *Manor of Brayton* acquired by Thomas Salkeld (d. 1627) of Whitehall and later by Wilfrid Lawson (d. 1710), in whose family it remained until 20th century, Brayton Hall becoming their principal seat (demolished 1940). ***Economy:*** coal mining recorded 1681; rapid expansion in 1840s after construction of Maryport & Carlisle Railway (1845) and sinking of Brayton Domain Colliery on Lawson royalty by John Harris of Greysouthen (whose name is perpetuated by colliery village of Harriston) in 1850; last pit closed 1940. Other industries included creamery of West Cumberland Dairy Co. (established 1888; still in operation); agricultural implement making (1901); mattress manufacture (since 1974). ***Places of worship:*** medieval parish church of St Kentigern, a pre-Norman foundation; rebuilt 1846-7. Congregational chapel built 1827. Primitive Methodist chapel built 1866; replaced by a new building 1896. Christian Brethren chapel at Richmond Hill built 1886. Wesleyan Hall built 1929. ***Schools and other institutions:*** several unendowed schools in parish by 1815, one of which was in Aspatria village. National school built 1825 and girls' school built 1837. Board schools opened 1895 (now Richmond Hill Primary School). Infant school in colliery village of Harriston built 1879; used as temperance hall and for Sunday worship. Beacon Hill Secondary School built 1963-5. Agricultural college, founded by Lawsons in 1874 to promote scientific farming; closed 1925. Other public institutions included: Market Hall and Noble Temple, venue for concerts and lectures, both built 1872; free reading and news room (1873); Drill Hall (c.1914) and War Memorial Recreation Ground, c.1920. Institute and reading room at Harriston built 1892. Community Centre in Aspatria (in former Primitive Methodist chapel). Village hall in former school at Harriston; closed c.2015.

BACKBARROW *see* Holker, Upper

BAMPTON

Ancient parish in West ward, Westmorland.

Acreage: 10,925 acres [4,421 ha], including extensive unenclosed fell land on Bampton Common (6,890 acres [2,788 ha]), Knipe Moor (295 acres [120 ha]) and smaller commons at Butterwick Green (40 acres [16 ha]) and Bomby Green (14 acres [6 ha]). Bampton Sackwath common field (62 acres [25 ha]) enclosed 1846. ***Population:*** 600 in 1801; gradual decline during 19th and 20th centuries (except during construction of Haweswater dam, when migrant labour boosted figure to 1,011 in 1931), falling to 283 in 2001. ***Landownership:*** parish contained *manors of Bampton Patrick* and *Bampton Cundale*, named after their respective owners in early 13th century, Patrick de

Culwen (later Curwen), and Henry de Cundale. *Bampton Patrick* passed to Warwick family, from whom it was purchased 1772 by Edward Hasell of Dalemain who sold it to earl of Lonsdale. *Bampton Cundale* passed from Cundales to Cliburns, who held it until 1554; it later passed to earls of Lonsdale. *Thornthwaite* was hunting forest, held successively by lords of Naworth, Warwicks and earls of Lonsdale. *Knipe* became property of Wharton family before being bought by John Lowther c.1730. Thus, by early 19th century, earls of Lonsdale were dominant landowners. Manchester Corporation took possession of major part of parish in 1920s (purchasing 6,000 acres [2,428 ha] of common land and 400 acres [162 ha] of enclosed fields and farmhouses from Lord Lonsdale) to secure land for Haweswater reservoir and its catchment. ***Economy:*** hill farming (Bampton Sheep Association for improvement of mountain sheep formed 1876); small limestone quarries and lime-burning in 19th century. ***Places of worship:*** medieval parish church of St Patrick, rebuilt 1726; restored 1884-5. Wesleyan Methodist chapel at Bomby built 1877; closed 1993. Mission church for construction workers at Haweswater dam; building removed to Thirlmere 1942. ***Schools and other institutions:*** Bampton Grammar School founded by Thomas Sutton 1623; re-constituted as Bampton Endowed School 1877 on new site; closed 2005. School at Roughill endowed by Edmund Noble 1662; closed 1950. Grammar school at Measand (Mardale) endowed 1711 (plaque gives date of foundation as 1713); rebuilt as Mardale School 1883; closed 1932; building removed to Walmgate Head, Bampton, on flooding of Mardale 1937. Public libraries endowed 1710-1757. Bampton Memorial Village Hall opened 1950.

BARBON*

Chapelry and township in Kirkby Lonsdale parish, Lonsdale ward, Westmorland.

Acreage: 4,261 acres [1,724 ha]. ***Population:*** 242 in 1801, rising to a peak of 364 in 1861 (inflated by presence of railway labourers); in range 200 to 275 across 20th century; stood at 263 in 2001. ***Landownership:*** *manor of Barbon* held by Barbon family from late 12th century, descending by marriage to Lascelles family by 1239. In mid-15th century it passed by marriage of Margaret Lascelles and James Pickering to Pickering family, whose heir Francis Vaughan held it in later 16th century. Held by Sir Richard Shuttleworth of Gawthorpe by 1588, in whose family it remains. Sir James Kay-Shuttleworth built Barbon Manor as shooting lodge 1863. ***Economy:*** predominantly agricultural until coming of railway in 19th century (station opened 1861; closed 1954). Coal mining on Barbon Moor, recorded 1692 to mid-19th century. Quarrying and lime-burning on Barbon High Fell in 19th century. Garden furniture manufacture from 1999. ***Places of worship:*** chapel of ease, recorded 1571; said to have been 'lately rebuilt' 1692; rebuilt again 1815. New church of St Bartholomew built on adjacent site 1892-3. Wesleyan Methodist chapel built 1888 (converted to dwelling c.2009). ***Schools and other institutions:*** school taught in chapel in early 18th century; school attached to chapel built when chapel was rebuilt 1815. Replaced by new school on different site, built by subscription 1867; closed 1982. Reading room built 1884; closed 1960s (converted to dwelling). Village hall built 1926 (included lending library); still in use.

BARDSEA *see* Urswick

BARROW-IN-FURNESS

Industrial town covering much of 'bierleys' (or townships) of Hawcoat and Yarlside in Dalton-in-Furness parish, Lonsdale hundred, Lancashire North of the Sands.

Acreage. Hawcoat bierley covered 7,544 acres [3,053 ha] and Yarlside 4,011 acres [1,623 ha]. Borough of Barrow-in-Furness (created MB 1867 and County Borough 1889) was extended 1872 to include Walney Island (q.v.), Sheep Island, Piel Island, and Foulney Island, and again 1881 to incorporate Rampside chapelry.

Landownership. Hawcoat and Yarlside were divisions of *manor of Dalton or Plain Furness*, part of lordship of Furness, granted to Furness Abbey on its foundation 1127, which reverted to Crown on Dissolution, 1537. In 1662 manor of Plain Furness was granted to George Monck, duke of Albermarle (d. 1670), passing by descent to George Brudenell, duke of Montague (d. 1790), whose daughter Elizabeth married Henry, 3rd duke of Buccleuch and Queensberry (d. 1812), through whose family it descended to 21st century. Site of Furness Abbey (an estate later known as 'The Manor') leased to John Preston of Preston Patrick 1549 and passed by descent to Cavendish family, dukes of Devonshire, of Holker Hall.

Origins and growth of the town. Until 1850s Hawcoat and Yarlside were rural farming areas, with traditional coastal industries of fishing and salt-making (memory of which preserved in place-name Salthouse). Piel Island was minor port with customs house until 1760. Rampside had become sea-bathing resort by 1820s and Barrow was main port for transport of iron ore from Furness mines by 1830s. Population of hamlet of Barrow estimated at around 150 in 1843; it had reached 661 by 1850.

Local economy was transformed by coming of Furness Railway 1846 and rapid expansion in exploitation of local iron ore deposits. Haematite Iron and Steel Works at Barrow founded 1859; taken over 1866 by Haematite Steel Co. (founded 1864). New harbour constructed 1863 (with Devonshire and Buccleuch docks opened 1867 and 1873, graving dock 1872 and Ramsden dock 1879). Naval Armaments Co. established 1888; absorbed by Vickers Maxim Co. 1897, so that Barrow became famous for naval shipbuilding, with ancillary activities such as ship-repair and rope-making. Other major industries included flax and jute works (established 1870; partly destroyed by fire 1892; closed c.1930; demolished 1948) and chemical wood-pulp and paper works at Salthouse (founded 1888; closed later 20th century). Barrow's population rocketed across 1860s, to stand at 17,992 by 1871, and continued to rise rapidly across later 19th century: town contained 56,625 inhabitants in 1901. Core of town, laid out by Sir James Ramsden in 1856, took form of grid of wide streets. Migrants to Barrow came from many parts of Britain and Ireland, a strong Scottish presence being reflected in tenement-style housing on Barrow Island.

Both industrial output and population peaked during First World War (town's population stood at 67,244 in 1921) but long period of decline set in from 1920s. Population fell across 20th century, partly as result of movement from town centre to new estates on Walney Island (q.v.), to stand at 49,577 in 1981 and 47,794 in 2001. Ship-building witnessed a revival during Second World War and from 1945 during post-War 'tanker boom'. From 1960s Barrow shipyard began to specialise in submarine building, ensuring its survival. Other founding industries fared less well: ironworks closed 1963; steelworks 1983; and closure of last remaining foundry (David Caird Ltd) 1988 finally brought an end to Barrow's iron and steel industry. New post-War industries included Lister's woollen yarn spinning plant at Roose (opened 1947; closed 1990), Kimberly Clark Paper Mill (1967; still in production) and gas terminals for Morecambe Bay gas field.

Places of worship. Chapel of ease at Rampside built or rebuilt 1621; rebuilt 1840 with additions 1866 (porch and vestry) and 1892 (chancel). School at

Newbarns, built 1843, used for worship until St George's Church, St George's Square, was built in new town (consecrated 1861; extended 1867). Further Anglican churches of St James, Hartington Street, built 1867-9, and St Paul, Abbey Road, built 1871 (western end completed c.1965 and 2005-06). Four new prefabricated churches opened 1878 when town's population was growing rapidly: St Matthew, Harrogate Street (rebuilt 1965-8); St Mark, Rawlinson Street, (extended 1882-3); St Luke, Roose Road (rebuilt 1963-4) and St John, Barrow Island, (rebuilt 1934-5). St Perran's, Roose, established as mission church under St Luke's. To cater for town's expansion after 1945, new Anglican churches of St Aidan's, Newbarns, built 1952 and St Francis, Ormsgill, 1955.

Roman Catholic. Church of St Mary of Furness, Duke Street, built 1866-7, with additions 1888 (steeple) and 1894 (baptistery). St Patrick's, Barrow Island, built 1877; rebuilt 1933. School chapel of Sacred Heart, Roose Road, built 1902. Further RC churches opened at Newbarns (Holy Family) 1951, and Ormsgill (St Pius X) 1957.

Nonconformist. Barrow's rapid growth saw chapels for numerous denominations, reflecting diversity of migrants. Wesleyan Methodist chapel, Hindpool Road (built 1862; closed 1935), followed by further Wesleyan chapels at Hartington Street (1874; closed 1951); Roose Road (1875; closed 1976); Abbey Road (1876; rebuilt 1902); Greengate Street (1876; replaced by new building 1904; known from 1953 as Central Methodist Church; closed c.1996; demolished), and Stonedyke (1877; closed 1991). United Methodist Free Church, Storey Square, opened 1894 (closed 1952). Mission rooms also established at Hawcoat (closed 1989) and Piel. Primitive Methodist chapels in Forshaw Street (1866; closed 1953); Hartington Street (1874), and Marsh Street (1875; closed 1953). Methodist New Connexion chapel (Christ Church), Abbey Road, built 1875; became United Methodist Church in 1907; destroyed in blitz 1941; remaining assets used to build Beacon Hill (Christ Church) Methodist Church, Holyoak Avenue, 1956. Baptist Church, Abbey Road, built 1873; destroyed in blitz 1941; replaced by new building on new site near Barrow Park 1953. Congregationalist chapels at Hindpool Road (built 1857; enlarged 1863; closed 1931) and Ainslie Street (built 1873). Trinity Presbyterian church, School Street, opened 1875; closed 1971. In 1991 Abbey Road Methodist and Ainslie Street Congregationalist churches merged to form Trinity Church Centre, using Abbey Road building. Spiritualist church, Buccleuch Street, built 1893; Salvation Army Citadel, Abbey Road, built 1910.

Synagogue opened in Abbey Road 1902; moved to Crellin Street 1918 and to School Street 1925; closed 1974.

Cemetery with Episcopalian, Dissenting and RC mortuary chapels opened 1873; crematorium opened 1962.

Schools. Curate at Rampside chapel taught children in 1720s; small schoolroom near chapel, recorded mid-19th century, had gone by 1900. National school built at Newbarns 1843; closed and demolished 1877. School at Rabbit Hill opened 1849 by Furness Railway Co. for children of its workers; now St George's Primary School. Board schools opened 1873 (initially in temporary buildings) in Holker Street, Cambridge Street, Barrow Island, Hawcoat, Rawlinson Street, and Roose. Higher Grade School, Abbey Road, opened 1879, moving to Duke Street 1888, becoming Alfred Barrow School. School of Science and Art, operated privately from 1877, taken over by council 1891 and housed in new Technical School building, Abbey Road, from 1903. Roman Catholic convent school, Nelson Street, established 1897; moved to Holker Street 1905 and to Crosslands 1929, where it became known as Our Lady's Convent School. By 1924 there were 16 council schools in Barrow, and a further three Roman Catholic schools. New school built at Risedale 1926, and grammar school for boys opened 1930, followed by girls' grammar school 1932. Primary schools established on new council estates at Greengate (1950), Ormsgill (1951) and South Newbarns (1953) and new St Paul's CE School 1957. St Aloysius' RC Secondary School opened 1953. New primary schools built at Yarlside and Hawcoat (1972) and St Mary's RC Primary was replaced by two new schools: St. Pius X, Schneider Road, and Holy Family, Newbarns, 1974. Roman Catholic preparatory school, forerunner of Chetwynde School, opened at Crosslands 1945.

Technical School expanded to create new Central College of Further Education, which moved to new building in Howard Street 1954. Technical School split from College again 1964, moving to new building in Thorncliffe Road. Reorganisation following introduction of comprehensive system 1979 created four comprehensive schools: Grammar School and, briefly, Risedale sites became Parkview School; Technical School became Thorncliffe School; St Aloysius' RC School merged with Our Lady's Convent School to become St Bernard's Catholic High School, and Alfred Barrow School retained its name. New Sixth Form College, Rating Lane, also opened 1979. Alfred Barrow, Thorncliffe, and Parkview Secondary Schools amalgamated 2009 to form Furness Academy. Further education provided at Furness College on new site at Channelside, opened 2012.

Other institutions. Market and town hall built 1866; new Town Hall, Duke Street, completed 1887. Other public buildings in centre of new town included Working Men's Club and Institute, built 1870-1, and Ramsden Hall, built as public bath 1871; converted to public hall 1886.

Workhouse opened at Roose 1880; separate accommodation for children in Cottage Homes, Roose Road, from 1905. Workhouse became Roose Hospital for geriatric and elderly care, 1948; closed 1993; building subsequently demolished. Temporary hospital opened in private house 1866; replaced by North Lonsdale Hospital, School Street, opened 1875; closed 1984 (subsequently demolished) when new Furness General Hospital, Dalton Lane, opened. Other hospitals included Isolation Hospital, Devonshire Road (opened 1882; closed c.1984), Smallpox Hospital at Rakesmoor (1903), Risedale Maternity Hospital (opened 1921; closed c.1984) and Infant Clinic, Abbey Road (opened 1934).

Library opened in temporary building 1882; transferred to newly opened Town Hall 1887. New purpose-built library opened in Ramsden Square 1922, which also housed town's museum until 1994 when relocated to Dock Museum, on site of redundant Graving Dock. Branch libraries at Salthouse and Victoria School (opened 1940) and Barrow Island (1944).

Theatre Royal, Albert Street, opened 1864; became Her Majesty's Theatre 1952; closed 1968. Music halls: The Alexandra, Forshaw Street, (1866; converted to Regal cinema 1931) and Alhambra Palace, Cavendish Street (1872; converted to Odeon, later Classic, cinema 1937; closed 1976). Other cinemas (of which there were eight in 1950s) included Ritz cinema, on corner of Holker Street and Abbey Road, opened 1936; closed and demolished 1999; replaced by Apollo cinema, Cornerhouse Park. Live performances held at Forum 28 Arts Centre, Duke Street.

Barrow Public Park opened on 45-acre [18-ha] site in Abbey Road 1901; laid out by Thomas Mawson 1904-07 and extended 1920; leisure centre built there 1991, and park restored to its original design 2005.

BARTON

CP in Barton ancient parish, West ward, Westmorland, comprising townships of High Barton and Low Winder. Barton Fell added to CP 1935 (previously divided between Sockbridge and Barton CPs).

Acreage: Barton CP contained 2,797 acres [1,132 ha], divided between former townships thus: High Barton: 2,569 acres [1,040 ha]; Low Winder: 301 acres

[122 ha]. Barton Fell covered 1,718 acres [695 ha]. ***Population:*** fluctuating in range 250-350 across 19th and 20th centuries, with peak of 388 in 1921; stood at 232 in 2001. ***Landownership:*** *manor of Barton* passed from Lancaster family (barons of Kendal) to Multons of Gilsland, then by marriage to Dacres, before being sold to Sir Christopher Musgrave of Edenhall in late 17th century; later purchased from Musgraves by Hasells of Dalemain. ***Economy***: predominantly agricultural until 19th century; focal point was Pooley Bridge, where fish market was held (had ceased by 1860) and two livestock fairs (one for sheep, cattle and horses in September; one for tups (rams) in October); both still held 1885. Tourism increasingly important by later 19th century: paddle steamers calling at pier at Pooley Bridge from 1859; hotels, lodging houses and pleasure boats by 1885. ***Places of worship***: medieval parish church of St Michael; most recently restored 1904. St Paul's chapel of ease, Pooley Bridge, built by subscription 1867-8. ***Schools and other institutions:*** free endowed grammar school, founded 1648 by Gerard Langbaine, provost of Queen's College, Oxford; became public elementary school in new premises c.1873 (closed; converted to dwelling). Parkin Memorial Hall, Pooley Bridge, built 1911.

BASSENTHWAITE

Ancient parish in Allerdale below Derwent ward, Cumberland.

Acreage: 6,915 acres [2,798 ha], including 3,630 acres [1,469 ha] of commons and stinted pasture, enclosed in 1771. Divided into two constablewicks, Highside and Lowside (or Hawes). Bassenthwaite and Skiddaw Common (1,018 acres [412 ha]) remains unenclosed fell grazing. ***Population:*** estimated at 530 in 1688. Rose from 450 in 1801 to peak at 557 by 1851; had fallen back to 445 by 1901; then remained fairly steady across 20th century to stand at 412 in 2001. ***Landownership:*** *manor of Bassenthwaite* was divided between two daughters of Adam de Bassenthwaite in early 14th century, one half descending through Irton family to Leighs of Isel and thus to Lawsons; other moiety through Martindale family until forfeited to Crown on attainder of Roger Martindale c.1570, after which it was granted to earl of Derby. Derby share passed to John, Lord Ashburton in 1714 on his marriage to Henrietta Stanley; he enfranchised tenants 1715. Other landowners included Highmores of Armathwaite (sold in 18th century to Fletcher-Vanes) and Speddings of Mirehouse. ***Economy:*** predominantly agricultural. Woollen cloth manufacture ('Skiddaw grey') and blue slate quarries in 17th century; several fulling mills. Antimony mine opened in 1820s at Robin Hood, but proved uneconomic;

had closed by late 1840s. ***Places of worship:*** medieval parish church of St Bega beside lake near Mirehouse (rebuilt c.1800; restored 1874). Hawes (or Halls) chapel, closer to village (Bassenthwaite Halls), built as chapel of ease 1471; rebuilt 1805. Replaced by new church of St John, nearby, 1878. Primitive Methodist chapel opened 1868; bought by Wesleyans 1882; extended 1930s; remains in use. ***Schools and other institutions:*** school near Hawes chapel in early 19th century; forerunner of schoolroom attached to chapel, dated 1862. Board school built 1874; now Bassenthwaite Primary School. Reading room and library, adjacent to Methodist chapel, built 1874 (closed; converted to dwelling). Village halls: Parish Room, built 1908 for Skiddaw Humane and Friendly Society; former Hawes chapel used as church room.

BAYCLIFF *see* Aldingham

BEAUMONT

Ancient parish in Cumberland ward, Cumberland. CP enlarged by absorbing Grinsdale and Kirkandrews-on-Eden CPs 1934.

Acreage: 1,628 acres [659 ha]. Beaumont Moor enclosed 1680 or 1690. Common fields at Corn Holme and Lee Holme enclosed 1781. ***Population:*** estimated at 140 in 1688; 219 in 1801; peaked at 323 in 1821, then declined steeply after 1861 to stand at 188 in 1931 (last census year for which separate figure available). ***Landownership:*** *manor of Beaumont* held by Brun family, lords of Bowness, whose female heirs were bought out by Dacres, barons of Burgh, in 14th century; descended with Burgh barony thereafter. ***Economy:*** soil rich; mostly arable farming in later 17th century, with fishing and fowling on marsh. Jetties on River Eden at New Sandsfield. Brick and tile works operated 1830s to 1870s. Reservoir for Carlisle Canal (1823) at Monkhill Lough. ***Places of worship:*** medieval parish church of St Mary; restored 1887-9; has served Kirkandrews-on-Eden as well since 1692. Wesleyan Methodist chapel at Monkhill built 1858; rebuilt 1904; still in use. ***Schools and other institutions:*** schoolhouse but no master in 1814 – parish was served by master at Kirkandrews.

BECKERMET ST BRIDGET

Ancient parish in Allerdale above Derwent ward, Cumberland; also known as Greater Beckermet. Parish included village of Calderbridge and hamlet of Sellafield.

Acreage: 5,160 acres [2,088 ha], including detached portion at Bleng Tongue (97 acres [39 ha]), in fells north of Gosforth. ***Population:*** estimated at 485 in 1688; rising from under 500 in 1801 to 680 in 1871, before declining to under 600 by 1901. Spiked temporarily at 1,019 in 1951 during construction of Calder Hall power station; then fell back to 620 in 1961. Stood at 385 in 2001. ***Landownership:*** village of St Bridget Beckermet was bond settlement held directly from lords of Egremont barony. Bulk of parish (stretching from Sellafield to Cold Fell and Stords Fell) formed *Calder Lordship*, held by Calder Abbey, which subsequently belonged to Lowther family. Calder Abbey was bought by John Tiffin 1730, passing through his daughter by marriage to Senhouse family and Thomas Irwin (d. 1877). ***Economy:*** largely agricultural until 20th century. Sandstone quarrying (Stakes bridge quarry). Windscale (from 1981 Sellafield) nuclear complex originated in Royal Ordnance Factory, Sellafield, manufacturing TNT during Second World War, which was chosen as site for Britain's first nuclear reactor pile 1947. ***Places of worship:*** medieval parish church of St Bridget, known as Old Church or Low Church; pre-Conquest crosses in churchyard suggest early foundation. New church of St Bridget at Calder Bridge built 1842. Ruins of Cistercian Calder abbey, founded 1134. ***Schools and other institutions:*** CE school in Beckermet opened 1857-8; closed 1965. School at Calder Bridge by 1860; rebuilt 1894; closed 1970. Old school at Calder Bridge had been converted into library and recreation room by 1900. Calder Bridge and Ponsonby village hall, at Calder Bridge, built c.1950.

BECKERMET ST JOHN

Ancient parish in Allerdale above Derwent ward, Cumberland, including hamlet of Carleton and modern settlement of Thornhill.

Acreage: 2,946 acres [1,192 ha], including commons at Brisco and Ullcoats, enclosed under Egremont enclosure award 1783, and smaller commons at Carleton Moor and Grange Brow (188 acres [76 ha]), enclosed 1816. ***Population:*** estimated at 430 in 1688. Remained usually under 500 until 1870s, when it began to grow, reaching 682 by 1891; rapid rise across 20th century as Thornhill village (built for mine workers in 1920s) grew, to stand at 1,076 by 1931 and 1,925 by 2001. ***Landownership:*** *manor of Beckermet* held by Fleming family of Rydal from 12th century. ***Economy:*** farming; village noted for bacon-curing in 19th century. Iron ore mining by late 19th century, with expansion in early 20th: mines at Beckermet (closed 1973), Ullcoats,

Florence (sunk 1914; closed 1968). Sandstone quarrying. ***Places of worship:*** medieval parish church of St John the Baptist; early origin suggested by survival of pre-Conquest sculpture; rebuilt 1810 and again 1878-9. Thornhill Mission opened 1932-3. Methodist chapel at Thornhill opened 1928; closed 1944. ***Schools and other institutions:*** parochial school was in Beckermet St Bridget (q.v.). New CE Primary School built in St John Beckermet 1965 and extended 1975. Thornhill Primary School built 1927.

BEETHAM

Township in Beetham parish, Kendal ward, Westmorland. CP enlarged by absorbing Farleton and Haverbrack CPs 1935.

Acreage: 5,170 acres [2,092 ha], including hamlets of Arnside (q.v.), Hale, Hazelslack & Storth, and Whasset. Commons (930 acres [376 ha]) enclosed 1821. ***Population:*** (figures include Arnside until 1901): rising from 668 in 1801 to 1,196 by 1891. After loss of Arnside, stood at 619 in 1901; rose with growth of Storth to stand at 843 in 1931 (last census year before further boundary change). ***Landownership:*** *manor of Beetham* was in hands of Beetham family until 15th century when it passed (by purchase or forfeit after battle of Bosworth) to Stanley family, earls of Derby. Passed to Cliffords, who sold it 1767 to Daniel Wilson of Dallam Tower (in Haverbrack, q.v.) in whose family it descended. ***Economy:*** farming, fishing and cockling, salt-making (as recorded in place-name Salt Cote). Kent estuary navigable up to Storth, where there were wharves in early 19th century; coastal trade declined after building of Arnside railway viaduct 1857 and resulting silting. Charter for market and fair granted 1311. Paper-making at Waterhouse mill from 18th century; continues as modern factory making specialist paper and cardboard. Limestone quarrying, lime-burning and gravel pits in 19th century. Garden centre established 1984. ***Places of worship:*** medieval parish church of St Michael (standing in Haverbrack township); restored 1873-5. Supposed medieval chapel of St John in Beetham. Mission room at Storth by 1897; became All Saints Mission Church (closed; converted to dwelling). Non-denominational chapel at Storth built early 20th century; purchased by Methodists 1935; now houses Storth Village Church (combined Anglican and Methodist). ***Schools and other institutions:*** ancient endowed grammar school; school built 1663; rebuilt at south end of village 1827; rebuilt again on adjacent site (on site of former toll house) 1905; now Beetham CE Primary School. Former (1827) building in use as reading room in 1911; purchased for use as community theatre (Heron Theatre) 1985. Storth school opened 1875; rebuilt on new site mid-20th century as Storth CE Primary School. Storth Institute (containing reading room and billiard room), replaced 1929 by village hall.

BELLBANK

Township in Stapleton parish, Eskdale ward, Cumberland (not to be confused with Bellbank township in Bewcastle parish (q.v.), which did not become a separate CP). Absorbed into Stapleton CP 1934.

Acreage: 1,446 acres [585 ha]. ***Population:*** fluctuating around 120-130 in first half of 19th century; then declining to 62 in 1931 (last census year for which separate figure available). ***Landownership:*** manorial status unclear; divided into small units of landownership by 19th century. ***Economy:*** largely agricultural; limekilns on banks of White Lyne and lead mine near Dodgsonstown in 1860s. ***Places of worship:*** none. ***Institutions:*** none.

BERRIER AND MURRAH

Township in Greystoke parish, Leath ward, Cumberland. Absorbed into Mungrisdale CP 1934.

Acreage: 2,552 acres [1,033 ha]. Commons enclosed, with others in Greystoke manor, 1796. ***Population:*** fell from 136 in 1801 to below 100 by 1901; standing at 86 in 1931 (last census year for which separate figure available). ***Landownership:*** part of barony of Greystoke. ***Economy:*** predominantly agricultural. Thomas Denton, writing 1688, noted a 'very good horse course' on moor at Berrier. Small scale coal mining before 19th century. Lime-burning in 19th century. ***Places of worship:*** none. ***Schools and other institutions:*** Berrier Endowed School built 1828; closed 1920s.

BEWALDETH AND SNITTLEGARTH

Township in Torpenhow parish, Allerdale below Derwent ward, Cumberland. CP enlarged by taking in part of Torpenhow & Whitrigg CP 1934.

Acreage: 1,607 acres [651 ha], in two parts, Snittlegarth forming a detached portion of 328 acres [133 ha]. ***Population:*** rising from 55 in 1801 to 95 in 1861; then declining to stand at around 40 from 1931 to 2001.

Landownership: in hands of Mulcaster family in 14th century; sold by Robert de Mulcaster to Robert de Highmore, in whose family it continued until sold with Armathwaite Hall estate (in Bassenthwaite) by Benson Highmore (d. 1767) to James Spedding; later sold to Sir F. Fletcher-Vane of Hutton-in-the-Forest and remained with his descendants in 1901. ***Economy:*** predominantly agricultural. ***Places of worship:*** Quaker meeting house on common, built 1687; closed 1790 (site marked by burial ground at 'Meetinghouse Close'). ***Schools and other institutions:*** none.

BEWCASTLE

Ancient parish in Eskdale Ward, Cumberland, containing four quarters or townships of Bailey, Bellbank, Bewcastle and Nixons.

Acreage: 28,563 acres [11,559 ha], divided between constituent townships thus: Bailey: 11,833 acres [4,789 ha]; Bellbank: 2,937 acres [1,189 ha]; Bewcastle: 6,204 acres [2,511 ha] and Nixons: 7,389 acres [2,990 ha]. Parish contained extensive stretches of moorland and hill common, of which Baileyhope stinted pasture (c.4,000 acres [c.1,600 ha]) was enclosed 1814; Highstone common (c.1,500 acres [c.600 ha]) 1815; Blacklyne common (c.3,500 acres [c.1,400 ha]) 1820; Arthurside stinted pasture (c.400 acres [c.160 ha]) 1835; and Greyfell common (3,112 acres [1,259 ha]) 1849. White Lyne Common (c.2,900 acres [c.1,175 ha]) remained unenclosed. ***Population:*** estimated at 420 in 1688. Rising from 917 in 1801 to 1,336 in 1831, then falling rapidly to c.600 by end of 19th century and declining further to 411 in 2001. ***Landownership:*** bulk of parish formed *manor of Bewcastle*, which was part of barony of Burgh by Sands c.1180; in crown hands c.1461; granted to Grahams of Netherby 1630. Bailey township lay in *manor of Nichol forest*, probably formed c.1604-20 by Cliffords, earls of Cumberland, who sold it to Grahams of Netherby 1628. Both manors remain in hands of Graham family. ***Economy:*** livestock farming (said in late 17th century to be 'very good land for herbage and meadow, but grows little corne'). Weekly market and two cattle fairs granted 1279. Lime, freestone and roadstone quarries; coal mining at Oakshawford in Bellbank township; lime-burning and small scale drain tile factory at Kinkry Hill, 1860-90s. Extensive afforestation, both Forestry Commission and private, in later 20th century. ***Places of worship:*** pagan worship suggested by name of Roman fort, *Fanum Cocidii* ('shrine of Cocidius'); early Christian site within former Roman fort, attested by Bewcastle cross of late 7th- or 8th-century date, forerunner of medieval parish church of St Cuthbert; partly rebuilt 1792. Kinkry Hill school used as mission room in mid-19th century. Presbyterian meeting house in Nixon's township, built 1788; demolished and replaced by new church in Bellbank township 1890-1; remains in use. Bewcastle House of Prayer, a modern religious community, established c.2012. ***Schools and other institutions:*** schoolmaster recorded 1664 and school recorded 1703. Three schools recorded 1818; two in 1828 and 1847, one of which was at Nook, built 1827. Two new schools built 1854-5, close to Bewcastle church and at Kinkry Hill in Bellbank township. Both closed when Board school opened in Bailey township 1878; now Bewcastle Primary School. Village hall (Roadhead Public Hall) opened 1908.

BIGGAR *see* Walney Island

BIGRIGG *see* Egremont

BIRKER AND AUSTHWAITE

Township in Millom parish, in Allerdale above Derwent ward, Cumberland. CP abolished 1934, when combined with most of Eskdale & Wasdale CP to create Eskdale CP.

Acreage: 8,213 acres [3,324 ha], including c.7,000 acres [c.2,830 ha] of unenclosed common fell land. ***Population:*** comparatively stable at around 100 inhabitants for most of 19th century (peak: 118 in 1851); declined steadily during early 20th century to stand at 52 in 1931 (last census year for which separate figure available). ***Landownership:*** *Austhwaite* (western part of township) granted before 1215 by lord of Millom to Benedict son of Ketel, whose descendants took surname de Austhwaite. It passed by marriage to Stanley family 1354, in whose hands it remained until bought by Lord Muncaster 1888. Pasture of *Brotherilkeld* (eastern section of township) granted to Furness Abbey 1242. ***Economy:*** hill farming; some copper mining; iron mining at Low Birker in later 19th century. ***Places of worship:*** none (served by parochial chapel of Eskdale). ***Schools and other institutions:*** none.

BLACKWELL *see* St Cuthbert Without

BLAWITH

Chapelry and township in Ulverston parish, Lonsdale hundred, Lancashire North of the Sands. Combined with Subberthwaite to form Blawith and Subberthwaite CP 1986.

Acreage: 2,998 acres [1,213 ha]. ***Population:*** 170 in 1811, rising to 193 in 1861 and gradually falling to 101 in 1981 (last census year for which separate figure available). ***Landownership:*** granted to William de Lancaster, baron of Kendal, c.1160; held jointly by manor of Ulverston and Conishead Priory from 1189; annexed to Duchy of Lancaster 1536; purchased 1736 by duke of Montagu, from whom it descended to dukes of Buccleuch. ***Economy:*** farming; charcoal burning, 12th to 20th centuries; small-scale iron smelting, 13th-14th centuries; small-scale slate quarrying 17th to 20th centuries; pill-box manufacturing, 19th century. ***Places of worship:*** chapel of ease, recorded 1577; rebuilt 1749; replaced 1863 by new church of St John the Baptist on new site close by; declared redundant 1988 and sold 2011. ***Schools and other institutions:*** school taught in chapel in early 18th century. Separate schoolhouse erected 1747; replaced 1859; closed c.1965 and converted to dwelling. Blawith and Nibthwaite village hall at Water Yeat built c.1920.

BLEATARN *see* Warcop

BLENCOGO

Township in Bromfield parish, Cumberland ward, Cumberland. Absorbed into Bromfield CP 1934.

Acreage: 1,770 acres [720 ha], including c.400 acres [c.162 ha] of common, enclosed 1848. ***Population:*** rising from 167 in 1801 to stand at around 220 from 1821 to 1861, before declining to 139 in 1931 (last census year for which separate figure available). ***Landownership:*** one of townships held of barony of Wigton. Granted by Crown to Sir Richard Cholmley during reign of Henry VII; held by Anthony Barwis 1578. Later came to Thomlinson family (Blencogo Hall appears to have been in their hands by early 17th century); belonged to George Reay 1901. ***Economy:*** predominantly agricultural. ***Places of worship:*** mission hall at Leegate opened 1888 (closed and converted to dwelling). ***Schools and other institutions:*** school, founded and endowed by George Dixon 1854; now used as Blencogo & District Village Hall.

BLENCOW, GREAT *see* Dacre

BLENCOW, LITTLE *see* Greystoke

BLENNERHASSET AND KIRKLAND

Township in Torpenhow parish, Allerdale below Derwent ward, Cumberland. CP abolished 1934 to create Blennerhasset & Torpenhow CP.

Acreage: 1,262 acres [511 ha]. ***Population:*** 204 in 1801, fluctuating at around 230 until 1861, then rising to peak of 503 in 1891 before falling back to 330 by 1931 (last census year for which separate figure available). ***Landownership:*** *manor of Blennerhasset and Upmanby* (the latter in Allhallows parish) was in hands of Mulcaster family by c.1280; then passed by marriage to Tilliols in early 15th century and from them to Moresbys. By later 17th century held by Sir Francis Salkeld of Whitehall and descended with Whitehall to Charltons of Hesleyside (Northumb.), who held it in 1901. *Manor of Kirkland* was held by convent of Rosedale (Yorks.), passing at Dissolution to Salkelds of Whitehall, with which estate it then descended. By later 19th century, Lawson family of Brayton were major landowners in township. ***Economy:*** largely agricultural; Blennerhasset Farm (now Mechi Farm) was model farmstead built 1863-4 by William Lawson. Market charter granted 1661, with market on Saturdays and fairs on Maundy Thursday and on 18 October. ***Places of worship:*** Independent (Congregational) chapel by 1704; rebuilt 1828 and again 1864 on land donated by Sir Wilfred Lawson; closed; converted to dwelling. Primitive Methodist chapel (tin tabernacle); now Blennerhasset Evangelical Mission. ***Schools and other institutions:*** school by 1860; became Board school; now Blennerhasset Primary School. Reading room and library built by Sir Wilfred Lawson 1890. Village hall built late 1940s by local unemployed men.

BLINDBOTHEL (otherwise Moorside)

Township in Brigham parish, Allerdale above Derwent ward, Cumberland. CP enlarged by absorbing Mosser and Whinfell CPs 1934.

Acreage: 1,261 acres [504 ha]; 1,000 acres [405 ha] of commons in Eaglesfield manor, which included those in Blindbothel township, enclosed 1815. ***Population:*** 82 in 1801 rising to 128 in 1851, then declining to 64 in 1931 (last census year before boundary changes). ***Landownership:*** part of Eaglesfield, one of Five Towns of honour of Cockermouth (q.v.). Consisted of small freehold estates by 16th century. ***Economy:*** primarily agricultural. ***Places of worship:*** none. ***Schools and other institutions:*** Paddle School, on former common land allotted at enclosure for establishment of school; now Eaglesfield CE Primary Academy.

BLINDCRAKE, ISEL AND REDMAIN

Township in Isel parish, Allerdale below Derwent ward, Cumberland. Absorbed Isel Old Park and Sunderland CPs and renamed Blindcrake CP 1934, thus uniting whole of ancient parish of Isel into one civil administrative unit.

Acreage: 4,251 acres [1,720 ha]. Commons in Isel parish, at Moota, Sunderland Common and in Isel Old Park, totalling 1,600 acres [647 ha], enclosed 1813. Land at Highmoors had been enclosed 1724 and 200 acres [81 ha] of Moota Common said 1688 to have been 'lately enclosed'. ***Population:*** rising from 188 in 1801 to 370 in 1851, then falling back to 233 in 1931 (last census year before boundary change). ***Landownership:*** *Redmain* was granted to Gisburn priory (Yorks.), passing after Dissolution to Curwens of Camerton from whom it was sold to Sir Wilfrid Lawson. Rest of ancient parish formed *manor of Isel*, which had been inherited by William de Multon by 1311. It passed by his daughter to Leigh family, who were responsible for building work at Isel Hall, through whom it passed to Thomas Leigh (d. 1573), whose wife Maud (née Redman) conveyed it by marriage to her third husband, Sir Wilfrid Lawson (d. 1632). The manor descended through Lawson family until 20th century. ***Economy:*** farming; limestone quarrying and lime-burning. Moota Quarry continues in operation. ***Places of worship:*** parish church lay in Isel Old Park township (q.v.). Minor name 'Chapel Guards' at Redmain may record early religious site there. Primitive Methodist chapel in Blindcrake, converted from dwelling house 1894; closed and converted back to dwelling. Reading room used for Anglican services. ***Schools and other institutions:*** endowed school between Isel Hall and Blindcrake, recorded 1814; rebuilt as National school 1836; closed (converted to dwelling). Lending library established c.1840. Blindcrake Mission and Reading Room created 1877 from two cottages; now village hall.

BOLTON (Cumberland)

Ancient parish in Allerdale below Derwent ward, Cumberland, comprising townships of Bolton High and Bolton Low, which were united to form Boltons CP 1887. Main centres of population were Boltongate (location of parish church), Bolton Low Houses and mining village of Mealsgate.

Acreage: 8,456 acres [3,422 ha], divided between townships thus: Bolton Low: 4,523 acres [1,830 ha]; Bolton High: 3,933 acres [1,592 ha]. Extensive commons, totalling 5,178 acres [2,095 ha], enclosed 1781. ***Population:*** rising from 695 in 1801 to 1,245 in 1831, then slow decline to c.900 by end of 19th century and to 501 in 1981; slight increase to 585 by 2001. ***Landownership:*** *manor of Bolton* was in hands of Lascelles family by early 13th century; male line failed c.1260. In Crown hands by 1318, having been forfeited by Sir Roger de Mowbray for his support for Robert Bruce. Returned to Mowbray family, but had passed to Nevilles of Raby by 1388, through whom it descended until 16th century, when it passed by marriage of Katherine, daughter and heir of John Neville, Lord Latimer, to Henry Percy, 8th earl of Northumberland, from whom it descended with other Percy estates in Cumberland to Wyndhams. ***Economy:*** farming; coal mining by 16th century (Bolton colliery supplied coal to Mines Royal from 1560s); had ceased by 1900. Extensive limestone quarrying and lime-burning in mid-19th century. Copper mine tried c.1825 but not very fruitful; tile works in mid-19th century. ***Places of worship:*** medieval parish church of All Saints in Boltongate, 'one of the architectural sensations of Cumbria' (Hyde & Pevsner). Quaker meeting house at Bolton Low Houses, built 1702; closed 1874; demolished 1904. Wesleyan Methodist chapel at Bolton Low Houses built 1817; rebuilt 1881; closed c.2000 (converted to dwelling). Mission room at Bolton Low Houses by 1929. ***Schools and other institutions:*** several unendowed schools in early 19th century. Parochial school at Boltongate built 1854; closed (now parish room, the Bell Institute); replaced by Boltons CE Primary School at Bolton Low Houses. Reading room built 1885.

BOLTON (Westmorland)

Chapelry and township in Morland parish, West ward, Westmorland.

Acreage: 2,791 acres [1,130 ha]. Commons (557 acres [225 ha], including 22 acres [9 ha] of open fields, enclosed 1813. ***Population:*** 324 in 1801; rising slowly to peak of 404 in 1881. Declined thereafter, falling to 262 in 1921;

recovered during later 20th century to stand at 416 in 2001. ***Landownership:*** held in 1314 by Ralph de Greystoke; passed through several families before being sold to Lowthers. Estate at Bewley granted to bishops of Carlisle c.1170; they built Bewley Castle as retreat in 14th century. ***Economy:*** predominantly agricultural. ***Places of worship:*** medieval chapel of ease, recorded 12th century; now church of All Saints; restored 1848. Primitive Methodist chapel built 1818; restored 1926; still in use. ***Schools and other institutions:*** school endowed 1721; rebuilt on same site 1856, and included reading room for township. Bolton Primary School now in modern building on different site. Eden Grove specialist residential school for children with learning and behavioural difficulties opened 1955; closed 2013. Memorial Hall built 1922.

BOLTON WITH ADGARLEY *see* Urswick

BOOTLE

Ancient parish in Allerdale above Derwent ward, Cumberland.

Acreage: 6,786 acres [2,746 ha], including 1,113 acres [450 ha] of common on Bootle Fell, enclosed 1857, part of which (705 acres [285 ha]) remains registered as common land. Coastal moorland at Hycemoor (460 acres [186 ha]) enclosed 1801; sand dunes at Eskmeals (955 acres [386 ha]) enclosed 1870. ***Population:*** estimated at 555 in 1688; 547 in 1801, rising to 901 in 1861, then levelling at around 800 for most of 20th century (peak: 1,007 in 1961). ***Landownership:*** *manor of Bootle* was held by Copeland family in 13th century but appears to have been divided between heiresses. In 17th century was in hands of Hudleston, Pennington and Senhouse families. ***Economy:*** formerly a small market town, with market charter granted 1347; market had ceased by mid-19th century, though cattle fairs and hiring fairs for servants continued then. Parish largely agricultural; tile works at Skelda Hill in mid-19th century (had ceased by 1900). Eskmeals (sand dunes at northern end of parish) was rabbit warren by 17th century and was acquired as naval gunnery range by Vickers of Barrow 1897. Royal Ordnance Factory established at Hycemoor 1941-2, with camp for 500 workers near Bootle Station. ***Places of worship:*** medieval parish church of St Michael; transepts added 1837 and tower built 1850s to 1882. Independent chapel, built 1780 for Countess of Huntingdon's Connexion; became United Reformed Church; restored 1993; has been independent Evangelical Church since c.1995. ***Schools and other***

institutions: grammar school at Hyce Moor endowed 1713; rebuilt 1847; closed 1991. School in Bootle, endowed by naval captain Isaac Shaw, built 1830; now Captain Shaw's CE Primary School. Bootle Union workhouse built 1856-7, replacing older workhouse in Bootle village. Main workhouse building demolished later 20th century; infirmary block converted to dwelling.

BORROWDALE

Chapelry and township in Crosthwaite parish, Allerdale above Derwent ward, Cumberland.

Acreage: 18,663 acres [7,553 ha], much of which remains unenclosed common fell, including Langstrath and Coombe Fells (3,957 acres [1,602 ha]). ***Population:*** rising from 342 in 1801 to 497 in 1901; stood at 438 in 2001 (increased seasonally by tourism). ***Landownership:*** *manor of Borrowdale* was granted out of forest of Derwentfells in 1211 to Furness Abbey and Fountains Abbey, who held it until Dissolution. Sir Wilfred Lawson of Isel Hall purchased it 1614 and it descended in Lawson family. ***Economy:*** livestock farming (monastic vaccary at Stonethwaite recorded 1302); sheep fair at Rosthwaite each September until late 19th century; revived 1989. Coppice woodland producing charcoal and hazel nuts (Borrowdale noted for latter in late 17th century). Lead and copper mining from 16th to 19th century. Slate quarrying at Quayfoot and Honister quarries. 'Black lead' (graphite or 'wad') mining from 16th century; formed basis of Keswick pencil industry; mine closed 1891. Tourism increasingly important from 1770s, with viewing station on Castle Crag; Bowder Stone and waterfall at Lodore became ports of call on circuit from Keswick. Hotels and other visitor accommodation by mid-19th century. Modern economy heavily reliant on tourist trade. ***Places of worship:*** chapel of ease (now St Andrew's Church) near Stonethwaite, recorded 1505; consecrated 1687; rebuilt 1824-5; chancel added 1873. Holy Trinity Church at Grange built by Margaret Heathcote; consecrated 1861. Wesleyan Methodist chapel at Grange built 1894. ***Schools and other institutions:*** school in west end of chapel in early 19th century; separate schoolroom built nearby 1825; now Borrowdale CE Primary School. School at Grange built in memory of Margaret Heathcote 1894 (now Climbers' Club hut). Rosthwaite Village Institute founded 1960s.

BOTCHERBY *see* St Cuthbert Without

BOTHEL AND THREAPLAND

Township in Torpenhow parish, Allerdale below Derwent ward, Cumberland.

Acreage: 3,389 acres [1,372 ha], including extensive commons: Bothel and Torpenhow Common (2,200 acres [890 ha]) enclosed 1811; Threapland Common (500 acres [202 ha]) enclosed 1813. ***Population:*** 313 in 1801, rising to peak of 495 in 1851, then settling in range 350-400 for much of later 19th and early 20th centuries. Fell to low point of 284 in 1971; then rose to stand at 438 in 2001. ***Landownership:*** *manor of Bothel* was in hands of Brun family by 13th century; divided between three heiresses in 14th century, who married into Harrington, Curwen and Bowet families. Harrington and Curwen shares had been acquired by Salkeld family of Whitehall by later 17th century, from whom manorial rights descended to Charltons of Hesleyside (Northumb.) in mid-18th century. *Manor of Threapland* passed in 1322 from Mulcastre family to Henry de Malton, whose heiress married Thomas Skelton. It descended through Skeltons until Lancelot Skelton of Armathwaite sold it c.1600 to John Salkeld, a younger son of Salkelds of Whitehall. ***Economy:*** predominantly agricultural. Freestone and limestone quarrying and lime-burning recorded in 19th century, but name Wharrels Hill ('quarry hill') implies quarrying from medieval period; wind farm there built 2007. ***Places of worship:*** Wesleyan Methodist chapel in Bothel built 1840; rebuilt 1890. Primitive Methodist chapel in Bothel built 1836 (now demolished). ***Schools and other institutions:*** school in Bothel by 1686, when it was endowed as free school. Rebuilt on new site 1989; now St Michael's CE Primary School. Library and reading room in village by 1900.

BOWNESS-ON-SOLWAY

Ancient parish in Cumberland ward, Cumberland, containing townships of Anthorn, Bowness (including village of Port Carlisle), Drumburgh (with hamlets of Easton and Glasson) and Fingland (including Whitrigg).

Acreage: 11,176 acres [4,523 ha], including extensive areas of former moorland and moss. Parish was divided between constituent townships thus: Anthorn: 3,034 acres [1,228 ha], including 2,000 acres [809 ha] of commons enclosed 1842; Bowness: 2,985 acres [1,208 ha], including 1,500 acres [607 ha] of commons enclosed 1810; Drumburgh: 2,964 acres [1,199 ha], including Glasson commons (400 acres [162 ha]), enclosed 1816 and Drumburgh commons and moss enclosed 1890; and Fingland: 2,194 acres

[888 ha], including Whitrigg common and marsh, enclosed 1837 and 1850. Some mossland remains registered as common land, including Drumburgh Moss (247 acres [100 ha]); Whitrigg Marsh and Moss (302 acres [122 ha]) and peat moss at Bowness (97 acres [39 ha]). ***Population:*** estimated at 830 in 1688; rising from 825 in 1801 to peak of 1,488 in 1841, reflecting growth of Port Carlisle. In 1841 population was divided between townships as follows: Anthorn: 207; Bowness: 624; Drumburgh: 470; Fingland: 187. After subsidiary peak of 1,369 in 1881, numbers drifted down to around 1,000 by 1930s, apart from temporary (probably War-related) peak of 1,812 in 1951, to stand at 1,221 in 2001. ***Landownership:*** *manor of Bowness* held of barony of Burgh by Brun (de Feritate) family until death of Robert le Brun (c.1395), when divided between heiresses. Seems to have been reunited with barony by purchase by Thomas Dacre in 16th century, though part descended to Grahams of the Fold until 1633, when George Graham released his rights in Bowness and Cardurnock to Thomas, earl of Arundel, then lord of Burgh barony. ***Economy:*** predominantly farming and fishing (including traditional 'haaf net' fishing in Solway estuary) until growth of trading and coastal resort functions in early 19th century.

PORT CARLISLE (formely Fisher's Cross) developed as small port: harbour built 1819; canal opened 1823 but closed 1853 and replaced by Port Carlisle Railway, opened 1854 (closed 1932). Success proved to be short-lived after railway to Silloth opened 1856; trade had 'entirely collapsed' by 1873.

Bowness and Port Carlisle had developed as small coastal resorts by 1840s, with genteel bathing and lodging houses. Caledonian & North British Railway opened 1854, with Solway viaduct (closed 1921) giving direct link to Scotland. Despite contraction of transport links, parish still retained services such as shops, hauliers, lodgings and tennis club in 1938. Other industries included gravel extraction and brick- and tile-making in 19th century, peat extraction from late 19th century. Chemical works built at Whiteholme on Drumburgh Moss 1857, making sulphuric and other acids and distilling tar; in later 19th century (as Cumbrian Manure Works) making artificial manures; continued until mid-20th century. Nearby was an asphalt works in 1860s. Large naval air base (*HMS Nuthatch*) constructed on Cardurnock peninsula during Second World War, opening 1944; closed 1959; radio transmitter masts erected on site 1964. ***Places of worship:*** medieval parish church of St Michael in Bowness village; restored 1891. Mission rooms at Drumburgh, Longcroft and Cardurnock in early 20th century. Wesleyan Methodist chapel at Port Carlisle built 1861 (now Solway Methodist Church), and Wesleyan Home Mission chapel in Bowness built 1872 (closed). Primitive Methodist chapel at Glasson built 1844; closed 1966. Congregational chapel at Anthorn, erected by subscription in 1869 (closed). ***Schools and other institutions:*** Rev. Thomas Pattinson's charity (endowed 1785) supported schools at Bowness and Drumburgh and schoolmaster at Anthorn. Parish school at Bowness, enlarged c.1835 but closed after Board schools opened. Drumburgh school built 1834; re-built 1859. School at Fingland built 'a few years' before 1847; re-opened 1895. School at Easton had closed by 1860. Board schools built at Bowness (now Bowness Primary School) and Anthorn 1875. Village halls at Bowness (Lindow Hall, built 1910), Anthorn and Port Carlisle.

BOWSCALE

Township in Greystoke parish, Leath ward, Cumberland. Absorbed into Mungrisdale CP 1934.

Acreage: 1,262 acres [511 ha], consisting largely of unenclosed common fell land. ***Population:*** hamlet of four farms, containing around 30 inhabitants from 1841 to 1931 (census years for which separate figures available). ***Landownership:*** part of barony of Greystoke (q.v.) until sold to Hasells of Dalemain (before 1860). ***Economy:*** hill farming. ***Places of worship:*** none. ***Schools and other institutions:*** none.

BRACKENTHWAITE

Township in Brigham parish, Allerdale above Derwent ward, Cumberland. Absorbed into Buttermere CP 1934.

Acreage: 4,395 acres [1,779 ha], including c.3,300 acres [c.1,340 ha] of unenclosed common on Brackenthwaite Fell. ***Population:*** around 140 in early 19th century, declining from 1850s to stand at 89 in 1931 (last census year for which separate figure available). ***Landownership:*** part of forest of Derwentfells in honour of Cockermouth. *Manor of Brackenthwaite* was granted c.1160 to Waldeof son of Dolfin, ancestor of Moresby family. Christopher de Moresby exchanged it with Thomas de Lucy for half of Distington manor 1355, thus returning it to lords of Cockermouth (q.v.), in whose hands it remained until forfeited to Crown 1530. Granted with manor of Loweswater 1546 to Lord Grey and Bannister, passing with Loweswater to Lawsons of Isel 1624. John Marshall bought manorial lordship 1814 and, in 1824, an estate including many properties in Brackenthwaite, acquired by

Robert Fisher in 17th century, which descended to Bertie family in 1722. The National Trust and Balliol College bought that estate and lordship of manor 1937. ***Economy:*** hill farming. Annual sheep fair at Lanthwaite Green, recorded 1816. Scale Hill Hotel was posting-house with boats for hire, catering for tourists by 1829. ***Places of worship:*** chapel of St Mary Magdalene in Brackenthwaite, recorded 1453 and 1507 (location uncertain; may have been at Rannerdale). ***Schools and other institutions:*** none.

BRAITHWAITE *see* Above Derwent

BRAMPTON (Cumberland)

Market town and ancient parish in Eskdale ward, Cumberland. CP created in 1883 by union of townships of Brampton, Easby and Naworth.

Acreage. Brampton CP contained 6,466 acres [2,617 ha], combining townships of Brampton: 3,323 acres [1,345 ha]; Easby: 1,269 acres [514 ha] and Naworth: 1,874 acres [758 ha] (excluding Tindale Fell, a detached portion of Naworth township: see Midgeholme). Brampton commons (1,532 acres [620 ha]) were enclosed 1778.

Landownership. Caput of *barony of Gilsland* from 14th century, when Naworth Castle replaced Irthington as seat of barons. Gilsland was granted to Hubert de Vaux (or Vallibus) by Henry II 1157/8, passing through Vaux family to Matilda, heir of Hubert II de Vaux (d. c.1240), who married Thomas de Moulton. Barony then descended through Moulton family to Margaret, daughter of Thomas Moulton (d. 1313), who carried it to Dacres by her marriage to Ranulf Dacre. It then descended through their heirs to George Lord Dacre (d. 1569), last of male line. In division of Dacre estates, Gilsland was assigned to Elizabeth Dacre (c.1564-1639), wife of Lord William Howard, and descended through their heirs, the earls of Carlisle.

Origins and growth of the town. Brampton received grant of weekly market and annual fair 1252 and became market centre for Gilsland barony and wide area of north-east Cumberland. Thomas Denton (writing 1688) described it as a 'market well stored with all sorts of corn, especially rye, in regard that they grow more rye than other grain in Gilsland.' It was 'improved both in building and trade' and he estimated population to be 630. By 1801 population was 2,125, of whom 1,682 resided in Brampton township.

By 19th century, Brampton continued to serve wide rural hinterland and possessed usual range of retailers, trades and professional services of a market town, including banking, clockmaking and printing. Coal mining provided employment: Brampton Coal Staith (1799) linked town by waggonway to earl of Carlisle's collieries and quarries on Tindale Fell. Although Newcastle and Carlisle Railway bypassed Brampton, with station outside town at Milton (opened 1836), branch line (Brampton and Hartleyburn Railway) took passengers by horse-drawn 'Dandy' from Staith (renamed Brampton Town Station 1881) to Milton Station; line closed 1890, reopening briefly twice between 1913 and 1923. Two breweries were established (one in 1785; one by 1829); both continued into 20th century. Tannery and skin yard from at least 1790, continued throughout 19th century, supplying large number of boot and shoemakers: in 1850s there were 17 businesses employing c.80 shoemakers in Brampton. Textiles employed large numbers (c.200 in 1841), initially weaving checks and ginghams for Carlisle manufacturers. Numerous weaving shops and dye houses in town: one establishment of 13 rooms contained 22 looms in 1815. Scotch Tweed Mill, established 1865, was employing 150 by 1873; destroyed by fire 1875; rebuilt but did not flourish and closed within a few years. Other small-scale manufacturing included hat making and nail making and, during latter part of 19th century, mineral water manufacture. Population grew rapidly, rising to 2,754 by 1841 and peaking at 3,557 in 1871.

With decline of manufacturing, Brampton's population fell across late 19th and early 20th century to stand at 2,526 in 1931. Modernisation of housing stock to replace overcrowded lanes and yards began in 1930s under auspices of the Brampton Public Utility Society. From 1951 construction of Spadeadam rocket testing site (see Kingwater) brought large numbers of skilled technical staff to area, resulting in building of two housing estates in Brampton. Population rose to 4,033 by 1971. Although influx was short-lived, Brampton continued to grow slowly as dormitory for Carlisle. Small industrial estate developed at Townfoot from 1980s. In 2001 population stood at 4,361.

Places of worship. Medieval parish church of St Martin (in corner of Roman fort, one mile west of Brampton); replaced 1788 by new church in town, on site of Brampton Hospital (former almshouses); enlarged 1827 and rebuilt 1878 (noted for stained glass windows by Morris & Co. to Burne-Jones designs). Presbyterian congregation from 1660s; chapel built 1772; replaced 1854; continues in use as United Reformed Church. Wesleyan Methodist chapel built 1799; rebuilt 1836; replaced by new building on new site 1900;

still in use as Brampton Methodist Church. Zion Chapel (Independent) built 1818; closed and converted to flats. Primitive Methodist chapel, Back Street, built 1823; sold 1879 after opening of new chapel on Moatside; now Bethesda Evangelical Church. Moatside Primitive Methodist chapel built 1878; closed c.1932; used as Playhouse Theatre 1936 to 2009; demolished 2013. Roman Catholics worshipped in house from 1893 to 1957, when St Ninian's RC Church was opened in Ashmore, a house in Craw Hill; closed 2014.

Schools. Grammar school, recorded 1588; from c.1688 was held in Brampton Hospital. When hospital was replaced by new parish church 1788, school continued in one room for some time (recorded 1816) but had ceased to exist by 1840s. Infant school established 1825. National school built 1817, with separate school for girls 1832; replaced by new building on different site, erected 1857 for boys and girls. Acquired by School Board 1874 and enlarged 1875 and 1879 and became town's primary school; replaced by modern building on site of former workhouse; now Brampton Primary School. Presbyterian school held in former chapel from c.1855; continued until early 20th century. Private boarding and day schools in 19th century included Croft House Academy, established 1840 by Joseph Coulthard, renowned for high standard of education that attracted pupils from across the country. County Secondary School opened 1909 in building known as 'White House', becoming grammar school ('White House Grammar School') after Second World War. Irthing Valley Secondary Modern School opened c.1950. The two schools were linked in partially comprehensive scheme during 1970s; then combined on secondary modern school site as fully comprehensive William Howard School 1980.

Other institutions. Moot Hall in Market Place, built 1817 replacing 17th-century structure. Ground floor open for egg, butter and poultry market until 1896; upper floor court house for barony courts; cupola contained market bell. Almshouses known as Brampton Hospital erected at expense of earl of Carlisle 1685; site used for new church 1788. Parish workhouse to south of town recorded 1777; enlarged 1837 on establishment of Brampton Union. Replaced by new workhouse on north side of town; completed 1877 (now demolished). Wilson Homes, retirement homes named after Isaac Wilson who funded their construction, built adjacent to the Sands; opened 1930. Brampton & District War Memorial Cottage Hospital opened 1923; still in use. Mechanics' Institution and reading room in mid-19th century; in 1860 had library of 800 volumes. Working Men's Club and reading room established c.1858; continued into early 20th century. Public halls included

St Martin's, built 1895 adjacent to parish church; Central Hall in former Presbyterian church; and New Hall in redundant Tweed Mill, subsequently a cinema, but destroyed by fire 1960.

BRAMPTON (Westmorland) *see* Long Marton

BRANTHWAITE *see* Dean

BRAYSTONES *see* Lowside Quarter

BRIDEKIRK

Township in Bridekirk parish, Allerdale below Derwent ward, Cumberland. CP enlarged by absorbing Dovenby and Tallentire CPs 1934.

Acreage: 873 acres [353 ha] excluding Hames Hall area (171 acres [69 ha]), formerly detached portion of Dovenby, which was added to Bridekirk CP 1887; thereafter Bridekirk covered 1,044 acres [422 ha] until 1934 boundary change. Commons enclosed 1835. ***Population:*** in range 100-125 across early and mid-19th century, then drifting down from 139 in 1871 to 105 in 1931 (last census year before boundary change). ***Landownership:*** granted to Gisburn Priory (Yorks.) in 12th century. After Dissolution granted to Henry Tolson 1543, in whose family it descended; tenants enfranchised 1701. *Woodhall* was separate estate of Tolson family, which they sold 1707; had passed to Fisher family by mid-19th century. *Hames Hill* estate was owned by Senhouse family from 1655 until 19th century. ***Economy:*** predominantly agricultural; limestone quarrying and lime-burning in 19th century. ***Places of worship:*** medieval parish church of St Bridget (containing richly-decorated 12th-century font); replaced by new building adjacent to old church (parts of which still stand) 1868; extended 2010. ***Schools and other institutions:*** for parish school, see Dovenby.

BRIGHAM

Township in Brigham parish, Allerdale above Derwent ward, Cumberland.

Acreage: 1,759 acres [712 ha], including 500 acres [202 ha] of common land enclosed 1819. ***Population:*** rising from 238 in 1801 to 503 in 1831; steady across middle decades but rising again to 818 in 1891; fell back to around

750 until later 20th century, then rising from 1971 to stand at 1,065 by 2001. ***Landownership:*** one of Five Towns within honour of Cockermouth. *Manor of Brigham* had been divided by 13th century between heiresses, moieties being held by Robert de Yanwith and Gilbert de Huthwaite c.1269. Yanwith share descended by marriage to Twinham family by 1278, who continued to hold it in 14th century; by 17th century it had passed to Fletchers of Moresby. Huthwaite moiety descended to Swinburns, who sold it to Goodwin Wharton 1699. It was purchased by Wilfred Grisedale 1727, and descended through his daughter to Lucocks of Cockermouth, and to Joshua Lucock (Bragg) of Lorton Hall before 1804. Tenants of both moieties had been enfranchised before 1688. *The Fitz* estate belonged to Senhouse family from 1670 to 1990. ***Economy:*** farming, overlain by industry in 19th century, when village also became a middle-class suburb of Cockermouth. Extensive quarrying and lime-burning in 19th century. Cockermouth & Workington Railway (in operation 1847 to 1966) had stations at Brigham and Broughton Cross and coal depot. Fitz mill (on Derwent in north-east corner of township), recorded by 1774, was flax thread mill in mid-19th century; briefly used for wool spinning in 1890s; closed by 1895. Caravan park at Low Brigham. ***Places of worship:*** medieval parish church of St Bridget, of pre-Conquest origin (contains fragments of Anglo-Saxon sculpture); restored 1863-76. Wesleyan Methodist chapel built 1856; replaced by new chapel on adjacent plot 1882-3; still in use. ***Schools and other institutions:*** schoolhouse built by inhabitants by early 18th century; perhaps on site of parochial school which stood at south-east end of village in mid-19th century. Replaced by National school between Brigham and Broughton Cross, opened 1875; now St Bridget's CE Primary School. Colliery Rescue Station built 1914 by Cumberland Coal Owners Association to coordinate rescues across West Cumberland coalfield; closed 1951; building later used as police station. Brigham Memorial Hall erected 1924 in memory of men who fell in First World War.

BRISCO *see* St Cuthbert Without

BROMFIELD
(otherwise Bromfield, Crookdake and Scales)

Township in Bromfield parish, Allerdale below Derwent ward, Cumberland. Enlarged by absorption of Blencogo and Langrigg & Mealrigg CPs 1934.

Acreage: 3,028 acres [1,225 ha] before boundary changes. ***Population:*** 297 in 1801, rising to 411 in 1861 then falling to 324 in 1931 (last census year before boundary change). ***Landownership:*** township contained two manors. *Manor of Bromfield* was granted to Holme Cultram Abbey by Adam son of Thomas de Brunfield. After Dissolution, it passed ultimately to Osmotherleys of Langrigg Hall and from them in 1735 to Barwise family, through whom it passed with Langrigg Hall until sold 1876 to Joseph Bowerbank. *Manor of Crookdake* belonged to Musgrave family from 15th century until death of William Musgrave in 1664; then passed through his daughter, Anne, wife of Sir John Ballantine, to Ballantine (later Ballantine-Dykes) family, who continued to hold it in 19th century. ***Economy:*** predominantly agricultural; township was crossed by Maryport & Carlisle Railway (opened 1845), with station at Leegate, and by Solway Junction Railway (opened 1869). ***Places of worship:*** medieval parish church of St Kentigern (or Mungo); restored 1861. ***Schools and other institutions:*** endowed free grammar school founded 1612; restored 1861 and 1898, becoming public elementary school; closed.

BROUGH SOWERBY

Township in Brough parish, East ward, Westmorland.

Acreage: 1,097 acres [444 ha], including small area of common land, Brough Sowerby Common (35 acres [14 ha]). ***Population:*** from peak of 180 in 1821, fluctuated in range 130-155 through rest of 19th century, declining to 78 in 1971 and rising since, to stand at 127 in 2001. ***Landownership:*** as Brough under Stainmore (q.v.). ***Economy:*** predominantly agricultural; vegetable oil refining in 21st century. ***Places of worship:*** Primitive Methodist chapel built in iron 1897; replaced by new building 1972. ***Schools and other institutions:*** none.

BROUGH-UNDER-STAINMORE

Township in Brough parish, East ward, Westmorland, including Church Brough and Market Brough.

Acreage: 1,573 acres [637 ha]. Common land at Lowgill Field (25 acres [10 ha]), enclosed 1855. Shared pastures of Market Brough Intake, enclosed 1842, and Church Brough Intake (78 acres [32 ha], enclosed 1855. ***Population:*** rising from under 700 in 1801 to peak of 966 in 1831; 840 in 1861, then declining to 590 by 1901; steady at around 600 across earlier 20th century, falling to low point of 574 in 1971, then rising across late 20th century to stand at 680 in 2001. ***Landownership:*** castle and *manor of Brough* descended with barony of Appleby (q.v.). Brough castle built beside Roman fort of VERTERIS from c.1100; much repaired by Lady Anne Clifford 1659, but allowed to decay thereafter as manor was administered from Appleby Castle. Wharton family property in Brough regarded as part of their manor of Wharton. ***Economy:*** attempts to found new town at Chruch Brough in 12th century failed as commercial activity had migrated to Market Brough by 1200. Market charter granted 1330, granting weekly Thursday market and four-day fair in September. Fairs flourished following increase in cattle droving in early 17th century, but market had dwindled by 1670s. Brough was significant stop on coaching road from London to Glasgow, with blacksmiths, wheel wrights and innkeepers serving travellers' needs. Transport services remained important in 20th century, with haulage firms and bus and coach company, deriving from garage established 1920. ***Places of worship:*** medieval parish church of St Michael (in Church Brough); extensively restored 1880. Chapel at Market Brough founded 1506, dedicated to Blessed Virgin and St Gabriel; did not survive. Primitive Methodist chapel built 1825; rebuilt 1877; restored following damage by lightning 1880; now demolished. Wesleyan Methodist chapel built 1803; enlarged 1905; remains in use. Baptist chapel built late 1840s; replaced by larger building 1850; refurbished 1883 but closed 1916; building sold to parish council c.1926. ***Schools and other institutions:*** grammar school recorded 1506 in association with chapel at Market Brough; re-established in reign of Edward VI. Replaced by new Board school 1875; now Brough Community Primary School. Former school building converted into Public Hall (used by Temperance Society (established 1834) in later 19th century). Former Baptist chapel converted into Memorial Hall c.1926.

BROUGHAM

Ancient parish in West ward, Westmorland. CP enlarged by absorbing Winderwath, formerly detached portion of Cliburn CP, 1935.

Acreage: 6,226 acres [2,520 ha] before boundary changes; 6,729 acres [2,723 ha] after addition of Winderwath (503 acres [203 ha]). Brougham Moor and Sandwath enclosed 1776; Whinfell (formerly hunting forest of barons of Westmorland), enclosed 1842. ***Population:*** rising from 167 in 1801 to 296 in 1881; then declining to 243 in 1931 (last census year before boundary change). ***Landownership:*** *Brougham castle* built by Robert de Veteripont, baron of Westmorland, in early 13th century, next to site of Roman fort of BROCAVUM. Castle descended with barony (see Appleby), Lady Anne Clifford rebuilding it 1651 and dying there 1676. *Manor of Brougham* held by Brougham family until 1608; manor house, Brougham Hall, bought by Lady Anne Clifford who renovated parts of it. Passed on her death to her agent, James Bird, who extended it and enlarged park c.1686, sweeping away nearby village. Brougham Hall repurchased by younger branch of Brougham family 1726, who held it until 1934. ***Economy:*** predominantly agricultural; quarrying on Whinfell. Whinfell park or 'forest' well stocked with red and fallow deer in later 17th century. Extensive forestry plantation on Whinfell by mid-19th century; holiday park (Oasis Holiday Park, now Center Parcs) in plantation, opened 1997. Brougham Hall used as testing site for military equipment, with prisoner of war camp nearby, during Second World War. RAF landing ground at Hornby Hall, 1941-5. ***Places of worship:*** medieval parish church of St Ninian ('Ninekirks'), an ancient foundation rebuilt by Lady Anne Clifford 1659-60; porch added 1841. St Wilfrid's chapel, adjacent to remains of Brougham Hall; rebuilt by Lady Anne Clifford 1658; restored in 19th century. ***Schools and other institutions:*** reported 1818 that no school needed as 'there are no poor in this parish'. Board school built at Woodside 1876; closed c.1980.

BROUGHTON (Cumberland)

CP in Bridekirk parish, Allerdale below Derwent Ward, Cumberland, created 1898 from bulk of townships of Great and Little Broughton. Broughton Moor CP created at same date, covering mining settlement on former moorland. This entry covers both CPs.

Acreage: Great Broughton township: 2,073 acres [839 ha]; Little Broughton township: 1,026 acres [415 ha]. After separation of Broughton Moor CP (1,736 acres [703 ha]), Broughton CP contained 1,360 acres [550 ha]. Common field at Little Broughton enclosed 1702. Common land on Broughton Moor (1,070 acres [433 ha]) enclosed 1829. ***Population:*** rising from 650 in 1811 to 2,267 in 1891 (last census year before Broughton Moor was separated). In 1901 Great and Little Broughton had a population of 1,334, which rose to 1,727 by 2001. Broughton Moor's population stood at 920 in 1901 and remained fairly stable until 1971, declining thereafter to 726 by 2001. ***Landownership:*** in hands of Percy family, earls of Northumberland and barons of Allerdale, by 15th century.

Granted to Sir Thomas Wharton 1530; bought back by Charles, duke of Somerset, from trustees of duke of Wharton 1739, descending thereafter with rest of Cockermouth Castle estate (see Cockermouth). ***Economy:*** noted for weaving and bleaching of linen cloth in later 17th century ('the finest, whitest and firmest cloath in the north of England', according to Thomas Denton). Limestone quarrying and lime-burning at Broughton Craggs from 17th century. Coal mining recorded from 1437; colliery at Broughton Craggs by later 17th century; expansion of mining in 19th century, especially on Broughton Moor from 1860s (Buckhill colliery opened 1873; closed 1932); extensive opencast coal working late 20th century. Brick and tile manufacture on Broughton Moor in 19th and early 20th centuries. Royal Navy Armaments Depot ('the Dump'), covering 1,050 acres [425 ha], from 1938 to 1992. ***Places of worship:*** chapel of St Lawrence, by River Derwent close to boundary with Ribton, in use mid-17th century but disused by 1688. Anglican chapel of ease (Christchurch) built between the two villages 1856. Anglican church of St Columba at Broughton Moor, built 1905. Broughton noted for Puritan nonconformity in 17th century: Baptist Chapel built 1672; still in use. Quaker meeting house at Little Broughton built 1687; replaced by new building 1742; closed 1982. Wesleyan Methodist chapels at Great Broughton (built 1846; now village hall) and Broughton Moor (built 1903; still in use). Primitive Methodist chapels at Great Broughton (rebuilt and enlarged c.1880; closed and sold 2015) and Broughton Moor (built 1861; closed 1969 and converted to dwelling). ***Schools and other institutions:*** almshouse and schoolhouse built and endowed by Joseph Ashley 1735. School rebuilt 1846; now Broughton Primary School. School at Broughton Moor built 1878; now Broughton Moor Primary School. Mission Hall, built 1882. Thomas Paisley Memorial Hall, containing billiard and reading rooms and library, built 1890; derelict by 2013. Village hall at Little Broughton.

BROUGHTON EAST (or Field Broughton)

Township in Cartmel parish, Lonsdale hundred, Lancashire North of the Sands, containing settlements of Wood Broughton, Field Broughton and Aynsome. Hamlets of Grange, Cart Lane and Kents Bank were transferred to Grange-over-Sands CP on its creation 1894. For those areas, see Grange.

Acreage: 3,425 acres [1,386 ha] before boundary changes; 1,905 acres [771 ha] after loss of territory to Grange CP. Common land in township enclosed under Cartmel enclosure award 1809. ***Population:*** 353 in 1811 rising to 1,719 in 1891 (reflecting growth of Grange as holiday resort); reduced to 218 in 1901 after creation of Grange CP; falling to 173 in 2001. ***Landownership:*** contained several small estates: *manor of Hampsfield* (or Hamsfell) held by family of that name until 14th century. In hands of Thornbugh family by 1521; sold by William Thornburgh to Robert Curwen of Cark; descended with Cark Hall thereafter (see Holker, Lower). *Wood Broughton* was held by Waleys (or Walles) family from 13th century to 16th; in 17th and 18th centuries it seems to have belonged to Knipe family. ***Economy:*** largely agricultural; limestone quarrying and lime-burning in 19th century. Analytical and technical agricultural testing laboratories established at Aynsome 1898; moved to Kents Bank, Grange, 1954. ***Places of worship:*** chapel of ease of St Peter at Field Broughton built 1731 and consecrated 1745; rebuilt 1892. ***Schools and other institutions:*** parish rooms at Field Broughton given to community by Hibbert Family 1921.

BROUGHTON-IN-FURNESS

Chapelry and market town in Kirkby Ireleth parish, Lonsdale hundred, Lancashire North of the Sands.

Acreage: 7,298 acres [2,953 ha], including 3,226 acres [1,306 ha] of commons and moss land enclosed 1847. ***Population:*** rising from around 1,000 in early 19th century to 1,375 by 1831; then fell gradually to stand at 1,117 by 1901 and 995 in 1951 with steeper decline across later 20th century to 529 in 2011. ***Landownership:*** *manor of Broughton* held by Broughton family from 12th century until forfeited by Sir Thomas Broughton who supported Lambert Simnel's revolt 1487. Granted to earl of Derby c.1490; sold to Edward Leigh 1653/4, who sold it to Roger Sawrey 1658; it then descended in his family, who rebuilt Broughton Tower 1744, until bequeathed to James Cookson 1881. ***Economy:*** market centre for surrounding rural hinterland: charter said to have been granted later 16th century. Described by Thomas West 1774 as 'a weekly market on Friday and a fair for all sorts of merchandise on the 1st day of August. The principal commodities are woollen yarn spun by the country people'. Market Hall built 1766. Market day changed from Friday to Wednesday c.1830; fairs noted for cattle and sheep by 19th century. Wool spinning had declined by mid-19th century, by which time hoop, brush and basket making were principal trades. Slate quarrying, notably at Broughton Moor in 20th century. Lead, copper and iron mining, recorded 18th and 19th centuries. ***Places of worship:*** medieval chapel of ease of St Mary Magdalene; new nave built 1873-4; tower replaced 1900. Wesleyan Methodist chapel built

1837; rebuilt 1875. Baptist chapel at Scroggs built 1701; closed c.1823 (when meetings started at Ulpha); deserted by 1842. For church of Holy Innocents at Broughton Mills, see Dunnerdale & Seathwaite. ***Schools and other institutions:*** endowed school founded 1650s by Leonard Towers of London. School building on Church Street (19th century) replaced by modern Broughton CE Primary School at Kepplewray. Schoolhouse at Aulthurst Side, built by subscription 1828; closed 1947. Broughton Tower used as residential school for children with special needs 1947 to 1997. Victory Hall built 1929; enlarged and refurbished 1999-2002. Reading room (now village hall) at Broughton Mills.

BROUGHTON MOOR *see* Broughton (Cumberland)

BURGH BY SANDS

Ancient parish in Cumberland ward, Cumberland, including townships of Boustead Hill, Burgh Head, Burgh West End, Longburgh, and Moorhouse and salt marshes of Burgh Marsh.

Acreage: 6,170 acres [2,497 ha], divided between constituent townships thus: Boustead Hill: 676 acres [274 ha]; Burgh Head: 866 acres [350 ha]; Burgh West End: 785 acres [318 ha]; Longburgh: 682 acres [276 ha]; and Moorhouse: 1,687 acres [683 ha]. Commons at Longmoors, Moorhouse, and Boustead divided 1686-1708; Moorhouse Green (50 acres [20 ha]) enclosed 1842. Burgh marsh (1,655 acres [670 ha]), common to all townships in parish, enclosed 1848 (much of it (1,162 acres [470 ha]) remains registered as common land). ***Population:*** estimated at 540 in 1688; rising from 780 in 1801 to 1,083 in 1851; then declining steadily until 1931, before rising again to stand at 1,143 in 2001. ***Landownership:*** caput of *barony of Burgh,* held by Morville family in later 12th century, passing on death of Hugh de Morville (d. 1202) through his daughter Ada to Moultons of Gilsland, and from them by marriage to Dacres in early 14th century. On death of George Lord Dacre in 1569 barony was allotted to his sister Anne, wife of Philip Howard, earl of Arundel, and descended with Howards until Henry Howard, duke of Norfolk, sold it to Sir John Lowther 1685. ***Economy:*** fertile soils and extensive arable land noted 1688, with grazing and wildfowling on marshes and fisheries in Solway estuary. Old Sandsfield was traditional port for Carlisle, taking vessels of up to 60 tons, until destroyed by storm 1796; ford (Stony Wath) and ferry across Solway nearby. Chemical works (manufacturing copperas and vitriol) at West End by 1847 (removed to Drumburgh 1857 – see Bowness-on-Solway). Tannery at Thurstonfield in mid-19th century. ***Places of worship:*** medieval parish church of St Michael, with defensive west tower of mid-14th century date; restored 1881. Quaker meeting house at Moorhouse, built 1681; rebuilt 1733; closed 1913, then used by Methodists until c.1967. Methodist chapel at Thurstonfield built 1861; still in use. ***Schools and other institutions:*** grammar school until 1786, when vicar converted it into barn; subsequently used as vestry. Schools at Burgh and Moorhouse under benefaction of Thomas Pattinson (d. 1785); that at Burgh built by parishioners on West End Green by 1829. New school built south of village 1870 and transferred to Board 1873; enlarged 1896; now Burgh by Sands Primary School. Subscription library recorded 1860. Public Hall erected 1894, as limited company.

BURNESIDE

Chapelry in Kendal parish, Kendal ward, Westmorland, comprising townships of Strickland Ketel and Strickland Roger, each of which remains a separate CP. Burneside village straddles boundary between the two CPs.

Acreage: Strickland Ketel township: 2,349 acres [951 ha]; Strickland Roger township: 3,210 acres [1,299 ha]. Commons enclosed 1816 (Strickland Ketel) and 1853 (Strickland Roger). ***Population:*** combined population of Strickland Ketel and Strickland Roger townships rose from 519 in 1801 to 770 by 1851; then rose steeply in later 19th century as mill village of Burneside grew, reaching 1,141 by 1891. Further growth across 20th century lifted population to 1,309 by 1931 and over 1,600 by 1980s; it stood at 1,637 in 2001. ***Landownership:*** Strickland ('Stercaland' in Domesday Book) appears to have been pre-Conquest forerunner of Kendal as centre of large estate covering much of southern Westmorland. Post-Conquest *manor of Strickland* was divided into two, suffixes referring to medieval owners. Strickland Roger later descended with Marquis fee of barony of Kendal (q.v.). *Manor of Burneside* was held from Roger de Lancaster by Gilbert of Burneside 1348; then passed to Bellinghams in 14th century. Bought by Thomas Braithwaite (d. 1610) of Ambleside and descended through Braithwaite family until sold by Richard Braithwaite (d. c.1750) to Thomas Shepherd of Kendal. All three manors had passed to Lowther family by 19th century. Cropper family acquired considerable landholding in 19th century, centred on Tolson Hall and Ellergreen. ***Economy:*** rural economy transformed by industry based on water-powered mills on River Kent. Paper mill at Cowan Head built by Thomas Ashburner 1753; Roger Wakefield (d. 1756) owned corn, fulling, frizing and sickle mills at Burneside. Three-storey woollen mill built at Burneside 1761 by John Wakefield; converted to paper mill 1833 by Hudson, Nicholson & Foster. Paper mills at Cowan Head and Burneside leased 1845 to James Cropper, from Liverpool, who enlarged and modernised them before assuming full control 1854: firm, still bearing his name, prospered with coming of railway in 1847 and Burneside mill continues as manufacturer and exporter of specialist papers. Mill at Cowan Head closed 1981. Mills on River Sprint in Strickland Roger township included bobbin mill at Sleddalefoot and woollen mill at Sprint Mill in 19th century. ***Places of worship:*** chapel of ease at Burneside, said to have been rebuilt 1602; rebuilt 1823–8 and again (in part) 1881; now parish church of St Oswald. Chapeltree Wood, near Hundhow, said to record site of early chapel. ***Schools and other institutions:*** school by chapel built 1717; rebuilt 1787 on endowment by Rev. Alen Fisher. Rebuilt as National school on new site adjacent to Burneside paper mill 1855 (extended 1885); moved to present site 1963; now St Oswald's CE Primary School. Bryce Institute opened in 1897, with reading & billiard rooms; still serves as village hall.

BURTHOLME

Township in Lanercost parish, Eskdale ward, Cumberland, enlarged by addition of part of Banks township in 19th century.

Acreage: 2,082 acres [843 ha] before boundary change; increased to 2,683 acres [1,086 ha] after absorbing part of Banks township. ***Population:*** rising from 279 in 1801 to peak at 347 in 1851; then falling to 177 in 1991. ***Landownership:*** part of initial endowment of Lanercost Priory, passing at Dissolution to Thomas Dacre of Lanercost. On failure of heirs, reverted to Crown in 1716 and leased to, later purchased by, earl of Carlisle. ***Economy:*** predominantly agricultural; quarrying and lime-burning in 19th century. Archaeological tourism increasingly important across 20th century; visitors attracted by wealth of historic sites, notably Hadrian's Wall and ruins of Lanercost Priory. ***Places of worship:*** nave of Lanercost Priory used as parish church of St Mary Magdalene since Dissolution. Primitive Methodist chapel at Calees, near Banks, erected 1874; closed 2004. ***Schools and other institutions:*** two schools in mid-19th century; that at Island recorded 1847; rebuilt 1862; now Lanercost CE Primary School. Library in school building, recorded 1901.

BURTON-IN-KENDAL

Township and small market town in Burton-in-Kendal parish, Lonsdale ward, Westmorland. CP enlarged by absorbing Dalton CP 1986.

Acreage: 1,473 acres [596 ha], including c.189 acres [76 ha] of common land enclosed under Dalton, Burton and Holme enclosure award 1815. ***Population:*** rising from 548 in 1801 to peak of 796 by 1841. Gradual decline thereafter, reaching low-point of 460 in 1921, then recovering to stand at 646 in 1961. ***Landownership:*** *manor of Burton* passed on death of Anthony de Burton (d. 1320) to his sister, wife of Sir Thomas Bethom, with whose descendants it remained until death of Sir Edward Bethom (d. 1472), when it passed to his niece, Agnes, wife of Sir Robert Middleton of Leighton. Descended from Middletons to Oldfield family of Cheshire in late 17th century. Passed by sale to Thomas Benison of Hornby; then, in 1770, to Thomas Pearson and in 1790 to William Atkinson (d. 1825), in whose family it remained until 20th century. ***Economy:*** weekly market and two fairs established 1661 by Sir George Middleton, lord of manor. By later 18th century had become most significant corn market in Westmorland, wealth of town boosted by its position on main north-south route, the Heron Syke Turnpike, which led to growth of coaching inns and carriers. Completion of Kendal-Lancaster canal (1819) reduced turnpike's importance (despite wharf at Burton) and coming of railway (1846) rendered Burton a backwater and led to decline in market and trading function. Burton had become desirable place to live by 19th century. Outside the town: farming; limestone quarrying, especially on Clawthorpe Fell; tannery in mid-19th century. ***Places of worship:*** medieval parish church of St James (said 1778 to be dedicated to St Helen); restored 1844 and 1872. ***Schools and other institutions:*** grammar school endowed 1657 by John Hutton of Farleton; rebuilt 1817; had closed by 1851. National school erected 1817; restored 1887. Morewood Memorial Infant School (named in memory of Rev. Robert Morewood, vicar of Burton), opened 1867; demolished 1967 when new school (now Burton Morewood Primary Academy) built nearby. Memorial Hall built 1956.

BUTTERMERE

Township in Brigham parish, Allerdale above Derwent ward, Cumberland. CP enlarged by absorbing Brackenthwaite CP 1934.

Acreage: 7,021 acres [2,842 ha], including 170 acres [69 ha] detached portion at Swinside (transferred to Lorton CP 1887). Buttermere Fell, estimated at 1,640 acres [664 ha], remains unenclosed common land. ***Population:*** 74 in 1801, rising to 136 in 1821, then falling back to 78 in 1851; stood at 126 in 1931 (last census year before boundary change). After inclusion of Brackenthwaite, combined population was 257 in 1971, but had dropped to 127 by 2001. ***Landownership:*** part of manor of Braithwaite and Coledale, within forest of Derwentfells (see Above Derwent). Birkness (or Burtness), that part of Buttermere township west of lake, was in manor of Loweswater (q.v.). Gatesgarth (retained in demesne in 13th century) was freehold estate of dukes of Norfolk in 18th century, bought by John Marshall 1815. Marshall estate bought by Balliol College and the National Trust 1934-7. ***Economy:*** livestock farming: Gatesgarth was demesne cattle farm ('vaccary') in late 13th century and major sheep farm, noted for its herdwick flock, from 19th century. Plumbago ('black lead' or 'wad') mining at Bowderbeck in 17th century. Slate mining on Honister Crag from 17th century, becoming large business from late 18th century; closed 1986; mine reopened 1997. Tourism from later 18th century, when Buttermere became part of circuit from Keswick, enhanced by fame of Mary Robinson (b. 1778), the 'Beauty of Buttermere'. Accommodation supplied by Fish Inn; Queen Victoria Inn (now Bridge Hotel; opened by 1847) and Buttermere Hotel (recorded by 1898; now youth hostel). ***Places of worship:*** chapel of ease (described as 'very ancient chapel' 1714); rebuilt 1840 and extended 1884. ***Schools and other institutions:*** school built c.1866; closed 1950. School building became village hall during 1990s.

CALDBECK

Ancient parish in Allerdale below Derwent ward, Cumberland, divided into four townships: High Caldbeck, Low Caldbeck, Caldbeck Haltcliff (which included village of Hesket Newmarket) and Mosedale (q.v.), which became separate CP.

Acreage: ancient parish covered 16,160 acres [6,540 ha], including 6,501 acres [2,631 ha] of undivided fell, common to townships of Caldbeck High, Caldbeck Low and Caldbeck Haltcliff. Caldbeck High township covered 2,116 acres [856 ha]; Caldbeck Low 1,453 acres [588 ha] and Caldbeck Haltcliff 2,333 acres [944 ha]. Common field at Nether Row (37 acres [15 ha]) and stinted pasture at Ox Park (73 acres [30 ha]) enclosed 1839. ***Population:*** estimated at 955 in 1688. Rose from 1,171 in 1801 to 1,608 in 1851, then fell to 613 in 1981, before picking up to 714 in 2001.

Landownership: *manor of Caldbeck* was retained by lords of Allerdale barony, passing from Lucys to Percys, until 1546, when granted to Thomas Lord Wharton. Descended in Wharton family until early 18th century when purchased by Charles, duke of Somerset, and reunited with former Percy estates. *Manor of Hesket*, held by Bewley family by early 16th century, was sold by Thomas Bewley to William Lawson 1630; Sir Wilfrid Lawson built idiosyncratic Hesket Hall in later 17th century. ***Economy:*** hill farming. Lead and copper mining since medieval period, with surges of activity in later 16th and 17th centuries (under Mines Royal) and in 19th; last barytes mine closed 1980s. Coal mines (at Lanthwaite) and slate quarrying (at Ratten Row) recorded 1688. Caldbeck village became centre of textile industry with spinning (recorded 1829) and woollen mill. Bobbin mill (c.1847) had one of largest water wheels in England. Weaving and bobbin manufacture remained major occupations in 1900. Other industries included brewing and lime burning in 19th century. Hesket became market centre, probably in 18th century, giving rise to name 'Newmarket'; market had faded away by later 19th century. Brewery at Hesket opened 1987. ***Places of worship:*** medieval parish church of St Kentigern; tower added 1727; restored 1880 and 1932-3. Quaker meeting houses at Whelpo, near Caldbeck village (built 1698; closed 1849) and Gillfoot, near Hesket Newmarket (built 1728 for holding business meetings; regular worship there from late 18th century to 1913). Wesleyan Methodist chapels at Caldbeck (built 1832; rebuilt 1863; still in use) and Hesket Newmarket (built 1839; became Free Evangelical church 1993). ***Schools and other institutions:*** endowed school had lost its endowment by 1815 and 'fallen down' by 1821. Said to be eight 'small day schools' in parish in 1830. National school at Upton built 1851; now Fellview Primary School. Board school at Howbeck, near Hesket Newmarket, built 1874, replacing an older school nearby; closed and converted to dwelling. Parish Hall at Upton, built 1928.

CALTHWAITE *see* Hesket-in-the-Forest

CAMERTON

Township in Camerton parish, Allerdale below Derwent Ward, Cumberland. CP enlarged by absorbing Ribton CP 1934.

Acreage: 787 acres [318 ha], including Camerton Common (425 acres [172 ha]) enclosed 1830. ***Population:*** 71 in 1801, rising as coal mining expanded to 245 in 1891; then falling back to 203 in 1931 (last census year before boundary change). ***Landownership:*** *manor of Camerton* descended with Seaton (q.v.) through Curwen family until sold by Joseph Curwen to Ralph Cooke of Kendal 1719; then descended through Cooke family (who rebuilt Camerton Hall 1833) until 1860s. ***Economy:*** salmon fishing in River Derwent (recorded 1688); coal mining: colliery described as 'rich' 1688. Mining expanded after opening of Cockermouth & Workington Railway 1847. Fireclay brick works important in later 19th and early 20th century. Camerton Colliery closed 1932. ***Places of worship:*** medieval parish church of St Peter; rebuilt 1694 and again 1796. ***Schools and other institutions:*** infant school opened 1897. Isolation hospital (Derwent Joint Smallpox Hospital) on northern edge of township in 1920s; used as store during and after Second World War; since demolished.

CARGO AND STAINTON *see* Stanwix

CARLATTON

Extra-parochial township in Eskdale ward, Cumberland.

Acreage: 1,462 acres [592 ha]. ***Population:*** 50 in 1801; 29 in 2001. ***Landownership:*** part of honour of Penrith (q.v.); thus in hands of kings of Scots 1242-1295 and in gift of English Crown thereafter. Held by earl of Carlisle by 1688 and remained in hands of Howard family in 20th century. ***Economy:*** predominantly agricultural. ***Places of worship:*** medieval church given to Lanercost Priory by Robert, son of Hubert Vaux, lord of Gilsland; last vicar recorded 1386. Site of church is not now known. ***Schools and other institutions:*** none.

CARLETON (near Carlisle) *see* St Cuthbert Without

CARLISLE

Cathedral city and county town of Cumberland.

Administrative history. This entry covers four ancient territories: city of Carlisle itself and townships of Rickergate, Botchergate and Caldewgate, which together formed Carlisle CP, created 1904, covering 2,025 acres [819 ha]. Component territories are: ***City of Carlisle,*** which consisted of civil divisions of Eaglesfield Abbey (5 acres [2 ha] of extra-parochial land), St Mary Within

(33 acres [13 ha]) and St Cuthbert Within (38 acres [15 ha]). ***Rickergate*** township in St Mary's parish contained 358 acres [145 ha] before gaining part of Stanwix 1894 and thereafter containing 400 acres [162 ha]. ***Botchergate*** township in St Cuthbert's parish covered 415 acres [168 ha]; Botchergate CP, created 1894, extended beyond bounds of township and covered 561 acres [227 ha]. ***Caldewgate*** township in St Mary's parish covered 1,700 acres [688 ha], part of which (988 acres [400 ha]) was absorbed into Carlisle; in 1894 remainder became Belle Vue CP, much of which was, in turn, absorbed into Carlisle 1912.

Roman and medieval origins (to c.1300). Carlisle's origins lie in Roman settlement of LUGUVALIUM. Fort was established 72-3 AD on site now partly overlain by medieval castle. Civilian settlement or *vicus* grew up to east of fort, covering much of area later enclosed by medieval city wall (though it is unclear whether Roman settlement was walled). By 3rd century Luguvalium was important military base on Roman frontier and also a centre of administration as principal town of *Civitas Carvetiorum* (self-governing territory of tribal group known as *Carvetii*). It appears to have been prosperous and a town in which a Romanised population survived into 5th century, well after end of Roman administration. Some sort of organised community existed 685, when, according to Life of St Cuthbert, Cuthbert visited Carlisle and met reeve; a 'fountain' was still functioning. Carlisle and a territory of 15 miles radius were granted to St Cuthbert by king of Northumbria before 685: it is likely that an Anglo-Saxon monastic community was placed there, probably in vicinity of later cathedral. Carlisle is said to have been destroyed by Danes c.875 and to have lain desolate until after Norman conquest. Archaeological evidence, however, points to some continuing occupation through 10th and 11th centuries. Re-birth of Carlisle and building of walls and castle are traditionally attributed to William II (Rufus) of England, who captured 'the land of Carlisle' in 1092, dispossessing native ruler, Dolfin, and fortifying town. Creation of medieval city probably dates from reign of Henry I, who founded priory 1122, embarked on building city walls and stone keep of castle, and in 1133 made Carlisle a bishop's see. Claims of Scottish crown to Carlisle region posed a continuing threat until 1235 (city was besieged by William, king of Scotland, 1173-4 and again by his successor, Alexander, 1216). Carlisle received its first charter from Henry II c.1158, granting merchant guild and other privileges. Walled city, covering 57 acres [23 ha], appears to have developed rapidly and there is evidence of medieval settlement in extra-mural suburbs of Rickergate, Botchergate and Caldewgate (though last was devastated by flooding 1484). Suburbs were used for distribution and treatment of wool and hides and some citizens grew cereals and pastured their livestock outside city walls. By 1237 city had become largely self-governing community. Wool was woven and dyed; leather was tanned and both commodities were exported to Ireland. Servicing castle garrison and priory and administering a large territory as shire town and cathedral city complemented industrial and trading economy. Henry III granted second charter 1251, to replace the first which had been lost by fire, as indeed was second charter in 1292, in a fire which destroyed much of city. Edward I granted new charter 1293 but also began to develop Carlisle as major military base in preparation for invasion of Scotland.

Border city (c.1300- c.1750). From outbreak of wars with Scotland in 1296 until well into 18th century, Carlisle's character was shaped by its proximity to Scotland and its role in guarding the Border. It remained a military city on a defended site: royal castle on its north-west edge and 16th-century Citadel at south-east end were connected by strong city wall, enclosing core of built-up area. Population estimated to have been c.1,500 in 14th and 15th centuries, sporadically swollen by military presence. Garrison provided townspeople with economic opportunities (if causing friction at times); some fine houses were erected in mid-14th century by well-to-do clergy and merchants and as townhouses for local gentry families. However, continuing instability of Border region in later 16th century, outbreaks of plague (particularly in 1598), and disruption caused by Civil War siege of city in 1645 combined to hold back Carlisle's economy. By late 17th century, some evidence of increasing wealth. Carlisle served a wide region: despite its comparative isolation and poor transport links, its role was more than that of market centre for Solway lowlands. Thomas Denton noted 1688 that corn dealers ('badgers') from Northumberland came to Carlisle market and went home laden with corn and grain and that city's three fairs included three-day horse fair in June and 'great beast fair' in August. City was a focal point in the cross-Border cattle trade which gathered pace after union of crowns. Eight trade guilds (merchants, tanners, skinners and glovers, butchers, smiths, weavers, tailors and shoemakers) continued to thrive. Population was probably around 5,000 by late 17th century (in 1688 Denton estimated 5,060 inhabitants in St Mary's and St Cuthbert's parishes combined), making it one of largest towns in north-west England. As well as trading function, it continued to house an increasingly wealthy clerical, military and legal elite by virtue of being centre of administration, as royal garrison, county town and cathedral city. Despite this, progress appears to have stalled by early 18th century. Travellers' comments on Carlisle paint a bleak picture: Defoe visited in 1724 and reported that 'the city is strong, but small, the buildings old, but the streets fair ... There is not a great deal of trade here either by land or sea, it being a mere frontier', while in 1759 a merchant from Bristol called it 'small deserted dirty, poorly built and poorly inhabited'. Seeds of change were visible, however, in textile 'factory house with looms' established on site of fulling mill and dye house at Abbey Mill by Deulicher brothers 1724.

Industrial and railway city (c.1750-1914). Improved transport links (completion of military road to Newcastle 1750 and spread of turnpike roads) from mid-18th century enabled Carlisle to share in industrial revolution. It grew significantly across later 18th century, population of city and its suburbs exceeding 10,000 by 1801, with greatest growth occurring in new industrial areas outside city walls. Industrial base initially centred on water-powered textile mills specialising in finishing of linen cloth and in calico printing. Calico printing started 1761 and gingham, check and bleaching establishments set up shop. Cotton spinning mills were established from 1790s, largest being steam-powered New Mill or Slater's Mill (1802) and Shaddon Mill (1836). Although print works closed in early 19th century, dyeing and cloth finishing ('beetling') expanded. Tanning, brewing (on industrial scale from 1756), hat-making (from 1760s), printing (from 1799), iron and brass founding (from 1804), soap-boiling (by 1794), and biscuit-making (Carr's biscuit factory established 1834) all added to industrial mix. Major transformation took place in 1813, when city walls were finally demolished, heralding final transition of Carlisle from insular enclosed community to fully-fledged modern city. From 1804 corporation lit and paved the streets; from 1819 it was lit by gas. Industry remained dependent on road transport until canal to Port Carlisle was dug 1823. On eve of railway age, Carlisle was a thriving regional centre: Carlisle gained its first bank 1787 and its first newspaper 1798. There were three banks: Carlisle Old Bank was a private bank but Carlisle City & District Bank (established 1836; amalgamated with London & Midland Bank 1896) and Carlisle & Cumberland Banking Co. (established 1835) issued their own notes. *Carlisle Journal*, established 1798, was first of plethora of newspapers, most of which were short-lived. Those which endured were the *Journal* itself, *Carlisle Patriot* (first issued 1815); *Carlisle Examiner and Northern Advertiser* (1857-1870); and *Carlisle Express* (founded 1861), which merged with *Carlisle Examiner* 1870 to form *Carlisle Express and Examiner*; which, in turn, merged with *Carlisle Journal* 1913. *Carlisle Patriot* amalgamated with *East Cumberland News* and became *Cumberland News*, which absorbed *Carlisle Journal* 1968.

Railway connected Carlisle to Newcastle 1838 and city soon became a hub in expanding railway network, being served by seven different lines. Maryport railway opened 1845; that to Lancaster 1846. Southern section of Caledonian Railway opened 1847, with through services to Glasgow and Edinburgh from 1848; Glasgow and South Western Railway followed 1851. Port Carlisle canal was replaced by railway to Silloth 1856 and Midland Railway's Settle-Carlisle line opened for freight traffic 1875. Citadel station, built 1847-8, became hub of railway network. Carlisle grew rapidly across 19th century, swelled by Scottish and Irish immigrants: by 1851 city had over 25,000 inhabitants; by 1901 over 45,000. By 1860s textile industry was in decline and number of handloom weavers was dwindling rapidly but railways gave employment in their own right and also opened up markets for firms that were to come to prominence in second half of 19th century. Impact of railways on employment was significant: in 1917 over 14% of all households were headed by a railway worker and in 1921 over 20% of male employment was in the transport and communications industries. Carlisle's industrial base diversified: Hudson Scott's printing works developed tin-plate decoration and, from 1886, manufacture of tin boxes; number of foundries increased and specialised engineering, such as crane manufacture (from 1858), developed; John Laing set up small building firm 1848, which was to become one of country's leading building and civil engineering contractors. On eve of First World War Carlisle was major industrial city, as well as continuing to serve as county town and seat of an expanded diocese.

Modern Carlisle (since 1914). In first half of 20th century, railways, engineering and metal trades remained major industries. Textile industry, in severe recession at beginning of century, made a limited recovery during inter-war years, but dwindled after 1945, though some specialised activity continued, such as Linton Tweeds, which took over Shaddon Mill 1912. Some major firms closed: last crane built at St Nicholas engineering works 1987; Victoria ironworks closed 1998. Losses were counteracted to some extent by growth of new firms, including Kangol, Pirelli and Nestlé. In 1927 new power station was erected at Willowholme and connected to national grid. Electric trams ran from city centre to suburbs from 1900, but by 1931 were judged to cause congestion and were taken out of service. Decline in railway transport in mid-20th century reduced Carlisle's role as transport hub; two of its lines closed in Beeching cuts: Carlisle-Silloth line (1964) and Waverley line (1965). Redevelopment of city centre from 1960s radically altered townscape: Civic Centre built 1960-3; inner ring-road opened 1971; increasingly derelict 19th-century lanes and courts redeveloped as The Lanes shopping precinct 1984.

Places of worship.

City: finds of Northumbrian age sculpture suggest Anglo-Saxon monastic site encompassing site of cathedral and St Cuthbert's church. Church on cathedral site re-founded as Augustinian priory 1122, becoming see of diocese of Carlisle 1133. Parish church of St Mary was within nave of Cathedral until 1870, when new church of St Mary built on east of cathedral close; demolished 1954. St Cuthbert's Church, almost certainly a pre-Conquest foundation, rebuilt 1778 on site of previous structure. Within walls of medieval city were also St Alban's chapel, possibly also pre-Norman foundation, which became chantry and was dissolved 1549, and two friaries, of Franciscans (or Greyfriars), founded 1233, and Dominicans (or Blackfriars), also established in 1230s. Nonconformist places of worship in city centre included: Quaker meeting house, at first in Abbey Close, 1653 to c.1660; re-established in city 1693; new meeting house, Fisher Street, built 1702; replaced by one on another site in Fisher Street 1776; in turn replaced by present meeting house on former burial ground at end of Fisher Street, 1963. English Presbyterian chapel (formerly known as 'old Scotch' Presbyterian chapel), Fisher Street, built 1730s, when congregation removed from their former chapel near west wall; rebuilt 1894; demolished 1986. Chapel in Annetwell Street (demolished by 1901), used before 1817 by Lady Glenorchy's Connexion; then by Independents, until Congregational chapel (now United Reformed Church), Lowther Street, built 1843. Church of Scotland, Chapel Street, built 1834. First Methodist chapel built in Fisher Street 1786; replaced by new chapel on site of present Methodist Central Hall, Fisher Street, 1817. Other Methodist chapels included United Methodist Chapel (formerly Wesleyan Association chapel), Lowther Street, erected 1837. Baptists met in Abbey Street from 1809 and rented former Methodist chapel, Fisher Street, after 1817; Berian Baptists, built chapel in Tower Street 1818. Roman Catholic church, West Walls, built 1799; replaced by church in Chapel Street 1824. Other places of worship included Salvation Army barracks, Annetwell Street, replaced by new Citadel, Abbey Street, 1972; several groups of Christian Brethren by 1900; Elim Pentecostal Church, which met in building in West Walls from 1927 and moved to former Anglican church of St Paul 1979; Elim Free Church established after a secession 1939; and Seventh-day Adventist Church, opened 1944 in hotel room in West Walls; moved to former Jehovah Witness meeting place; purchased property in Finkle Street, 1982.

Rickergate: new Anglican district churches of St Paul, Lonsdale Street, (since 1979 Elim Pentecostal Church), built 1869-70 and St Aidan, Warwick Road, built 1899-1902. Roman Catholic church of St Mary & St Joseph, Warwick Square, built 1891-3; English Presbyterian church (now St George, United Reformed Church), Warwick Road, built 1862-3; Primitive Methodist chapel, Cecil Street, built 1852; demolished 1966; Baptist church, Aglionby Street, built 1887-8.

Botchergate: new Anglican district churches of Christ Church, Botchergate (erected 1828-30; closed 1938; demolished 1953); St John, London Road (built 1867) and St Stephen, James Street (built 1865; demolished 1964). Congregational (formerly Evangelical Union) chapel, Cecil Street, built 1860. Wesleyan Methodist chapel, Union Street, opened 1891; closed 1990; demolished 1991. Unitarian church, James Street, built 1889; closed. Mission church on Blackwell Road by 1901; St Herbert's church, Currock, dedicated 1932.

Caldewgate: medieval chapel of Holy Trinity, Caldewgate, recorded from 14th century (located in Paddy's Market area); its name preserved in new (Anglican) Trinity Church, completed 1830 as chapel of ease to St Mary's; spire removed 1947; demolished 1982. Other Anglican district churches of St James, St James' Road, built 1865-7; St Barnabas, Newtown, built 1935-6. Primitive Methodist chapel opened in Willowholme, 1826; moved to Cecil Street, 1852. Wesleyan Methodist chapel, Caldewgate, opened 1865. Methodist Church, Wigton Road, built 1929. Church of Christ chapel, Denton Street, built 1881; demolished 1966. Congregational chapel (now Carlisle Christian Fellowship), Charlotte Street, built 1860-1. St Bede's RC Church, Silloth Street, built 1866 as school and chapel; new church of St Bede, Wigton Road, built 1959.

Schools.

City: Carlisle Grammar School (usually called High School until 1880), recorded as early as 1186; new school built in West Walls 1806; replaced by new building there 1832. Moved to Strand Road 1883 (see below, Rickergate). Central School for boys and girls, a large structure at West Walls, built 1812 on ground given by corporation. Fawcett School (boys and girls), West Walls, opened 1851; took over buildings of Central School some time before 1884. St Patrick's Schools for boys and girls, Spring Gardens, founded 1825 for education of children of all denominations; enlarged 1844; moved to new building 1908; became part of Cardinal Newman School (see below, Rickergate) 1968. British or Lancastrian School, for boys and girls, originated in Lancastrian School established 1811 in Watergate. British school, Mary Street, built 1834; closed 1885; demolished 1973. High School for Girls,

Castle Street, opened 1884; removed to Lismore Place 1909 (see below, Rickergate). Academy of Arts, Finkle Street, established 1823, was forerunner of College of Art, Finkle St, founded 1856; moved to Tullie House (see below, Other Institutions) 1893, and from there to Stanwix (q.v.) 1951.

Rickergate: Carlisle Grammar School, moved to new premises on Spring Garden Lane 1883 and to Strand Road (name changed to Trinity School) 1968. Carlisle and County High School for Girls, Lismore Place, opened 1909; became mixed school (St Aidan's School) 1970; closed 2008. Cardinal Newman RC School, opened 1968. Further education classes previously held at Tullie House, moved to new Technical College (now Carlisle College), Victoria Place, 1954.

Botchergate: Christ Church National School, completed 1830; St John's National Boys' school, Close Street, erected 1869. Board schools: Brook Street Board schools, built 1892. School of Industry, Botchergate, built 1806; active 1847. St Martin's College took over City General Hospital site, Fusehill Street, after closure 2000; now campus of University of Cumbria.

Caldewgate: mixed school in Shaddongate opened 1836, adjoining cotton factory of Peter Dixon and Sons, for 'facilitating the education of the offspring of the operative classes'. Shaddongate Adult School and Reading Room founded 1857. Holy Trinity School, Caldewgate, built 1842; enlarged 1895; closed 1915. Caldewgate Ragged and Industrial School (also known as 'Head's School'), Shaddongate, opened by philanthropist George Head Head 1851; had closed by 1891. St Bede's RC School, opened 1866 as school chapel mission; new buildings 1883-4; enlarged 1889; old school closed 1962. Newcastle Street and Kendal Street Elementary School, for both boys and girls, built 1875. Board schools: Caldewgate School opened 1872; enlarged 1885; Caldewgate Junior School closed 1988, when combined with Caldew Lea School. Ashley Street Elementary School, for boys and girls, built 1895; closed 1988. Denton Holme schools (later Robert Ferguson schools), Morley Street, built 1879; senior mixed school, East Dale Street, built 1904; closed 1970. Belle Vue Primary School opened 1970.

Other Institutions. Carlisle Castle: royal fortress since c.1100; keep begun 1122 by Henry I and completed by David I of Scotland; used as prison for border reivers and others in the turbulent 15th and early 16th centuries, when rule of law broke down across border region. Became military garrison in 1830s, with barrack blocks in outer bailey. Citadel built 1541-3 as independent fortress at vulnerable southern end of city walls; re-constructed

1804-1822 to form entrance to city. Assize courts, originally held in Guild hall, moved to Citadel 1811; western tower housed Crown Court, connected to goal by underground passage. Purpose-built gaol built close to Citadel 1688, replacing castle as county gaol. New gaol on Citadel site completed 1827; closed 1922; pulled down 1931. New Crown and County Courts, Earl Street, built 1990-2. Town Hall, housing corporation and mayor's court and city sessions, built 1668-9 on site of medieval town hall; extended 1717 and 1825. Ground floor occupied by shops. Guildhall (or Redness Hall), built by Richard de Redeness after fire of 1392 (dated to 1396-1407); each of eight guilds had a room as meeting place. Guildhall Museum opened 1978. Covered market, replacing street market around market cross ('Carel Cross') which closed 1927, originated when butchers' shambles were moved to Old Bluebell Lane 1799, to which butter and egg market were also moved 1854. New covered market opened 1889; refurbished from 1989. Civic Centre, at foot of Lowther Street, built 1960-3.

Leper hospital of St Nicholas, Botchergate, probably founded in 12th century; survived Dissolution to be destroyed during siege of Carlisle 1645. Dispensary, founded 1782 in Abbey Street; moved into part of Tithe Barn 1825; new dispensary opened in Chapel Street 1855; absorbed into National Health Service 1948; sold 1962. House of Recovery (with the object of curing and preventing contagious diseases) established 1820 in building near Collier's Lane; moved to Crozier Lodge, beside Cumberland Infirmary, 1847; remained a fever hospital until at least 1910. Cumberland County Infirmary, Newtown Road, built 1830-2; extended 1870-4, 1908-13; new block opened 2000. Home for Incurables (Strathclyde House), Wigton Road, opened 1885; closed 1991. Cumberland and Westmorland Home and Workshop for the Blind established in West Tower Street 1872; moved to new premises in Lonsdale Street, built 1878-9; moved to Petteril Bank 1950s. Parish workhouses for St Mary's parish built in Devonshire Walk 1780s; Caldewgate Workhouse, for St Cuthbert's parish, built in Coal Fell 1829-30. Both closed and replaced by Carlisle Union Workhouse, Fusehill Street, built 1863-4, part of which became City General Hospital (later Carlisle Maternity Hospital) from 1937. It closed 2000; site was taken over by St Martin's College (now University of Cumbria). St Mary's Home for Penitents (for 'reception and reformation of fallen women'), Coal Fell, opened 1872; laundry built 1894; home closed c.1920 but laundry continued; home re-opened at Coledale Hall, Belle Vue, 1926.

Roots of Tullie House Library, Museum and Art Gallery can be traced back to several separate institutions in early 19th century. Carlisle Subscription

Library, established 1768; moved into new premises on corner of English Street/Devonshire Street 1831; amalgamated with public library 1896. Mechanics' Institute established 1824 in Castle Street; subsequently housed in the Athenaeum, Lowther Street, built 1840, also containing large lecture hall, exhibition hall and museum. Institute moved to Fisher Street 1850. In 1891 Mechanics' Institute and Library were presented to corporation. Lord Street Reading Room, established 1848 in John Street, Botchergate; opened 1851. Its library was offered to corporation 1891 on condition city adopted Public Libraries Act. Museum founded by Carlisle Literary and Philosophical Institute 1835 was later transferred to the Athenaeum; collection was seized 1860 for non payment of rent and given to corporation who housed it in Finkle Street where it fell into disuse. Taken over by committee of working men 1874 and re-opened 1877. Tullie House was presented to Corporation 1890 and opened in as Museum and Public Library 1893, with an attached School of Art. In 1986 library moved to Globe Lane; museum redeveloped on Tullie House site; re-opened 1991.

Macready Theatre, Blackfriars Street, built 1813-4; but had fallen into disuse by 1850s. Three theatres by 1860s: Theatre Royal, Swifts Row; Bijou Theatre in Mechanics' Hall, Fisher Street; Matchbox Theatre, on The Sands (opened 1858; used by Salvation Army from 1880; demolished 1892). Her Majesty's Theatre, Lowther Street, built 1874 as Victoria Hall; re-named 1879; closed 1963. Star Music Hall, Rickergate, opened 1877. Palace Theatre/Cinema opened as music hall 1906; closed as theatre 1932; reopened as cinema same year; reopened as The Studios 1970; demolished 2004. Star Picture Hall, opened 1912; changed name to the Rex 1932; became bingo hall 1962. City Cinema, English Street, opened 1915; closed 1960. Lonsdale Cinema opened 1931; closed 2006.

Races held regularly from mid-18th century on course at the Swifts, beside River Eden, with a grandstand built 1839; moved to Blackwell 1904 (see St Cuthbert Without). The Sauceries and the Bitts, open spaces beside Eden became public parks and Botchergate recreation ground, Fusehill Street, opened 1892. Local radio station, BBC Radio Carlisle (later renamed Radio Cumbria), began broadcasting from Carlisle 1973.

CARTMEL (village) *see* Allithwaite, Lower, and Holker, Upper

CARTMEL FELL

Chapelry and township in Cartmel parish, Lonsdale hundred, Lancashire North of the Sands.

Acreage: 4,959 acres [2,007 ha]. Commons enclosed under Cartmel parish enclosure award 1809. ***Population:*** 322 in 1801, rising to 371 in 1821; gradual decline thereafter to 268 in 1901; stood at 309 in 2001. ***Landownership:*** largely held by customary tenants of Cartmel Priory. *Manor of Burblethwaite* held by Knipe family from 14th to 17th centuries; by Robinsons of Fell Foot in 18th century; acquired 1827 by Thomas Atkinson of Kendal. Other estates included Thorphinsty, held by Hutton family from 16th to 18th centuries. ***Economy:*** farming; woodland industries (charcoal-burning; potash burning) important from late-medieval to 19th centuries. Iron smelting: bloomery forge at Burblethwaite in early 18th century. ***Places of worship***: chapel of ease of St Anthony built c.1504; extended c.1520; restored 1911. ***Schools and other institutions:*** school taught by reader of chapel; endowed 1695 with bequest of £200. Small schoolhouse built 1871; replaced 1877 by new schoolhouse at Shanty Beck; closed 1971; now village hall. Richard Frankland's nonconformist academy housed briefly at Hartbarrow 1685.

CASTERTON*

Township in Kirkby Lonsdale parish, Lonsdale ward, Westmorland.

Acreage: 4,324 acres [1,750 ha], including 2,200 acres [890 ha] of common fell enclosed 1816. ***Population:*** c.200 in later 17th century. Rose from 266 in 1801 to 623 in 1841 after establishment of boarding school, which continued to inflate census figures: in 1991 only 265 of 574 inhabitants usually resident. ***Landownership:*** *manor of Casterton* was part of Richmond fee of barony of Kendal (q.v.). Leased 1669 to Wilson (later Carus-Wilson) family, who remained large landholders until end of 19th century. ***Economy:*** largely agricultural. Coalmining from later 17th century; had ceased by mid-19th century. Lime-burning from 17th century. Tourism from later 20th century: caravan park established 1960s. ***Places of worship:*** chapel of St Columba at Chapel House mentioned 1356; closed 1531; demolished by 1692. Holy Trinity church erected 1831-3; chancel added c.1860; restored 1891. ***Schools and other institutions:*** William Carus-Wilson's boarding school for clergy daughters (famous from *Jane Eyre*) moved from Cowan Bridge 1833; closed

2013, becoming junior department of Sedbergh School. Second girls' boarding school, the 'Servants' School', also established by Carus-Wilson, moved to Casterton 1837; re-founded as Endowed Church School 1880s; merged with Clergy Daughters' School 1921. Village school built 1841; closed 1972. Village hall (former mission hall) given to village 1913.

CASTLE CARROCK

Ancient parish in Eskdale ward, Cumberland. CP enlarged by absorbing Geltsdale 2003.

Acreage: 3,031 acres [1,227 ha], formerly divided into two quarters, Town and Outside Quarters; included 1,900 acres [769 ha] of common land on Castle Carrock Moor and Fell, enclosed 1805. ***Population:*** estimated at 360 in 1688. Rising from 252 in 1801 to peak of 383 in 1831, then declining to 271 in 1901. Secondary peak of 346 in 1921; then declining to stand at around 290 for much of later 20th century; 303 in 2001. ***Landownership:*** granted to Eustace de Vallibus in 12th century, whose descendants took surname Castle Carrock. Divided between heiresses; part subsequently came to Musgrave family, who by 1688 had sold part to earl of Carlisle and part to Dacre family. ***Economy:*** largely agricultural; lime-burning in 19th century. Reservoir supplying water to Carlisle opened 1909. ***Places of worship:*** medieval parish church of St Peter; rebuilt 1826-28; 'Normanised' 1888-9. Independent (Congregationalist) chapel built 1853; closed and converted to dwelling. Primitive Methodist chapel built 1899; had closed by 1965. ***Schools and other institutions:*** parish school in village, recorded 1818, built with profits from land granted at enclosure of commons. Rebuilt on new site 1874; now Castle Carrock Primary School. Watson Institute (now Watson Hall), a tower-like structure containing a public hall, library and reading room, built 1897 and given to village by James Proctor Watson.

CASTLERIGG, ST JOHN'S & WYTHBURN

Township in Crosthwaite parish, Allerdale below Derwent ward, Cumberland. CP lost territory to Keswick 1894-9 and 1934.

Acreage: 19,699 acres [7,972 ha] before boundary changes, reducing to 19,589 acres [7,928 ha] after creation of Keswick CP. Further exchanges with Keswick 1934 resulted in net loss of further 107 acres [43 ha]. Township contained c.7,000 acres [2,833 ha] of common land on fells in manor of Castlerigg and Derwentwater, enclosed 1849. Similar acreage of fell commons remains unenclosed in Wythburn. ***Population:*** in range 550-650 for most of 19th century, surging to 1,650 in 1891, when Thirlmere reservoir under construction. Decline across 20th century (partly reflecting loss of territory to Keswick CP), standing at 407 in 2001. ***Landownership:*** contained major part of three manors: for *manor of Castlerigg & Derwentwater*, see Keswick. *Manor of Wythburn* was held by Brathwaites of Warcop in 17th century and sold to George Fletcher of Hutton-in-the-Forest, descending to Fletcher-Vanes. *Manor of Legburthwaite* was held by Leathes family until late 1870s, when family's Dalehead estate was sold to Manchester Corporation for Thirlmere reservoir scheme. ***Economy:*** predominantly hill farming. Forestry developed on Derwentwater estate in Castlerigg in 18th century and around Thirlmere reservoir, supplying Manchester from 1894, in late 19th and 20th centuries. Tourism grew in importance from late 18th century. For Threlkeld granite quarry (which lay in this township), see Threlkeld. ***Places of worship:*** township comprised two chapelries of St John's-in-the-Vale and Wythburn. Both chapels of ease recorded 1554; St John's rebuilt 1845; Wythburn rebuilt 1640 and again 1740, with chancel added 1872. ***Schools and other institutions:*** school beside chapel at St John's by 1814; rebuilt 1848; closed 1948 (now diocesan youth centre). School attached to Wythburn chapel by 1833; closed 1937. Thirlmere Recreation Hall, Legburthwaite, opened 1942 (building had formerly been mission church for construction workers at Haweswater).

CASTLE SOWERBY

Ancient parish in Leath Ward, Cumberland, comprising five townships or 'bounds': Bustabeck, Howbound, Rowbound, Southernby and Stockdalewath.

Acreage: 8,605 acres [3,482 ha], divided between 'bounds' thus: Bustabeck: 1,743 acres [705 ha]; Howbound: 1,674 acres [677 ha]; Rowbound: 1,178 acres [477 ha]; Southernby: 1,130 acres [457 ha]; and Stockdalewath: 1,534 acres [621 ha]. Sowerby pasture and other commons, totalling c.5,000 acres [c.2,000 ha], enclosed 1769. ***Population:*** estimated at 705 in 1688; rising from 515 in 1801 to peak of 1,015 in 1841; then steadily declining to 337 in 2001. ***Landownership:*** part of honour of Penrith (q.v.), with which it descended. ***Economy:*** farming: noted for its corn, especially oats, which was said to be 'the best in the county' in later 17th century; woodland on former common land, with charcoal-burning, perhaps in medieval period. Coal mining at Hewerhill by later 17th century; mine there in mid-19th century; mining had ceased by 1900. Limestone and free stone quarrying; brick- and tile-making in Bustabeck Bound in mid-19th century. ***Places of worship:*** medieval parish church of St Kentigern in Howbound township. Said to be remains of chapel called Lady Chapel at 'Birksceugh' in early 19th century. Chapel of ease at Raughton Head, now church of All Saints, consecrated 1678 (possibly replacing earlier building); rebuilt 1761. Quaker meeting house at Sowerby Row built 1713; closed 1787. For Stockdalewath Methodist chapel, see Dalston. ***Schools and other institutions:*** school at Sowerby Row endowed 1751 by John Sowerby; closed c.1963. Schoolmaster at Raughton Head chapel in late 17th century; school adjoining chapel yard built 1744; rebuilt 1806; dismantled and materials reused on new site 1857; now Raughton Head CE Primary School. Village hall at Millhouse, rebuilt 1998-9.

CATTERLEN

Township in Newton Reigny parish, Leath ward, Cumberland. CP enlarged by addition of Newton Reigny CP 1934.

Acreage: 1,561 acres [643 ha], including 417 acres [172 ha] of commons, enclosed 1819. ***Population:*** 92 in 1801, rising to 150 in 1871, falling to 98 in 1931 (last census year before boundary change). ***Landownership:*** *manor of Catterlen* held by Vaux family until 1657, when it passed by marriage to Richmonds of Highhead, who held it until 1774. It then passed by will to Isabella, wife of Henry Curwen of Workington, who sold it to duke of Norfolk, from whom it passed to Howards of Greystoke. ***Economy:*** agricultural. ***Places of worship:*** supposed site of abandoned church. Primitive Methodist chapel built 1867. ***Schools and other institutions:*** school in Catterlen for whole of Newton Reigny parish, endowed by Isabella Miller (d. 1736); rebuilt 1835; closed before 1924 and converted to dwelling.

CHURCH CONISTON (formerly Coniston Fleming)

Chapelry and township in Ulverston parish, Lonsdale hundred, Lancashire North of the Sands. Enlarged by addition of Monk Coniston (see Hawkshead and Monk Coniston with Skelwith) to form Coniston CP 1894.

Acreage: 7,424 acres [3,004 ha] before enlargement, including 1,097 acres [444 ha] of common land on Tilberthwaite Fell, enclosed 1858. After addition of Monk Coniston 1894, Coniston CP covered 10,427 acres [4,220 ha]. ***Population:*** rising from 338 in 1801 to 587 in 1831; then jumping to 1,148 in 1841. Rose to a peak of 1,324 in 1861 (numbers swelled by railway builders) and then gradually declined to 962 by 1921 (last census year for which separate figure available). ***Landownership:*** *manor of Coniston* was granted to William de Lancaster c.1160 and was granted by Gilbert fitz Reinfred to Gilbert son of Bernulf c.1200, from whom it descended to Adam de Urswick, passing by marriage of his daughter to Richard le Fleming c.1250 and descending through Flemings of Rydal until 20th century. *Manor of Tilberthwaite* granted to Pennington family by early 13th century; was subdivided by 1730. ***Economy:*** livestock farming and woodland industries. Charcoal burning (which continued into 20th century) and iron smelting from medieval period. Bloomery forge from c.1675 to c.1750. Copper mining from late 16th century; major expansion in 19th century; closed 1942. Limestone and slate quarrying recorded later 17th century. Major expansion of slate quarrying and mining, particularly in Tilberthwaite, in 19th-20th century. Furness railway opened 1859; closed 1962. Tourism increasingly important from 19th century. ***Places of worship:*** chapel of ease recorded 1578, granted parochial rights 1586; rebuilt as St Andrew's Church 1819; chancel added 1891. Baptist chapel at Bowmanstead erected 1837, enlarged 1841; used by Plymouth Brethren c.1894 to 1903, then reverted to Baptists; sold to private owners after 1950. Primitive Methodist chapel established 1859 by Cornish miners; had been converted into Masonic Hall by 1914. Wesleyan Methodist chapel established 1875; still in use. Roman Catholic mission from 1866; church of Sacred Heart established 1872; still in use. Plymouth Brethren continued to meet in houses 1903 to 1972. ***Schools and other institutions:*** school taught in chapel until c.1790. Two small schools recorded 1818. New boys' school established 1854 (old school building became Mechanics' Institute); school closed and building became Diocesan Youth Centre 1970; closed 2014. Coniston Primary School built on new site north of village in later 20th century and secondary school (John Ruskin School) built to east of village. Institute and library, Yewdale Road, built 1878; enlarged 1898 and further enlarged 1901 by addition of Ruskin Museum, which was renovated and extended 1999.

CLAIFE

Township in Hawkshead parish, Lonsdale hundred, Lancashire North of the Sands.

Acreage: 4,457 acres [1,804 ha], including 1,500 acres [607 ha] of common land, enclosed 1799. Township contained two divisions: Upper Claife (Wray area): 2,077 acres [841 ha]; and Lower Claife (containing Near and Far Sawrey): 2,501 acres [101 ha]. ***Population:*** rising from under 400 in early 19th century to 563 in 1901 (peak: 627 in 1891); then falling across 20th century to stand at 298 in 2011. ***Landownership:*** part of Furness Abbey's estate in Furness Fells. Curwen family of Workington held extensive estate on Claife Heights by early 19th century. James Dawson, Liverpool surgeon, built Wray Castle 1840-7. ***Economy:*** farming and woodland industries, particularly charcoal burning. Iron smelting at Cunsey forge from 1618; bloomery forge superseded by blast furnace 1711-12; closed 1750. Tourism increasingly important from 18th century; Claife was western terminus of Windermere ferry with viewing station nearby. ***Places of worship:*** church of St Margaret of Antioch, Low Wray, said to have been built 1845 (porch sundial dated 1856). St Peter's Church, Far Sawrey, built 1866-72. Quaker burial ground at Colthouse from 1658; Quaker meeting house built nearby 1688; extended 1978; still in use. ***Schools and other institutions:*** school at Far Sawrey founded 1775; closed 1971; now used as village hall ('Braithwaite Hall'). School at High Wray built early 1840s; closed 1931; now village hall. Claife poor house, beside Esthwaite Water on boundary between the two divisions, mid-19th century. Sawrey Institute, opened as reading room 1884.

CLEATOR

Ancient parish in Allerdale above Derwent ward, Cumberland. Became Cleator Moor UD 1894, reverting to CP 1934, when name changed to Cleator Moor CP.

Acreage: 2,947 acres [1,193 ha], including c.400 acres [162 ha] of common land on Cleator Moor and c.1,200 acres [486 ha] on Dent Fell, enclosed 1825. ***Population:*** estimated at 330 in 1688; doubled from 362 in 1801 to over 763 by 1841; then surged across second half of 19th century as result of industrial development, attracting many immigrants from Ireland. New town of Cleator

Moor laid out on former common from 1880s. By 1861 population stood at 3,995; it peaked at 10,420 in 1881. It then declined, to 8,120 by 1901 and to low point of 6,411 in 1951, after which it rose to 7,686 by 1971; in 2001 it stood at 6,939. ***Landownership:*** *manor of Cleator* thought to have been possession of St Bees Priory; tenants enfranchised before Dissolution. ***Economy:*** iron smelting beside River Ehen from late 17th century: first blast furnace in Cumbria built at Cleator 1694; also steel and bar-iron works (closed 1799) and spade forge. Whitehaven Haematite Iron & Steel Co. established blast furnaces 1842. Major 19th-century iron ore (haematite) mines at Todholes, Crossfield, Crowgarth, Jacktrees and Montreal, most of which closed early 20th century. Coal mining began 1788; expansion in mid-19th century; most had closed by early 20th century. Limestone quarrying and lime-burning in 19th century. Flax mill established 1800; rebuilt 1859; became Kangol hat factory, established by Jacques Spreiregen 1938; closed 2009. Engineering works, Birks Road, founded 1870; in operation 1968. Brewery built 1873; closed 1977. Closure of heavy industries between World Wars led to establishment of industrial estates with some light manufacturing industry. ***Places of worship:*** medieval parish church of St Leonard; largely rebuilt 1841 and reconstructed 1900-03. Cleator Moor became separate ecclesiastical district 1869, served by new parish church of St John (consecrated 1872). Mission church at Wath Brow built 1881. Quaker meeting house at Crossfield built 1677; closed c.1746. Roman Catholic mission church of St Bega from 1853; present church (Our Lady of the Sacred Heart) consecrated 1872. Wesleyan Methodist chapel built 1862; destroyed by fire 1932; new church built 1934; still in use. Presbyterian church built 1865; rebuilt 1893; closed. Primitive Methodist chapel built 1892; closed 1965. Other nonconformist chapels included Congregational chapel, built 1876 and United Methodist Free Church (by 1900). ***Schools and other institutions:*** British school established at Cleator flax mill c.1850; became council school 1910; since closed. Montreal CE Primary School, Cleator Moor, originated 1866 in grand building, abandoned because of subsidence 1946, when school moved to prefab buildings at Wath Brow; replaced by new building 1986. Seniors moved to Ehenside Secondary School, Cleator Moor, 1958; demolished 2008. Roman Catholic school at Cleator (founded 1853); closed on merging with St Patrick's RC Primary School, Cleator Moor (founded 1873); still open. Town Hall, built 1877. Library at Cleator, recorded 1860. Public library at Cleator Moor, adjoining Town Hall, established 1894; forerunner of Carnegie Library, built 1906. Community hall (Jubilee Room), Church Street, Cleator, built 1899.

CLIBURN

Ancient parish in West ward, Westmorland. Detached portion of parish at Winderwath was transferred to Brougham CP 1935.

Acreage: 1,890 acres [758 ha], including detached portion at Winderwath (503 acres [204 ha]). Approximately 603 acres [c. 244 ha] of common land enclosed 1820 and common field at Cliburn Ling (72 acres [29 ha]) enclosed 1867. ***Population:*** rising from 157 in 1801 to 367 in 1861 (attributed to workers laying railway line), then falling gradually to 204 in 2001. ***Landownership:*** *manor of Cliburn* comprised two divisions: Cliburn Talebois and Cliburn Hervey, named after their owners; subsequently held as one manor by Cliburn family from 15th century until sold 1654. Manor and demesne bought from Edward Lee by Sir John Lowther 1667 and descended in that family. Detached portion at *Winderwath* was separate estate, passing from Cliffords to Thomas Braithwaite before 1612; it then passed to Wyvill family who held it from before 1658 until they sold it 1787 to Hasells of Dalemain, who held it until mid-19th century. ***Economy:*** predominantly agricultural: lands said to be 'dry, sandy and barren' in later 17th century. ***Places of worship:*** medieval parish church of St Cuthbert; restored 1849, followed by major restoration 1886-7. Medieval chapel at Winderwath (recorded 1343). Methodist chapel built 1852; closed 1974; converted to dwelling. ***Schools and other institutions***: free school endowed 1807; school built 1809, then rebuilt on new site 1877; closed 1983. Original school building used as reading room until Second World War. Village hall (a building relocated from Haweswater where it had been used for workers building dam) replaced by new village hall, opened 2002.

CLIFTON, GREAT

Township in Workington parish, Allerdale above Derwent ward, Cumberland.

Acreage: 998 acres [404 ha]. Clifton Common (900 acres [364 ha]), serving both Great Clifton and Little Clifton (q.v.), enclosed 1817. ***Population:*** 268 in 1801; comparatively steady until mid-19th century, when it surged from 374 in 1851 to 609 in 1861, as result of opening of collieries. Continued to rise to peak of 1,635 in 1961, since when it declined, to stand at 1,101 in 2001. ***Landownership:*** vill of Clifton, comprising Great Clifton, Little Clifton and Stainburn, was one of Five Towns in honour of Cockermouth. *Manor of Clifton* was in hands of Eaglesfield family, from whom it passed by marriage in early 16th century to Bardseys and on death of Nicholas Bardsea (d. 1586), to his daughter Elizabeth, wife of Lancelot Salkeld of Whitehall, through whose family it descended until sold to Sir James Lowther in 18th century. ***Economy:*** coal mining by 1635 (colliery said in 1688 to be worth £100); expanded in 18th and 19th centuries. Main pits were in north of township: Lowther Pit, with coke ovens at Coldfitz Wood, and William Pit. Mining continued until after nationalisation: Clifton colliery closed 1959. Opencast mining by 1970s. Pottery at Crossbarrow in mid-19th century; closed by 1900. Sandstone quarry at Brackenbarrow mid-19th century; closed by 1900. ***Places of worship:*** for medieval chapel of ease, see Little Clifton. Wesleyan Methodist chapel erected 1845; rebuilt 1892; replaced by modern building on new site. ***Schools and other institutions:*** Board school for Great and Little Clifton built near William Pit 1883 (now a dwelling); replaced by Derwent Vale Primary school, built on new site 1995. Co-operative Society Hall used for public meetings and workmen's reading and games room, recorded 1938.

CLIFTON, LITTLE

Township in Workington parish, Allerdale above Derwent ward, Cumberland, including village of Bridgefoot.

Acreage: 1,089 acres [441 ha]. ***Population:*** 166 in 1801; had risen to 476 by 1861, attributed to opening of collieries; peaked at 599 in 1871; declined across 20th century to stand at 391 in 2001. ***Landownership:*** see Great Clifton. ***Economy:*** Clifton iron furnace (first coke-fired blast furnace in Cumbria) erected 1723 by Messrs Cookson; closed 1781. Coal mining, with collieries in 19th century at Harrygill, Reelfitz and Westray (ceased by 1900). Forges at Bridgefoot (sickles and edge tools) and Reelfitz (spade manufactory) in mid-19th century. Brickworks near Lowther Pit in late 19th century. ***Places of worship:*** medieval chapel of ease serving both Little and Great Clifton townships. Had become 'very ruinous' and partly collapsed by 1717; rebuilt, perhaps 1734; again 1858, as St Luke's church, when ecclesiastical parish of Clifton was formed; rebuilt again 1900. Wesleyan Methodist chapel at Bridgefoot built 1867; rebuilt 1875. ***Schools and other institutions:*** National school, Chapel Brow, built c.1860; closed and converted to dwelling. Reading room established 1896; by 1903 the Allerdale Coal Company had provided reading and recreation room at Chapel Brow, known as William Fletcher Memorial Institute.

CLIFTON (Westmorland)

Ancient parish in West ward, Westmorland.

Acreage: 1,782 acres [721 ha], including c.125 acres [50 ha] of common land, enclosed 1816. ***Population:*** rising gradually from 219 in 1801 to 461 in 2001. ***Landownership:*** *manor of Clifton* given by Sir Hugh de Morville to Gilbert Engaine in reign of Henry II; carried by marriage of Eleanor daughter of Gilbert Engaine to William de Wybergh of St Bees 1365; descended in Wybergh family until c.1920. ***Economy:*** Wetheriggs brick and tile works near Clifton Dykes established 1760; pottery production started 1855; closed as pottery c.2008. Freestone and limestone quarries. Clifton Hill racing stables built 1912. Sawmills established 1963; acquired by A.W. Jenkinson Forest Products 1994. ***Places of worship:*** medieval parish church of St Cuthbert; restored and chancel rebuilt 1849. Methodist chapel built 1885; sold 1986 and converted to dwelling. ***Schools and other institutions:*** schoolmaster funded by Mary Scott's legacy 1764; school housed in building at Town End. New school erected 1874; restored and enlarged 1901; modern building of Clifton Primary School adjacent. Reading room, now village hall.

COCKERMOUTH

Chapelry, borough and market town in Brigham parish, Allerdale above Derwent ward, Cumberland. Became UD 1894 (until 1974). For the Goat area, transferred from Papcastle CP 1934, see Papcastle.

Acreage. 2,425 acres [981 ha] before boundary changes, including 1,500 acres [607 ha] of common moorland, enclosed 1832. After gaining the Goat area (133 acres [54 ha]) from Papcastle CP, and losing 518 acres [210 ha] to Setmurthy CP in 1935, Cockermouth UD comprised 2,040 acres [826 ha] CP gained further land to west of town from Brigham CP in early 1990s.

Landownership. From 12th century Cockermouth Castle was caput of *barony of Allerdale* and *honour of Cockermouth* (comprising the Five Towns south-west of town and forest of Derwentfells, the land between rivers Cocker and Derwent). Honour of Cockermouth descended through heirs of Waldeof, lord of Allerdale in early 12th century, to William Fitz Duncan (d. c.1153). After death of his widow Alice de Rumilly (d. 1186-7), it was partioned between daughters Cecily, wife of William le Gros, earl of Aumale, and Mabel wife of Reginald de Lucy. Aumale share descended to William de Fortibus III (d. 1259), earl of Aumale, and then to his widow Isabella (d. 1293), when it reverted to Crown. Lucy share descended via Alice de Lucy, wife of Alan de Moulton, whose heirs took surname Lucy, to Anthony de Lucy (d. 1343), who was granted Aumale share 1323, thus reuniting estate. His granddaughter, Maud (d. 1398), settled her estates on her second husband Henry Percy, 1st earl of Northumberland, thereby bringing Cockermouth to Percy family, in which it descended (with some breaks when confiscated to Crown) to Joceline, 11th earl of Northumberland (d. 1670), whose daughter Elizabeth married Charles Seymour (1662-1748), 6th duke of Somerset. Their son Algernon, 7th duke of Somerset, became 1st earl of Egremont (1749) but died childless 1750, the estate passing through his sister Katherine, wife of Sir William Wyndham, to Charles Wyndham (d. 1763), 2nd earl of Egremont, since when it has descended in Wyndham family.

Origins and growth of the town. Cockermouth was a seigniorial borough, established, probably in later 12th century, at foot of castle. Core of town appears to have been Market Place to east of River Cocker; a planned extension, the wide, bowed Main Street to west of Cocker, had been laid out by early 13th century. It was market centre for a wide agricultural hinterland (market charter granted 1221), acting as central cog in economy through which rural produce was distributed and in which industrial processes were concentrated. Rental of c.1270 records a thriving borough, its wealth based on woollen cloth and leather trades, and lordly income also deriving from salmon fishery in River Derwent. By 1670s, Sir Daniel Fleming could call Cockermouth 'the best Market Towne in this part of the county', with many fine buildings and 'no small reputation'. Its weekly market, fortnightly cattle fairs and Whitsuntide and Martinmas hiring fairs drew custom from a wide area. Water power provided by rivers Cocker and Derwent and two becks, drove corn and fulling mills from 12th century. Population may have been around 1,000 in later 13th century and was estimated at 965 in 1688. Water power led Cockermouth's transformation into industrial centre in 18th and 19th centuries, adding mills and factories to its market function. Paper mill at Simonscales by 1760s (a flax mill in mid-19th century); corn mill at Rubby Banks became textile mill by mid-19th century. By later 19th century, mainstays of town were flax and woollen mills and agricultural service industries (implement making; auction mart). Prosperity and industrial growth were reflected in increase in population to 2,652 in 1785 and 2,865 in 1801, with doubling to 5,775 in 1851. In first half of 19th century there were over 40 industrial sites: 5 corn millers, 7 tanners, 14 textile manufacturers, 2 dyers, 4 hat manufacturers, 5 nail makers, 3 brick makers, 4 brewers and maltsters, as well as spectrum of smaller businesses typical of a market town, from clog makers to clockmakers and tinsmiths to tailors. Arrival of railways (from Workington 1847; from Penrith and Keswick 1864) saw town expand to south of its medieval confines, with new housing on The Moor. Two larger businesses founded in 19th century survive: Mitchell's Auction Co. (established 1849) and Jennings Castle Brewery (moved from Lorton (q.v.) 1887). For industrial history of the Goat, see Papcastle. Population levelled at around 5,300 from 1861 to 1951 (dipping under 5,000 in 1921 and 1931). Cockermouth again flourished in later 20th century, initially as dormitory town, but increasingly as tourist centre, growing steadily from 5,235 in 1951 to 7,877 in 2001.

Places of worship. Medieval chapel of ease on site of All Saints Church (confirmed by names Kirkgate, Kirkwent, Kirkcroft, recorded in late medieval period). Dedication changed from St Mary to All Saints by 1385. Rebuilt 1711 (retaining 14th-century tower); destroyed by fire 1850 and rebuilt again 1852-4; remained parochial chapel under Brigham until 1865. Other medieval religious sites included chapel of St Helen, at head of St Helen's Street, recorded 1270 and surviving into late 15th century, and St Leonard's chapel at west end of town, also recorded 1270. In 1865 a second Anglican church, Christ Church, opened to serve western end of town. Independent (Congregational) church established 1651, meeting first in members' houses, then, from 1687, in converted house on The Sands; chapel built 1719; rebuilt 1735; present church (now United Reformed Church) adjacent to older chapel, built 1850. Quaker meeting house at head of Kirkgate built 1688; rebuilt 1781 and again 1884. Earliest Methodist chapel in High Sand Lane (now Victoria Hall); new Wesleyan chapel built 1841 (now Town Hall); replaced in turn by Lorton Street Methodist Church, built 1932. Primitive Methodists rented High Sand Lane chapel 1841, buying it 1851; moved to vacant National school in New Street 1885. High Sand Lane chapel used by Salvation Army in 1901. Roman Catholic services held in barn behind Sun Inn at foot of Kirkgate until St Joseph's RC Church, Crown Street, built 1856. Cemetery on Lorton Road, with two mortuary chapels, laid out 1856.

Schools. Grammar school prior to 1546; first recorded building alongside All Saints Church, erected 1676. School of Industry 'for the gratuitous education of 30 poor girls' established 1809. National school, New Street, built 1845; became a Board school 1870. National school moved to new premises in Kirkgate 1869, becoming All Saints National School; moved again to new premises on Slatefell Drive 1973; now All Saints CE Primary School. British school, Market Street, built 1847. Fairfield Girls' Board School built 1876, and Fairfield Boys' Board School opened 1884, replacing New Street school.

New building on adjacent site opened 1971; now Fairfield Primary School. School adjacent to St Joseph's RC Church from 1877; moved to new premises on The Level 1967; now St Joseph's RC Primary School. Cumberland County Industrial School (where boys sent by courts for vagrancy /crime were taught trades in addition to basic education) built near Lorton Road 1881; closed 1921. Cockermouth Secondary School opened in former Industrial School premises 1929, becoming Grammar School 1958 to 1984. Derwent Secondary Modern School, Castlegate Drive, opened 1958; amalgamated with grammar school to become comprehensive Cockermouth School 1984. Former grammar school site closed 1991.

Other institutions. Widows' Hospital in Kirkgate endowed by Thomas Leathes 1760. Dispensary established 1785. Nursing home or small hospital had been established at Harford House by 1905; Cottage Hospital, Isel Road, built 1915; replaced by new community hospital on adjacent site, opened 2014. Cockermouth Poor Law Union Workhouse, Gallowbarrow, opened 1843, serving some 50 townships; closed 1929. (Union School educated workhouse children until 1887 when it was moved to Flimby). Mechanics' Institute with library and reading room established 1845. Wordsworth Institute, Main Street, providing lectures and adult classes, founded 1882; moved to Christ Church rooms c.1921. Book club, founded 1785, had become Cockermouth Library by 1847, its collection forming nucleus of Carnegie Public Library, Main Street, opened 1904. Public meeting places included Public Hall (built 1874), Royal Assembly Rooms and Drill Hall (built 1886). Public park, Harris Park, laid out 1894. Cinema, the Grand Theatre, built 1913-14; closed 1978. Kirkgate Centre (cinema and arts centre) established in former All Saints National School building 1995.

COLBY

Township in Appleby St Lawrence parish, East ward, Westmorland. Colby CP, created 1894, covered larger area than township.

Acreage: 1,181 acres [478 ha] before boundary changes; after 1894 CP covered 1,402 acres [567 ha]. Colby Moor (360 acres [146 ha]) and Colby Pasture enclosed 1854. ***Population:*** around 140-150 in period 1821-51; 178 in 1861; had fallen back to 121 by 1891 (last census year before boundary changes). Population of Colby CP fell to 79 by 1961 but had risen to 120 by 2001. ***Landownership:*** held by Colby family 12th century to c.1400, when passed by marriage to Warcops who held it until c.1660, when most of tenants enfranchised. ***Economy:*** predominantly agricultural; small-scale quarrying and lime-burning mid-19th century. Tourism by early 21st century: caravan park at Hawkrigg Farm. ***Places of worship:*** mission room ('iron church') built later 19th century; in use 1938. United Methodist chapel built 1874; closed; rebuilt as dwelling 2002. ***Schools and other Institutions:*** village hall established 1990s (building formerly a school classroom elsewhere).

COLTON

Parish (carved out of Hawkshead parish 1676) in Lonsdale hundred, Lancashire North of the Sands, including the settlements of Bouth, Finsthwaite, Haverthwaite, Nibthwaite and Rusland.

Acreage: 14,322 acres [5,796 ha], embracing seven constituent divisions: Bethecar Moor: 1,392 acres [563 ha]; Colton East: 2,939 acres [1,189 ha]; Colton West: 1,709 acres [692 ha]; Finsthwaite: 2,258 acres [914 ha]; Haverthwaite: 1,256 acres [508 ha]; Nibthwaite: 2,551 acres [1,032 ha]; Rusland: 2,219 acres [898 ha]. ***Population:*** rising from 1,516 in 1801 to peak of 2,008 in 1851, then dropping back to 1,648 by 1901 and falling across 20th century to 765 in 2001. ***Landownership:*** part of Furness Abbey's lordship of Furness Fells; thus part of manor of Hawkshead (q.v.). ***Economy:*** pastoral farming, woodland industries and ironworking predominated. Furness Abbey had stock farms (some termed 'herdwicks') in late-medieval period. Coppice woodland, yielding charcoal for iron smelting and other woodland products, from medieval period to 20th century. Iron-working sites include Backbarrow (bloomery forge c.1685; replaced by blast furnace built 1711, which continued to use charcoal until 1926; furnace ceased production 1967); Nibthwaite (furnace-forge, operating 1736-c.1850), Stony Hazel, Rusland (bloomery forge, operating 1718-43). Bobbin manufacture at Nibthwaite (2 mills, one built 1834) and Stott Park (built 1835; production ceased 1971). Gunpowder making at Bouth from 1862 to 1928. Other industries using woodland products included tanning (at Nibthwaite and Rusland) and basket and hoop-making. Furness railway opened Newby Bridge branch through Haverthwaite 1869; closed 1965. Tourism increasingly important across 20th century: modern attractions include lake cruises from Lakeside; Haverthwaite & Lakeside steam railway (established 1973); Stott Park Bobbin Mill; Lakes Aquarium. ***Places of worship:*** chapel recorded 1530; consecrated 1578; gained parochial status 1676; extended 1721; restored 1889-90; now parish church of Holy Trinity. St Peter's church, Finsthwaite, built 1724; rebuilt 1873-4. St Paul's church, Rusland, built 1745; rebuilt 1866-69. St Anne's church, Haverthwaite, built 1824-5. Baptist chapel at Tottlebank built c.1750; extended 1864 to accommodate schoolroom. Quaker meeting house at Rookhow built 1725 for monthly business meetings; stable block converted to hostel 1992. Methodist chapel, Backbarrow, built 1850; closed 1991 and rebuilt as Chapelstone Cottages 1999. ***Schools and other institutions:*** preaching schoolmaster at Colton chapel endowed by Adam Sandys of Bouth 1662; schoolhouse built 1745; closed c.1969. School built at Finsthwaite 1724, at same time as chapel and taught by curate; endowed 1729; rebuilt 1874; closed 1969; building subsequently used as Lakeside & Finsthwaite Village Hall. National school built at Rusland 1840; became Satterthwaite & Rusland CE Primary School; closed 2006. School at Haverthwaite, c.1871 to 1969. Older schools replaced by Leven Valley CE Primary School later 20th century. Reading rooms at Rusland (built 1890; still in use), Bouth (built c.1910 by gunpowder works for their employees; now village hall) and Oxen Park (built by subscription c.1903).

CONISTON *see* Church Coniston; Hawkshead &
Monk Coniston with Skelwith

COPELAND (barony) *see* Egremont

CORNEY

Ancient parish in Allerdale above Derwent ward, Cumberland. Absorbed into Waberthwaite CP 1934.

Acreage: 4,440 acres [1,797 ha], including detached portion of 81 acres [33 ha] at Grange, in Waberthwaite, and 1,342 acres [543 ha] of unenclosed common on Corney Fell. Corney Moor enclosed 1818. ***Population:*** estimated at 480 in 1688; rising from 222 in 1801 to a peak of 292 in 1831, then declining steadily to 185 in 1931 (last census year for which separate figure available). ***Landownership:*** *manor of Corney* thought to have passed by marriage from de Corney family to Penningtons of Muncaster in 13th century; they continued to hold manor of Corney and Middleton Place in 20th century. ***Economy:*** predominantly agricultural; noted for its cattle in mid-19th century. Sporadic small-scale iron mining recorded in mid-19th century. ***Places of worship:*** medieval parish church of St John the Baptist; vestry added 1847; restored 1882. Nunnery of Seaton or 'Lekeley' (on boundary with Bootle parish) founded late 12th century by lord of Millom; remained a very small, poor house. ***Schools and other institutions***: none.

COTEHILL *see* Wetheral

CRACKENTHORPE

Township in Appleby St Michael parish, East ward, Westmorland. CP created 1894 from part of Appleby St Michael parish not in Appleby MB.

Acreage: 1,357 acres [550 ha]. ***Population:*** in range 100-130 across century 1821 to 1921, then falling to stand at 77 by 2001. ***Landownership:*** *manor of Crackenthorpe* held by Machell family from 12th century until sold by Lancelot Machell (d. 1788) to earl of Lonsdale 1786. Bought back 1877 by Captain James Octavius Machell (d. 1902); sold out of family again 1928. Crackenthorpe Hall rebuilt 1685; extended 1880s. ***Economy:*** predominantly agricultural. ***Places of worship:*** site of St Giles' chapel at Chapel Hill. Primitive Methodist chapel ('tin tabernacle') 1901; in use until 1960s. ***Schools and other institutions:*** none.

CROGLIN

Ancient parish in Leath ward, Cumberland, comprising townships of Croglin and Newbiggin. Most of parish (except small area added to Cumrew CP) absorbed into Ainstable CP 1934.

Acreage: parish contained 7,112 acres [2,930 ha], divided between townships thus: Croglin: 4,986 acres [2,018 ha]; Newbiggin: 2,127 acres [861 ha]. Croglin High and Low commons and Croglin and Newbiggin townfields (5,900 acres [2,431 ha] in all) enclosed 1815. ***Population:*** estimated at 345 in 1688; rising from 169 in 1801 to peak at 362 in 1831; then dropping back to 190 in 1931 (last census year for which separate figure available). ***Landownership:*** *manor of Croglin* held by Hastings family from 12th century, passing by marriage to Wharton family c.1300; sold by trustees of duke of Wharton 1739 to duke of Somerset, owner of Egremont and Cockermouth Castle estates. *Manor of Brackenthwaite and Newbiggin* in hands of Dacre family by c.1500 and thus descending to earls of Carlisle. ***Economy:*** farming; grouse-shooting; coal mining, quarrying (limestone, freestone, and marble or 'porphyry') and lime-burning in 19th century; livestock fair. ***Places of worship:*** medieval parish church of St John, rebuilt 1878, closed 2012. Wesleyan Methodist chapels at Newbiggin (built 1847; rebuilt on new site 1867) and Croglin (built 1877). ***Schools and other institutions:*** school taught in church 1704; endowed 1723; rebuilt on common midway between Croglin and Newbiggin c.1770; rebuilt 1866; closed c.1962. Reading room and library recorded 1901. Village hall and reading room built by subscription 1937.

CROOK

Township in Kendal parish, Kendal ward, Westmorland. CP enlarged by addition of parts of several adjacent CPs 1935.

Acreage: 2,119 acres [858 ha] before boundary changes. Commons enclosed 1823. ***Population:*** rising from 179 in 1801 to peak of 279 in 1871 and 1881, then falling back to 230 by 1931 (last census year before boundary changes). Population of enlarged CP stood at 340 in 2001. ***Landownership:*** part of both Marquis and Lumley fees of barony of Kendal (q.v.). Crook Hall (formerly Thwatterden Hall) owned in mid-16th century by Myles Philipson and was in Philipson family for several generations. ***Economy:*** largely agricultural; woollen worsted manufacture and bobbin turning at Crook Mill by 1830, and machine works at Birks Mill in 19th century. ***Places of worship:*** medieval chapel of ease (14th-century bell); surviving tower of old church, dedicated to St Catherine, thought to date from 1620. Rebuilt on new site 1887. Quaker meeting house built 1711; demolished 1841, but burial ground remained. Independent (Congregational) chapel built 1866; closed 1966. ***Schools and other institutions:*** schoolhouse built 1795; rebuilt near Howes 1852; superseded by new school in centre of township 1873; closed 1984. Training colony for farm work (for epileptic boys) at Browhead (see Crosthwaite & Lyth). Village hall (Crook Memorial Hall) built 1926-7.

CROSBY-ON-EDEN

Ancient parish in Eskdale ward, Cumberland, comprising townships of High Crosby, Low Crosby and Walby. Brunstock was detached part of Crosby parish (see Stanwix). Crosby-on-Eden CP abolished 1934; part subsumed into Stanwix CP, part to Wetheral CP.

Acreage: ancient parish covered 2,855 acres [1,155 ha], of which 561 acres [227 ha] formed detached portion at Brunstock. Remaining 2,294 acres [928 ha] divided between constituent townships thus: High Crosby: 967 acres [391 ha]; Low Crosby: 647 acres [262 ha]; Walby: 523 acres [212 ha]. Common fields called Crosby Holme, the Tofts, Little Hill and the Willows enclosed 1772. ***Population:*** rising from 258 in 1801 to 337 in 1831; decline in mid-19th century, then rise to 398 in 1881. Fell sharply to 297 in 1891 and declined thereafter, to 238 in 1931 (last census year for which separate figure available). ***Landownership:*** part of barony of Linstock, granted to second bishop of Carlisle by Waldeof son of Gospatric, lord of Allerdale. ***Economy:*** mostly agricultural; modern golf course beside Eden. For RAF Crosby-on-Eden, Second World War military airfield, see Irthington. ***Places of worship:*** medieval parish church of St John the Evangelist; rebuilt 1854. ***Schools and other institutions:*** schoolhouse built 1807. Crosby-on-Eden CE Primary School built by subscription as National school 1844; enlarged 1882. Village hall at Low Crosby since mid-20th century.

CROSBY GARRETT

Ancient parish in East ward, Westmorland. For Little Musgrave, a detached portion of parish, see Musgrave.

Acreage: ancient parish (including Little Musgrave) contained 5,108 acres [2,067 ha], of which 3,900 acres [1,578 ha] constituted Crosby Garrett township. Part of commons, totalling 434 acres [176 ha], enclosed 1850; remainder of Crosby Garrett Common (2,033 acres [823 ha]) regulated 1884 under Commons Act 1876. ***Population:*** rising from 177 in 1801 to 245 in 1861. Temporary surge to 585 in 1871 as result of railway construction but fell back to 224 in 1881, since when decline across 20th century to low of 107 in 1971; stood at 112 in 2001. ***Landownership:*** *manor of Crosby Garrett* held by William de Soulby 1296; in possession of Musgraves by 1314 who held it until early 17th century, when it passed by marriage to John Davison of Co. Durham. Sold early 18th century to Thomas Gale of Whitehaven, who devised it to Rev. William Bird (d. 1746), rector of Crosby Garrett, whose grandson William (d. 1822), also rector, sold it. Held by Thompson family of Stobars Hall in 19th century. ***Economy:*** predominantly agricultural; quarrying and lime-burning. Mill on Scandale beck converted to hand-loom weaver's shed c.1760. Station on Midland Railway's Settle-Carlisle line opened 1876; closed 1952. ***Places of worship:*** medieval parish church of St Andrew; partly rebuilt 1866 and 1874; extensively restored 2010. Site of medieval chapel at Chapel Well by Scandale Beck. Mossgill Baptist chapel built privately 1815; described as 'Presbyterian meeting house' c.1830 but Baptist by mid-19th century; closed 1992. Wesleyan Methodist chapel built 1882; became United Chapel when Baptist chapel closed. ***Schools and other institutions:*** school endowed 1648; replaced by new, larger school 1865; converted into village institute and reading room when school was replaced by Board school 1894; school closed; now used as village hall.

CROSBY RAVENSWORTH

Ancient parish in West ward, Westmorland, including townships of Crosby Ravensworth, Maulds Meaburn and Reagill. Parts of Crosby Ravensworth CP were transferred to Orton and Shap Rural CPs 1935.

Acreage: 11,049 acres [4,471 ha], divided between constituent townships thus: Crosby Ravensworth: 3,752 acres [1,518 ha]; Maulds Meaburn: 4,256 acres [1,722 ha]; Reagill: 2,340 acres [947 ha] and parts of Birkbeck Fells township totalling 633 acres [256 ha], the remainder of which was in Orton parish. Cow Close (stinted pasture) and Grayber common field at Maulds Meaburn enclosed 1818. Common moorland at Reagill enclosed 1813 and at Maulds Meaburn (1,200 acres [486 ha]) enclosed 1858. Crosby Ravensworth Fell (5,540 acres [2,242 ha]), which was extra-parochial and common to townships of Birkbeck Fells and Crosby Ravensworth, remains unenclosed, as does Bank Moor (264 acres [107 ha]). ***Population:*** rising from 789 in 1801 to peak of 971 in 1851; then declining steadily to 538 in 2001. ***Landownership:*** *manor of Crosby Ravensworth* descended through Hastings family, passing to Threlkeld family by marriage of Isabel Hastings to William Threlkeld in early 14th century; then descended through Threlkeld family until early 16th century when passed by marriage to Pickerings. Thomas Pickering sold it to Sir John Lowther in early 17th century. *Manor of Maulds Meaburn* granted to John le Fraunceys 1242, whose grandson took surname Vernon; the family sold it to Sir John Lowther 1602. *Reagill* was granted to Shap Abbey c.1200, passing to Whartons at Dissolution; sold to Lowthers c.1730. ***Economy:*** predominantly agricultural: said in late 17th century to have rich soil and to be well wooded. Annual fair for livestock at Maulds Meaburn, established 1828 but had petered out by 1880s. Coal pits at Reagill in 18th century; quarrying and dressing of limestone and lime-burning in 19th century; stone dressing continued across 20th century. Spa at Shap Wells; hotel built 1833. ***Places of worship:*** medieval parish church of St Lawrence; rebuilt 1811-12 and restored in stages between 1848 and 1887. Chapelgarth in Reagill said to be site of medieval chapel. Wesleyan Methodist chapel in Crosby opened 1875; United Methodist chapel in Maulds Meaburn built 1878 (now private house). ***Schools and other institutions:*** Crosby Ravensworth school endowed 1630; rebuilt and endowment increased 1784. Endowed girls' school at Crosby Ravensworth built 1830. The two schools merged 1898 to become Crosby Ravensworth Mixed School. School at Reagill founded 1684; buildings enlarged 1881; closed 1963; now village hall. School on village green at Maulds Meaburn built 1834; closed 1963. The three village schools combined to form new Crosby Ravensworth CE Primary School, built 1963. Village halls at Crosby Ravensworth (built 1927), Maulds Meaburn ('tin hut', replaced by modern Village Institute 2011) and Reagill (in former school building).

CROSBY (near Maryport) *see* Cross Canonby

CROSS CANONBY

Ancient parish in Allerdale below Derwent ward, Cumberland, containing townships of Birkby, Crosby, Crosscanonby and Netherhall. Netherhall became separate CP 1894, and was abolished 1928 to create Maryport CP (q.v.).

Acreage: parish contained 2,889 acres [1,169 ha], divided between townships thus: Birkby: 734 acres [297 ha]; Crosby: 1,040 acres [421 ha]; Crosscanonby or Canonby: 603 acres [244 ha]; and Netherhall: 512 acres [207 ha]. Crosby and Birkby Commons (127 acres [51 ha]) enclosed 1853. Small areas of unenclosed common land survive along coast and foreshore. ***Population:*** Birkby contained c.100 inhabitants and Crosby c.200 in early 19th century. Crosby's population doubled in 1850s as result of expansion of coal mining, to stand at 506 by 1861. Population of whole parish continued to grow, reaching 1,257 by 1891 (last census year before separation of Netherhall). Modern CP contained in range 850-1,000 inhabitants across 20th century, standing at 1,054 in 2001. ***Landownership:*** *manor of Birkby* was part of demesnes of Allerdale barony and descended with honour of Cockermouth (q.v.). *Crosscanonby and Crosby* were granted to Carlisle Priory, passing at Dissolution to Dean and Chapter of Carlisle. *Netherhall* (house largely demolished 1979) was seat of Senhouse family from 16th century to 20th. John Senhouse (d. 1604), antiquary, displayed collection of altars from Roman fort of ALAVNA (part of his estate) at Netherhall. ***Economy:*** farming; fishery in Ellen estuary, recorded 1688; saltpans and coal mining from 17th century: saltpans leased 1634; in operation until c.1760. Colliery on Senhouse estate 1688; coal mining at Birkby and Crosby from early 18th century, with major expansion in 19th. Mining at Birkby ceased 1893 but was revived during Second World War; Birkby Colliery finally closed 1950. Fireclay and brick-making from 1862. ***Places of worship:*** medieval parish church of St John; restored 1879-80. Primitive Methodist chapels at Crosby (built 1863; closed and converted to dwelling) and Crosby Villa (built c.1885; closed 1969). ***Schools and other institutions:*** parochial school in Crosby by 1818. National school built there 1861, forerunner of Crosscanonby St John CE Primary School now in modern building on new site. Reading room and library established 1872. Crosscanonby Community Centre built 1983.

CROSTHWAITE AND LYTH

Chapelry and township in Heversham parish, Kendal ward, Westmorland.

Acreage: 7,958 acres [3,221 ha]; commons totalling 4,042 acres [1,619 ha] enclosed 1815. ***Population:*** 509 in 1801, rising to 761 in 1851 and 771 in 1881; gradual decline to 636 in 1931 and to 562 by 2001. ***Landownership:*** part of barony of Kendal (q.v.). ***Economy:*** largely agricultural. Quarrying and lime-burning in 19th century. Paper-making at Starnthwaite Mill 1709 to 1849; then bobbin-turning mid-19th century; had become sawmill by 1880s. ***Places of worship:*** medieval chapel of ease of St Mary granted parochial rights 1556; tower constructed 1626; church rebuilt 1878 and tower replaced 1885. ***Schools and other institutions:*** school taught in church before 1665, when George Cocke left legacy to build school; this was replaced by new school room 1822 and school was rebuilt on new site 1878; now Crosthwaite CE Primary School. Girls' school at Howe, opened 1872; closed 1946. Starnthwaite Home Colony founded 1892 on site of disused mill by Herbert V. Mills as a utopian community at Starnthwaite Mill and Browhead Farm, in Crook (q.v.); subsequently used as farm colony for epileptic boys (from 1902) and as approved school (from 1935), known as Starnthwaite Ghyll School from 1962. Argles Memorial Hall opened 1931; refurbished 2002-03.

CULGAITH

Township in Kirkland parish, Leath ward, Cumberland. CP enlarged by absorbing Blencarn & Kirkland and Skirwith CPs 1934.

Acreage: 2,837 acres [1,148 ha] before boundary change. Culgaith commons (1,587 acres [643 ha]) enclosed 1775. ***Population:*** rising from 254 in 1801 to 467 in 1871, then declining to stand at 344 in 1931 (last census year before boundary changes). ***Landownership:*** *manor of Culgaith* granted by Henry I to Adam son of Swein. Had passed to Sir Christopher Moresby (d. 1348) whose descendant, Lady Knevett, sold manor in later 16th century to Crackanthorpes of Newbiggin. ***Economy:*** predominantly agricultural until coming of railway in 1870s (Culgaith station opened 1880; closed 1970). Sandstone quarrying at Crowdundale, 19th century. Tile works and kiln opened 1836; in operation until mid-20th century. Butter factory (Vale of Eden Dairy Factory) by railway station opened 1887. ***Places of worship:*** chapel of ease of All Saints recorded 1456; rebuilt 1756 on same site; gothic windows inserted c.1896. Wesleyan Methodist chapel built 1830; closed 2013. ***Schools and other institutions:*** free grammar school, open to all children of Culgaith and Blencarn, endowed with land at time of enclosure 1775; superseded by new school built 1876, now Culgaith CE School. Victoria Institute (containing reading room and concert hall) built 1897; now village hall, housing 'The VIC' community cafe, opened 2009. Open air swimming pool, recorded 1923; filled in c.2010.

CUMDIVOCK *see* Dalston

CUMMERSDALE

Township in Carlisle St Mary's parish, Cumberland ward, Cumberland.

Acreage: 2,014 acres [815 ha] before loss of northern part of township to Carlisle MB (Caldewgate CP) 1894, which reduced Cummersdale CP to 1,798 acres [728 ha]. Further loss of territory to Carlisle 1951. Common land at Cummersdale and Carlisle Moor (1,350 acres [546 ha], straddling boundary with Carlisle) enclosed 1770. ***Population:*** rising from 382 in 1801 to 659 by 1851 and 909 by 1891 (last census year before boundary changes). Population of reduced CP stood in range 550-650 across 20th century. ***Landownership:*** part of bishop of Carlisle's barony of Dalston (q.v.). ***Economy:*** largely agricultural until 19th century; Dalston Estate Small Holdings (part of Land Resettlement

Scheme) from 1920s. Clay pits for brick-making on former common. Cotton dyeing and bleach works at Holme Head works, part of Ferguson brothers' factories, opened 1824; closed 1991. Cluster of water-powered mills on River Caldew near High Cummersdale: corn mill recorded from 1268; commercial flour-milling for Carlisle until 1905; demolished 1970s. Cotton spinning mill established 1780s by Forsters of Carlisle; operated by Fergusons from c.1830 and by John Dalton & Son from 1848; closed 1903; sold to Edmund Stead 1905 and converted to hand block printing department of Stead McAlpin; closed 1978. Print works built by Forsters 1801; bought by Thomas McAlpin 1835; now Stead McAlpin & Co. Dye works established by Ferguson brothers 1782; taken over by Lowthian and Parker 1849; purchased by Stead McAlpin 1901; closed 1911; demolished 1970s. Bacon curing works in High Cummersdale later 19th and early 20th century. Pirelli tyre factory built at Low Cummersdale 1966-9. ***Places of worship:*** St James' Church, St James Road, built 1865-7, serving Denton Holme district of Carlisle; daughter church, St James', High Cummersdale, converted from factory building mid-20th century. Mission hall (former school) at Newby from late 19th century; closed and converted to dwelling. St Luke's Church, Morton, built 1958-60. Church of Jesus Christ of Latter Day Saints (Mormons), Langrigg Road, built 1967. St Edmund's RC Church, built 1972-3; rebuilt 1981. Carlisle Cemetery, with Anglican and nonconformist mortuary chapels, opened 1855; expanded south of Fairy Beck late 19th century; crematorium dedicated 1956. ***Schools and other institutions:*** National school at Newby built 1833; had closed by 1890s. British school at Holme Head works built 1841; closed 1884. Factory schools at Cummersdale print works and cotton mill (from 1860s; closed c.1914; building used as joiner's shop; demolished 1993). Board school at High Cummersdale opened 1884; now Cummersdale Primary School. In outer suburbs of Carlisle: New Laithes Junior School opened 1963; adjacent Infant School opened 1965; Morton Secondary School (now Richard Rose Morton Academy) opened 1968. Embassy Club, High Cummersdale, converted from bacon-curing works 1920s as community building with reading room, dance hall, billiards and games rooms; now village hall.

CUMREW

Ancient parish in Eskdale Ward, Cumberland. CP enlarged by gaining part of Croglin CP 1934.

Acreage: 2,772 acres [1,122 ha] before boundary changes, divided between townships of Cumrew Inside (1,782 acres [721 ha]) and Cumrew Outside

(990 acres [401 ha]). Encompassed part of King Harry common (6,000 acres [2,428 ha] in total), straddling boundary with Cumwhitton, enclosed 1801. ***Population:*** estimated at 255 in 1688. Rose from 181 in 1801 to peak of 231 in 1821, falling thereafter to stand at 90 in 1931 (last census year before boundary changes). Population of enlarged CP stood at 85 in 2001. ***Landownership:*** part of barony of Gilsland (see Brampton). ***Economy:*** predominantly agricultural; sheep-rearing on fell commons important in 17th century. Lime-burning in 19th century. ***Places of worship:*** medieval parish church of St Mary; rebuilt 1890. ***Schools and other institutions:*** school (supported by quarter pence and 'whittlegate') recorded 1815. New school built, just outside parish boundary near Rising Sun Inn, 1845; closed c.1948.

CUMWHINTON *see* Wetheral

CUMWHITTON

Ancient parish in Eskdale ward, Cumberland, containing townships of Cumwhitton and Northsceugh with Moorthwaite.

Acreage: 5,824 acres [2,357 ha], divided between townships thus: Cumwhitton: 2,579 acres [1,044 ha]; Northsceugh with Moorthwaite: 3,245 acres [1,313 ha]. Common moorland at King Harry (c.6,000 acres [2,428 ha] in total, shared between Cumwhitton and Cumrew) enclosed 1801. Small common field at Moorthwaite (35 acres [14 ha]) enclosed 1832. ***Population:*** estimated at 310 in 1688; rising from 446 in 1801 to peak of 579 in 1831, then downward trend to 293 by 2001. ***Landownership:*** granted by Bavin family to Lanercost Priory c.1200; after Dissolution passed to Howard family and thus to earls of Carlisle. ***Economy:*** predominantly agricultural; some sandstone quarrying in 19th century. ***Places of worship:*** medieval parish church of St Mary; rebuilt in 18th century but retaining older masonry. Wesleyan Methodist chapel built 1891; now closed and converted to dwelling. ***Schools and other institutions:*** unendowed school built c.1819. Parish school at Sandy Syke endowed 1830s by Adam Hodgson; replaced by Board school at Piper's Stile, built 1876; closed c.1970. Public hall and reading room (now village hall) built 1901.

DACRE

Ancient parish in Leath ward, Cumberland, containing townships of Dacre, Great Blencow, Newbiggin, Soulby and Stainton.

Acreage: 7,466 acres [3,021 ha], divided between constituent townships thus: Dacre: 1,192 acres [482 ha]; Great Blencow: 674 acres [273 ha]; Newbiggin: 3,047 acres [1,233 ha]; Soulby: 843 acres [341 ha]; Stainton: 1,641 acres [664 ha]. Commons and common fields at Newbiggin, Stainton and Great Blencow enclosed 1775; Dacre and Soulby commons (1,480 acres [599 ha]) enclosed 1808. ***Population:*** estimated at 620 in 1688. Total parish population in range 900-1,000 for most of 19th and early 20th century (peak 995 in 1831; low point 863 in 1921); rising again from 1970s, to stand at 1,326 by 2001. Newbiggin (341 in 1841) and Stainton (305 in 1841) were most populous townships. ***Landownership:*** *manor of Dacre and Soulby*: held by Dacre family (seat at Dacre Castle) until 15th century, descending to Joan, daughter of Thomas Dacre (d. 1458), who married Sir Richard Fiennes, who took title Lord Dacre. From them it descended to Margaret, sister and heiress of Gregory Fiennes, Lord Dacre (d. 1594); she married Sampson Lennard and Dacre passed through Lennard family to Thomas Lennard, 5th Lord Dacre, earl of Sussex (d. 1715). In 1716, his widow sold it to Sir Christopher Musgrave, who sold it to Edward Hasell of Dalemain (see below). *Stainton, Newbiggin and Great Blencow* were part of barony of Greystoke (q.v.). *Dalemain* estate (in Newbiggin township), held in 12th and 13th centuries by de Morvilles and, from 14th century (or possibly earlier) by Layton family. After death of William Layton (d. 1675) it was sold by trustees of five heiresses to Sir Edward Hasell, since when it has remained in that family. ***Economy:*** predominantly agricultural; Great Blencow noted for mustard-growing in 17th century. Extensive limestone quarrying and lime-burning in 19th century, notably at Whinberry Quarry and Flusco (Newbiggin township) and at Redhills (Stainton township), where fossil limestone quarried for ornamental purposes. Cockermouth, Keswick & Penrith railway (opened 1865; closed 1970) served commercial lime works, with station at north end of Newbiggin village, named Blencow Station. Former lime works at Flusco (large-scale quarrying began 1922; closed 1960s) now municipal waste recycling depot. Snuff mill at Stainton, recorded 1861. Tourism increasingly important from later 20th century: caravan site at Waterfoot (in Soulby township); holiday chalet parks in Newbiggin; Rheged Discovery Centre (on site of former limestone quarries and kilns in Stainton township), opened 2001. ***Places of worship:*** medieval parish church of St Andrew, on site of Anglo-Saxon monastery, founded c.720, recorded by Bede. Supposed medieval chapel of St John at Stainton (recorded in place-name Kirk Rigg). School at Stainton used as mission room c.1900; school at Newbiggin used for church services. Wesleyan Methodist chapels at Newbiggin (built 1863; closed; sold 1971, now a dwelling), Dacre (built 1873), and Stainton (built 1877, still in use). ***Schools and other institutions:*** endowed grammar school, Great Blencow, founded 1577 by Thomas Burbank. Rebuilt 1795 to include boarding accommodation and master's house; in 1880, said to be 'free for the classics to all the world'; closed 1911. Endowed parish school, Dacre, built 1749; rebuilt 1834; now closed. School at Newbiggin built 1825; replaced by new school 1866; closed 1964. Endowed school, Stainton, founded 1759, with further endowment for 'four poor girls' 1832; rebuilt 1838 and again 1863. New school on new site built 1960; now Stainton CE Primary School. Village halls at Blencow and Laithes (built 1961); Dacre (20th century); Newbiggin (built 1956; rebuilt 2010-11) and Stainton (built 1919-20).

DALSTON

Ancient parish in Cumberland ward, Cumberland, comprising townships of Buckabank, Cumdivock, Dalston, Hawksdale, Ivegill or Highhead, and Raughton & Gaitsgill. CP enlarged by gaining part of Castle Sowerby CP and losing land to Skelton CP 1934.

Acreage: 12,459 acres [5,042 ha] before boundary changes, divided between constituent townships thus: Buckabank: 2,148 acres [869 ha]; Cumdivock: 2,194 acres [888 ha]; Dalston: 656 acres [265 ha]; Hawksdale: 2,989 acres [1,210 ha]; Ivegill or Highhead: 1,650 acres [668 ha]; and Raughton & Gaitsgill: 2,822 acres [1,142 ha]. Commons in manor of Dalston, totalling 2,500 acres [1,012 ha], enclosed 1807. ***Population:*** 2,120 in 1801, rising to 3,023 in 1831 before declining (attributed in 1841 to depressed state of manufacturing and in 1861 to cotton mill fire and depressed state of hand-loom weaving) to 2,004 in 1891 and 1,631 in 1931 (last census year before boundary changes). Population of enlarged CP stood at 2,643 in 2001. ***Landownership:*** *barony of Dalston*, almost co-extensive with parish, granted to bishop of Carlisle 1230 and remained in possession of the bishops thereafter, with their seat at Rose Castle in Hawksdale township. Several subsidiary manors within barony: *manor of Little Dalston* held by Dalston family from 13th century until sold by Sir George Dalston in 1761 to Monkhouse Davison, after whose death it was bought 1795 by John Sowerby, in whose family it remained until early 20th century. *Manor of Cardew* held by John Burdon c.1300; transferred to John Denton, in whose family it remained until sold by George Denton to Sir James Lowther 1686. *Manor of Highhead* acquired by L'Engleys family in early 14th century, passing by marriage to Ralph Restwold (d. 1383), descended through Restwolds until sold in 1540s to John Richmond, in whose family it remained until divided between heiresses in 18th century. ***Economy:*** agricultural economy overlain by industrial growth from 18th century. Dalston had Friday market for meat etc. in 19th century (had ceased by 1900). Quarrying in Cumdivock and lime-burning in Gaitsgill in 19th century. Fulling mill at Dalston by 1688, recorded in place-name Walk Mill. Forge at Dalston, established 1756, manufacturing spades and other agricultural implements. Cotton manufacture from late 18th century: first cotton mill established 1782; by mid-19th century there were three: Buckabank Mill, New Rookery Walk Mill and Low Green Mill. Linen manufacture from late 18th century; both flax mill and canvas manufacture in mid-19th century. Nestlé powdered milk factory, opened 1962 near site of Low Green mill. ***Places of worship:*** medieval parish church of St Michael in Dalston village; reconstructed 1749 and 1890. Medieval chapel in Rose Castle, rebuilt late 17th century. Highhead chapel recorded 1358; rebuilt 1682 and again 1836; closed 1974. Other townships gained Anglican churches in 19th century: Christ Church, Ivegill, built 1868; St Jude's, Gaitsgill, built 1869; demolished c.1970; All Saints, Cumdivock, built 1870-2. Wesleyan Methodist chapel at Dalston built 1825; used as National school by 1860. Wesleyan Association Methodist chapel opened 1851. United Methodist Free Church, Dalston, built 1881; enlarged 1903; still in use. Methodist chapel at Stockdalewath, built 1892; still in use. ***Schools and other institutions:*** *Cumdivock:* school at Chalkfoot built 1780 by parishes of Westward and Dalston; had closed by 1900. *Dalston:* grammar school founded in 17th century; rebuilt 1815; converted to elementary school in 1870; closed (by 1938 used as parochial hall). National school held in former Methodist chapel 1860; moved to new building on new site 1864; now St Michael's CE Primary School. Caldew School (secondary) built c.1960. Parish workhouse built 1827. Working men's reading room and library established 1848. Parochial library established 1874; by 1938 in possession of National school. Victory Hall built 1922. *Gaitsgill:* school recorded 1818; Primrose Hall built 1885 for estate workers by Lt. Col. J. Carleton Salkeld of Holm Hill; now used as village hall. *Hawksdale:* Board school opened 1879; closed and converted to dwelling; Lime House independent school moved from Wetheral to Holm Hill 1946. *Highhead:* reader at chapel endowed to teach children 1736. *Ivegill:* schoolmaster recorded 1688; small school built 1835; new school, now Ivegill CE Primary School, built 1873. Village hall, opposite school, built 1964.

DALTON (Westmorland)

Township in Burton-in-Kendal parish, only part of parish to lie outside Westmorland in Lonsdale hundred, Lancashire. Transferred to Westmorland 1895. Absorbed into Burton-in-Kendal CP 1986.

Acreage: 2,167 acres [877 ha], including 58 acres [24 ha] of common land, enclosed 1815. ***Population:*** in range 100-150 across 19th century, with peak of 155 in 1841; stood at 86 in 1981 (last census year for which separate figure available). ***Landownership:*** *manor of Dalton* held by Croft family from 12th century until death of James Croft c.1457. On division of Croft estates 1489 Dalton was assigned to his daughter Mabel, wife of Piers Legh of Lyme and descended in Legh family until death of Peter Thomas Legh 1797. Sold to Rev. Geoffrey Hornby 1803 and remained with Hornby family (later Penrhyn-Hornby and Mason-Hornby) to 21st century. ***Economy:*** agricultural; limestone quarrying and lime-burning in 19th century. ***Places of worship:*** none. ***Institutions:*** none.

DALTON-IN-FURNESS

Ancient parish in Lonsdale hundred, Lancashire North of the Sands, covering much of southern end of Furness peninsula, and divided into four quarters or 'bierleys': Hawcoat, Yarlside, Above Town, and Dalton. Hawcoat and the bulk of Yarlside became part of new town of Barrow-in-Furness in 1867 (q.v.); from 1881 Dalton CP (UD from 1894 to 1974) consisted of divisions of Dalton Proper and Above Town (which embraced townships of Ireleth and Lindal & Marton). This entry covers Dalton Proper and Above Town, i.e. the three townships of Dalton, Ireleth and Lindal & Marton, each of which was reincarnated as separate CP in 1987 and is treated separately below.

Acreage: ancient parish covered 17,962 acres [7,269 ha], of which 1,045 acres [423 ha] lay in Dalton Proper township and 5,362 acres [2,172 ha] in Above Town quarter, which comprised townships of Ireleth (4,196 acres [1,698 ha]) and Lindal & Marton (1,166 acres [472 ha]). Commons at High Moor, Low Moor, Stewner and Ireleth Marsh, Powka and Lindal Moor in Ireleth, Lindal and Marton (totalling c. 500 acres [c.200 ha]) enclosed 1831. ***Landownership:*** part of Furness Abbey's manor of Plain Furness (see Barrow in Furness). ***Population:*** ancient parish contained population of 1,954 in 1801, rising to 3,231 by 1841. Rapid expansion of iron ore mining resulted in surge in population, including creation of new industrial settlement at Askam. By

1871 (from which time Barrow-in-Furness was enumerated separately) population of Dalton had risen to 9,902 and it peaked at 13,349 in 1881. Decline of mining industry resulted in slow decline to 10,339 by 1931. It remained at 10,913 in 1981 but growth of new housing saw increase to 12,799 by 2001.

DALTON. Modern CP of Dalton Town with Newton, established 1987, covers much of earlier division of Dalton Proper, together with Newton, which was formerly part of Yarlside quarter.

Origins and growth of the town. Dalton was a borough and market town of medieval low Furness and seat of manorial government (Dalton Castle being court house and bailiff's residence). Market charter granted 1239. In 1670s Sir Daniel Fleming described Dalton as 'a good market for all kinds of provisions', with Saturday market and fair on 12 October. Market declined from 18th century and was superseded by Ulverston but hiring fairs at Dalton continued to 20th century. Iron-ore mining recorded from 13th century; this remained small in scale in early 19th century, when Dalton contained hand-loom weavers as well as tanning, malting and brewing industries. Town grew rapidly in second half of 19th century with major expansion of iron mining after opening of Furness Railway 1846. Major iron mines in Dalton Proper were at Crossgates, Elliscales, Longlands and Mouzel. Mining at Newton from 1850s; last mine closed 1944. Limestone quarrying and lime-burning in 19th century. Light industries after Second World War, including Furness Footwear and Blair's underwear factories (1948; closed 1983). By late 20th century, Dalton and surrounding villages had become dormitory settlements for Barrow. Some tourism, notably South Lakes Wild Animal Park, opened 1994.

Places of worship. Medieval parish church of St Mary, Dalton, rebuilt 1825-6 and 1830; replaced by new building 1885. St Margaret, Ulverston Road, built 1904; taken over by Roman Catholics 1981. Methodist chapel above stable in Skelgate by 1823; replaced 1864 by new chapel in Wellington Street, still in use as Dalton Methodist Church. Primitive Methodist chapel, Ulverston Road, opened 1868; replaced by new chapel, Cleator Street, 1883. Old chapel used by Church of Christ early 20th century and by Seventh Day Adventists since 1987. Cleator Street chapel closed 1959; converted to housing. Bible Christian chapel (later United Methodist Church), Broughton Road, built 1872; closed 1935; later demolished. Congregationalists used former grammar school, Beckside Road, until chapel built 1869; now United Reformed Church. Baptists met in former Methodist chapel in Skelgate from 1868; chapel built at Broughton Road 1872; rebuilt 1885; closed 1963; now a bakery. Other places of worship included Plymouth Brethren and Welsh Calvinists (1874) who used old grammar school in Beckside Road. Roman Catholic chapel of Our Lady of the Rosary, built 1879; demolished after move to former Anglican church of St Margaret, 1981. Cemetery with Anglican and nonconformist mortuary chapels opened 1860; one taken over by Greek Orthodox Church of St Simon and St Kentigern, founded 1997. Mission room at Newton opened 1900, later called St Barnabas' church; closed 1977 and converted to dwelling. Wesleyan Methodist chapel, Newton, opened 1891; closed 1965.

Schools. Dalton free school founded 1622; survived into 19th century; then refinanced and rebuilt by local entrepreneurs, re-opening as church school (The Green School) 1862; closed 1969; replaced by new St Mary's CE Primary School, Coronation Drive. Grammar school, Beckside Road, opened 1746; had closed by 1876. Board schools: boys' school, Broughton Road, and girls' school, Chapel Street, opened 1878 (latter now Chapel Street Infant and Nursery School). Roman Catholic school, Ulverston Road, opened 1880; replaced 1972 by Our Lady of the Rosary Primary School, Crooklands Brow. Junior mixed school, Nelson Street, opened 1884; became girls' primary school 1928. Dowdales mansion, purchased by Lancashire County Council, opened as Dowdales Central School (now Dowdales School) 1928. Primary schools in Nelson Street (girls) and Broughton Road (boys) closed 1980 and replaced by new George Romney Junior School, Cobden Street. Mixed elementary school at Newton opened 1877; now Newton Primary School.

Other institutions. Town Hall built 1884. Parish workhouse at Billingcote opened 1735; replaced by new workhouse at Goose Green 1826. Carnegie Free Library, Nelson Street, opened 1903. Empire cinema, Market Street, opened 1912; reopened as Roxy 1937; closed 1965. Electric Picture Palace in Co-operative Hall, Nelson Street, opened 1919, known locally as 'Co-op pictures'; closed c.1960. Leisure complex, including swimming pool, opened 1907.

IRELETH. Modern CP of Askam and Ireleth, established 1987, covers approximately same area as former township of Ireleth.

Economy: copper mine (disused by 1846); slate quarrying at Rebecca Hill. Major expansion of iron mining in mid-19th century at Roanhead (closed 1942) and Park mine (where what was then largest iron ore deposit known was discovered 1850; mine closed 1922). Ironworks established at Askam 1865; closed 1918 (site cleared 1933), limestone provided by new quarry at Goldmire. Brickworks established at Greenscoe 1845; still in operation. K Shoes factory at Askam (from1953; closed 1996) and British Cellophane factory at Sandscale (opened 1954; closed 1991). Wind farms at Far Old Park and Hare Slack Hill, built 1998-9. Population of Askam and Ireleth CP stood at 3,632 in 2001. ***Places of worship:*** medieval chapel of St Helen, near Chapel Hill, Haume (site recorded 1846). School at Ireleth licensed as chapel of ease 1637; building solely used as chapel after new school built 1778; rebuilt as St Peter's church on new site 1865. Wesleyan Methodist chapel built 1870; replaced by new church nearby 1909; now shared by Methodists and Anglicans. Zion Primitive Methodist chapel built 1870; closed 1985; now converted to dwelling. United Methodist chapel (corrugated iron building) opened 1878; closed and became Rankin Hall social club. Christian meeting house of Church of Christ built 1907; closed 1956 and became St Anthony's Roman Catholic Church, which closed 2009. ***Schools and other institutions:*** school at Ireleth, founded 1612 or 1624 by Giles Brownrigg of London; rebuilt 1778 and again 1862; now Ireleth St Peter CE Primary School. Senior mixed school, Askam, opened 1873; closed c.1973; now Askam & Ireleth Community Centre. Victoria Council School (mixed infants) opened 1887; rebuilt 1977, now Askam Village School. Temperance Hall, Ireleth, built 1872, used by Bible Christians in late 19th century; closed c.1907; renovated 1990s as community centre. Carnegie Library, Lord Street, Askam, opened 1904. Askam village hall (or Askam Lots Social Club) established c.1940s.

LINDAL and MARTON. Modern CP of same name covers much of former township, though western boundary does not coincide with historic boundary with Ireleth and Dalton Proper.

Economy: early mining attested by place-name Orgrave. Extensive limestone quarrying and iron mining by mid-19th century, with pits at Eure Pits, Lindal Cote, Lindal Moor and ironworks at Orgrave Mill, Poaka and Whitriggs (all ironworks closed before 1888). Further expansion of mining in second half of 19th century, especially on Lindal Moor. Mining declined in early 20th century; last mine to close was Margaret Mine at Henning Wood, Lindal, closed 1960. ***Places of worship***: temporary iron church at Lindal built 1875 (later used as drill hall; destroyed by gale 1903); replaced by St Peter's church, built 1885. Wesleyan Methodist chapels at Tarn Flatt, Marton (built 1866; closed 1960) and Lindal (built 1871; closed 1985). Christian meeting house, Lindal, built 1875; closed after Second World War. ***Schools and other institutions:*** school at Lindal built by Harrison Ainslie & Co. (iron mine owners) 1854; enlarged 1860; now Lindal and Marton Primary School. Buccleuch Hall built c.1875, used as Sunday school before becoming village hall. Reading room, Lindal, now a dwelling.

DEAN

Ancient parish in Allerdale above Derwent ward, Cumberland, which included townships of Dean, Branthwaite and Ullock, Pardshaw & Deanscales. CP enlarged by absorbing Eaglesfield CP 1934.

Acreage: 6,528 acres [2,642ha], divided between constituent townships thus: Dean: 2,265 acres [917 ha]; Branthwaite: 2,264 [916 ha]; Ullock, Pardshaw & Deanscales: 1,999 acres [809 ha]. Common land totalling 2,300 acres [931 ha], much of it on Dean Moor, enclosed 1815. ***Population:*** estimated at 620 in 1688. Rising from 678 in 1801 to 876 in 1841, then falling to 582 in 1931 (last census year before boundary changes). Population of enlarged CP rose steadily to stand at 1,077 by 2001. ***Landownership:*** one of the Five Towns in honour of Cockermouth. *Manor of Dean* descended with honour until granted to Sir Thomas Wharton 1530; repurchased by duke of Somerset (lord of honour of Cockermouth) 1738/9. *Manor of Branthwaite* descended in Branthwaite family until passing by marriage to Robert de Whitrigg c.1320. On death of Sir Thomas Whitrigg, it passed to his sister Joan, wife of Thomas Skelton (d. 1365) and descended through Skelton family until 1757. ***Economy:*** largely agricultural until 19th century. Limestone and red and white freestone quarrying; lime-burning. Coal mined on Dean Moor by 1430s; colliery there from 18th century to 20th; closed 1951. Opencast mining later removed remaining deposits. Tile works on Dean Moor in mid-19th century. Woollen mill at Branthwaite in early 19th century; paper mill at Branthwaite by 1829; disused by 1898. Linen thread manufactory at Deanscales early 19th century. ***Places of worship:*** medieval parish church of St Oswald. Quakers met in open on Pardshaw Crag from 1650s; meeting house built at Pardshaw Hall 1672 and replaced by present building 1729 (closed for weekly worship 1923; later used as hostel). Wesleyan Methodist chapels at Pardshaw (built 1823; renovated 1879), Ullock (built 1870) and Branthwaite (built 1884, for congregation meeting in school room erected 1823). ***Schools and other institutions:*** grammar school at Dean church stile endowed 1596 by John Fox, native of Branthwaite; closed later 19th century. National school at Dean built 1863; enlarged 1871; now Dean CE Primary School. Quaker schoolhouse at Pardshaw Hall built 1745; continued until early 20th century. Reading rooms at Branthwaite (established 1884) and at Wesleyan chapel at Ullock. Village halls at Dean (Kirkstile Community Hall, dated 1912) and Ullock (former Methodist chapel, given to community 1950).

DEARHAM

Township in Dearham parish, Allerdale below Derwent ward, Cumberland.

Acreage: 2,149 acres [870 ha], including 480 acres [194 ha] of common land enclosed 1827. ***Population:*** rising increasingly rapidly as coal mining expanded, doubling from 515 in 1821 to 1,037 by 1841 and rising to peak of 2,598 in 1891. Steady decline to 1,642 in 1971; then recovery to 2,028 by 2001. ***Landownership:*** *manor of Dearham* was divided into halves, one moiety, in hands of Barwis family, probably by 1400, passed by marriage of Grace Barwis (1609-1640) to Thomas Lamplugh of Ribton, in whose family it descended until 1722 when sold to Sir James Lowther. Other moiety was in hands of Thomas de Multon in 13th century, who granted it to Calder Abbey; it had been divided among freeholders by 1687. ***Economy:*** extensive coal mining from 18th century; several pits closed second half of 19th century; Crosshow and Townhead pits reopened 1895 to 1903. Revival during First World War: Allbright Drift continued to operate until 1950. Pottery at Whistling Syke built by Aaron Wedgwood 1708; his brother Jonathan Wedgwood established Dearham Pottery on site which continued in operation until c.1900. Brick-making in 19th century. ***Places of worship:*** medieval parish church of St Mungo (modern dedication), an early foundation, to judge by wealth of pre-Conquest sculpture; restored 1882. Mission room by church built 1891. Wesleyan Methodist chapel built 1839; rebuilt 1888; still in use. Primitive Methodist chapel built 1856; closed. ***Schools and other institutions:*** school built and endowed by John Christian 1715. Replaced by Board school on new site, built 1874; now Dearham Primary School. Library of Dearham General Improvement Society founded 1855. Temperance hall built 1876. Village hall.

DENT

Chapelry and township in Sedbergh parish, Ewecross wapentake, West Riding of Yorkshire.

Acreage: 20,890 acres [8,454 ha], including extensive areas of common moorland, enclosed 1859. Chapelry contained five divisions: North Lord's Land: 1,928 acres [780 ha]; South Lord's Land: 3,881 acres [1,571 ha]; Lenacre: 1,726 acres [699 ha]; Deepdale: 4,453 acres [1,802 ha]; and Kirthwaite: 8,901 acres [3,602 ha]. ***Population:*** 1,773 in 1801. After period of decline, reached peak of 2,096 in 1871, swelled by workers building

Settle-Carlisle railway; steady decline thereafter to 675 in 2001, including some who were not usually resident. ***Landownership:*** *manor of Dent* (comprising Dent town, North and South Lord's Land, Deepdale and Kirthwaite) held by FitzHugh family, passing by marriage on death of George 7th Lord FitzHugh (d. 1513) to Parr family. Passed to Crown c.1574 and sold 1629 to Sir Allan Apsley, who sold it 1670 to Richard Trotter of Dent and others in trust for the tenants. *Lenacre* belonged to Monteagle and Middleton families; Monteagle manor was enfranchised 1595-6. ***Economy:*** pastoral farming; quarrying and lime-burning; hand-knitting of stockings was widespread. Coal mining in Cowgill Head area from 18th century to c.1870. 'Marble' quarrying from late 18th century; marble mill at Stone House for cutting and polishing marble established 1820s; closed 1907. Dee Mill, Cowgill, a worsted and cotton mill in 1840s; had closed by 1892. Tourism increasingly important from 20th century. ***Places of worship:*** medieval parochial chapel of St Andrew; tower built c.1785; major restoration 1889-90. Church of St John the Evangelist, Cowgill, built 1837-8. Quaker meeting houses in the Loaning, Dent (bought 1682; rebuilt 1701; closed by 1800; sold to Methodists 1834) and Lea Yeat (built 1703; closed 1911). Inghamite chapel 1754; later Sandemanian, abandoned by 1780s. Independent congregation by 1809; built Zion Congregational chapel, Flintergill, 1835; now Dent Meditation Centre. Wesleyan Methodist chapels in Dent (former Quaker meeting house, from 1834; now Dentdale Methodist Church) and Dent Head (1856-1876). Primitive Methodist chapels at Dent (1841-1934) and Deepdale (opened 1889; closed c.1996). Dent Foot Methodist chapel. ***Schools and other institutions:*** school recorded pre-1560. Dent Grammar School endowed 1604; closed 1897. National school, Dent, built 1845; replaced by new school on adjacent site c.1968; now Dent CE Primary School. Old building became village hall (Dentdale Memorial Hall). National school at Cowgill (Kirthwaite) built 1866; closed 1968. Parish workhouse at Hall Bank opened 1733; continued to be used by Sedbergh Poor Law Union until new workhouse built at Sedbergh 1857. Dent Reading Room built 1880; now used as library.

DENTON, NETHER

Ancient parish in Eskdale ward, Cumberland.

Acreage: parish and township did not quite coincide, parish containing 4,880 acres [1,975 ha]; township 4,984 acres [2,017 ha]. Common moorland totalling 3,300 acres [1,336 ha] in Nether and Upper Denton enclosed 1800. ***Population:*** rising from 245 in 1801 to 354 in 1891; then falling to 318 in 2001. ***Landownership:*** part of barony of Gilsland (see Brampton). Held by Denton family until 1506 when exchanged with Lord Dacre for Warnell (in Sebergham). ***Economy:*** predominantly agricultural. Carrick's Dairy & Pure Milk Supply Co. Ltd (making butter & cheese) established at Low Row, beside station on Newcastle-Carlisle railway, c.1881. Bakery and laundry added in 1920s to service Carrick's cafes. Dairy ceased during 1950s, laundry closed 1965 and bakery 1968. ***Places of worship:*** nothing remains of medieval parish church of Nether Denton, except a sculpture, possibly 12th century. St Cuthbert's church built on site of ancient church 1868-70. Wesleyan Methodists met in Low Row schoolroom, erected 1850. Separate Wesleyan chapel and Sunday school built 1883; closed. ***Schools and other institutions:*** school recorded 1818; rebuilt 1874 as Board school at Low Row; closed. Library and reading room at Low Row built by subscription 1894. Taylor Institute, Low Row, built 1916; now village hall.

DENTON, UPPER

Ancient parish (also known as Over Denton) in Eskdale ward, Cumberland; it included part of the village of Gilsland.

Acreage: 1,039 acres (420 ha), including part of 3,300 acres [1,336 ha] of common land in Nether and Upper Denton, enclosed 1800. ***Population:*** rising from 85 in 1801 to 200 in 1921; then falling back to 92 in 2001. ***Landownership:*** held 1279 by Richard Stowland who conveyed it to John Witherington in whose family it descended until purchased by Lord William Howard in 17th century. ***Economy:*** largely agricultural. Tourism from 19th century, attracted by mention of Mumps Hall in Sir Walter Scott's *Guy Mannering* (1815), proximity of Gilsland Spa and, increasingly, of Hadrian's Wall and associated Roman remains. ***Places of worship:*** medieval parish church, dedication unknown; restored 1881 and 1930s; now disused. Methodist chapel in Gilsland built 1869; still in use. ***Schools and other institutions:*** children went to school in Haltwhistle in early 19th century until day school established in Upper Denton (by c.1830). School at Gilsland (now Gilsland CE Primary School) established before 1900.

DILLICAR

Township in Kendal parish, but in Lonsdale ward, Westmorland; includes hamlets of Lowgill and Beckfoot. Absorbed into Grayrigg CP 1986.

Acreage: 1,121 acres [454 ha], including Dillicar Common (500 acres [202 ha]), enclosed 1853. ***Population:*** rising from around 80 in early 19th century to 135 in 1851; peak of 205 in 1861 (numbers inflated by railway construction), then falling back to 161 by 1871; continued to fall, dropping to 47 in 1981 (last census year for which separate figure available). ***Landownership:*** hamlet (by 15th century termed manor) of Dillicar held of barony of Kendal (q.v.). Held by Robert Laurence (d. 1439) and descended in Lawrence family until mid-16th century. ***Economy:*** agricultural until arrival of railways: London & North Western Railway (1848) and Ingleton line (opened 1861) met at Lowgill, where there was station with associated employment. ***Places of worship:*** Primitive Methodist chapel at Beckfoot (building dated 1872); closed; converted to dwelling. ***Schools and other institutions:*** school at Railway Terrace, Lowgill, built by railway company 1869; closed 1948.

DISTINGTON

Ancient parish in Allerdale above Derwent ward, Cumberland.

Acreage: 3,064 acres [1,240 ha], including 1,137 acres [460 ha] of commons, enclosed 1768. ***Population:*** estimated at 410 in 1688. Rising from 724 in 1801 to 1,106 in 1851; then climbing steadily from 1871 to 2,837 in 1971, declining to 2,247 in 2001. ***Landownership:*** *manor of Distington* descended from Gilbert de Dundraw c.1189 via Moresby and Dykes family to Fletchers who sold to Sir John Brougham 1720 who, in turn, sold to Sir James Lowther 1737. Other landowners were Robertson Walkers of Gilgarran c.1805 to 1951 and Fisher family of Distington Hall. ***Economy:*** soil said in 17th century to be good for corn. Sandstone (freestone) quarrying at Barngill and Pica Quarries; limestone quarrying and lime-burning at Barfs. Coal mining recorded from 1614; numerous small pits which declined by mid-19th century. Revived when Oatlands Pit, a deep mine, opened 1880; closed 1934. Opencast mining in late 1950s and from 1989 to 1990s. Coke ovens in mid-19th century. Edge tools, hats and linen thread also manufactured in 19th century. Distington Haematite Iron works opened 1879; later High Duty Alloys (from 1940), Alcan and Pechenay; closed 2007. ***Places of worship:***

medieval parish church of Holy Spirit; rebuilt 1886. Wesleyan Methodist chapel, Main St, built 1830; still in use. Primitive Methodist chapel, Boot Brow, built 1838; closed c.1950; sold for housing. Presbyterian mission hall, Church Road; later used by British Legion. Mission hall at Pica built 1896; demolished c.1975. Crematorium on site of Distington Hall, opened 1974. ***Schools and other institutions:*** grammar school ('given' by William Fletcher but not endowed) and petty school recorded 1722. Parochial school built on common 1754. Board school, Main Street, built 1878; closed 2002, when community school built on site of previous Infant school (built 1961). Dyon Council School near Gilgarran (1886 to 1962) replaced Walker's schoolroom at Gilgarran. Museum at Distington, founded by Joseph R. Wallace, flourished 1850-1896. Victoria Hall opened 1866; Albert Hall, Common End, opened 1899; closed and converted to dwelling. Reading room opened 1901 and used as public library in 1980s; now disused. Community centre and library opened 1996.

DOCKER

Township in Kendal parish, Kendal ward, Westmorland.

Acreage: 1,373 acres [556 ha]. Lambrigg and Docker Fells (c.641 acres [259 ha]) enclosed 1886. ***Population:*** rising from 65 in 1801 to peak of 95 in 1831, then falling to lie in range 50-70 for most of 20th century; 55 in 2001. ***Landownership:*** granted by William de Lancaster to St Leonard's hospital, York, for grazing horses. After Dissolution granted to Richard Washington, who conveyed it to Duckets of Grayrigg, who sold it 1690 to Lowthers. ***Economy:*** farming. ***Places of worship:*** none. ***Institutions:*** none.

DOVENBY

Township in Bridekirk parish, Allerdale below Derwent ward, Cumberland. Absorbed into Bridekirk CP 1934.

Acreage: 1,764 acres [552 ha.], excluding detached portion at Hames Hill (171 acres [69 ha]), which became part of Bridekirk CP (q.v.) 1887. Dovenby and Papcastle commons enclosed 1842. ***Population:*** rising from 153 in 1801 to peak of 272 in 1861; then falling back to 163 by 1931 (last census year for which separate figure available). ***Landownership:*** *manor of Dovenby* held by Lamplugh family from early 15th century. Manor was mortgaged in 17th century but Dovenby Hall and demesnes descended in Lamplugh family and by marriage via Woodall and Brougham families to Mary Dykes (1774-1860) who married her cousin Joseph Ballantine-Dykes. Dovenby Hall was seat of Ballantine-Dykes family until 1930s. Set in parkland, it had private station, opened 1867, on Maryport & Carlisle Railway. ***Economy:*** largely agricultural; coal mining 1830 to c.1900. Dovenby Hall became headquarters of M-Sport rally car production 1998. ***Places of worship:*** Independent chapel (shown as Methodist on 1899 OS) built 1869; closed 1929; later converted to garage. ***Schools and other institutions:*** grammar school founded 1609; rebuilt 1678 and again 1708. Rebuilt 1845 on eastern edge of township, midway between Dovenby and Bridekirk; now Bridekirk-Dovenby School. Hospital for four widows established 1609; building had gone by 19th century. Reading room opened c.1900; later converted to dwelling. Dovenby Hall was used as mental hospital from 1930 to 1997.

DRIGG

Ancient parish in Allerdale above Derwent ward, Cumberland. Part of Drigg CP was transferred to Seascale CP 1901.

Acreage: 3,982 acres [1,611 ha]; parish was coterminous with township of Drigg and Carleton (Drigg lying north of River Irt and Carleton to south). Common moorland (1,500 acres [607 ha]) enclosed 1828. ***Population:*** estimated at 560 in 1688; in range 350-450 for much of 19th and 20th centuries. Rose to peak of 579 in 1891 (last census year before loss of territory to Seascale CP). Population of modern CP peaked at 563 in 1961. ***Landownership:*** *manor of Drigg* held successively by Stutevill and Wake families, lords of Liddel barony, in 12th and 13th centuries. It had been divided between heiresses by 1282; part came to Curwens of Workington, who sold manor to Sir William Pennington of Muncaster c.1597, in whose family it remained until 20th century. ***Economy:*** farming; fisheries and mussel beds (from which pearls were obtained in 17th century) in Irt and Mite estuaries (place-name 'Fishgarth' near Carleton Hall); salt-boiling suggested by name Saltcoats. Brick- and tile-making mid-19th century. Royal Ordnance Factory, producing TNT, established on coastal dunes 1940; site used as storage depot for radioactive waste since 1959. ***Places of worship:*** medieval parish church of St Peter; rebuilt on same site 1850. ***Schools and other institutions:*** school at Carleton endowed 1727; endowment lost and premises sold before 1901. Another school, supported by 'quarterage', recorded 1818. Endowed grammar school at Drigg founded 1828; rebuilt 1878; closed and converted to dwelling.

DRUMBURGH *see* Bowness-on-Solway

DUFTON

Ancient parish in East ward, Westmorland.

Acreage: 16,848 acres [6,818 ha], including 11,468 acres [4,617 ha] of unenclosed common land on Dufton Fell. Act for enclosure of three fellside pastures (c.2,300 acres [931 ha]), 1827. Open field at Keisley Field (52 acres [21 ha]) enclosed 1859. ***Population:*** 392 in 1801, rising to 554 in 1831; decline thereafter, falling to 307 by 1901 and to 169 in 2001. ***Landownership:*** *manor of Dufton* held by Greystoke family by 1335; descending with Greystoke (q.v.) until granted on 99-year lease to Sir Christopher Clapham. Purchased by John Winder (d. 1699), whose son Williams Winder (1690-1766) demised it to Edward Milward, who sold it 1785 to earl of Thanet; descended thereafter with barony of Appleby (q.v.). ***Economy:*** sheep farming noted in later 17th century. Quarrying, lead mining and smelting employed many inhabitants in 19th century. Mines operated by London Lead Co., who built smelt mill south of village 1785 (replacing older mill); closed c.1873, by which time lead mining had ceased. Dufton Fell mine reopened for barytes 1882; closed 1897. ***Places of worship:*** medieval parish church of St Cuthbert; rebuilt 1775 (or 1784) and repaired 1853. Wesleyan Methodist chapel recorded 1803; rebuilt 1820; closed and converted to dwelling 1935. Primitive Methodist chapel built 1839; rebuilt on new site 1905; still in use as Dufton with Knock Methodist Church. ***Schools and other institutions:*** endowed school founded 1670; rebuilt by subscription 1824. New school built on village green 1858; closed and converted to dwelling. Village benefitted from patronage of London Lead Co., which remodelled village, provided library and reading room by 1820 and piped water 1858. Subscription library established 1846. Village hall built as Conservative and Unionist Club 1911.

DUNDRAW
(also known as Dundraw, Kelsick & Wheyrigg)

Township in Bromfield parish, Cumberland ward, Cumberland.

Acreage: 2,366 acres [958 ha], including c.600 acres [c.240 ha] of commons, enclosed 1822. ***Population:*** rising from 198 in 1801 to over 300 in the period 1821 to 1861, before declining to low point of 158 in 1961; then rising to

198 in 1981; stood at 167 in 2001. ***Landownership:*** part of barony of Wigton. *Manor of Dundraw* held by Gilbert de Dundraw in late 12th century and passed by descent to Crofton family, whose heiress Margaret de Crofton married Isold Brisco c.1400. Dundraw remained in hands of Briscos of Crofton until mid-20th century. ***Economy:*** predominantly agricultural. ***Places of worship:*** non-denominational mission hall, built 1898 (now converted to dwelling). Chapel or mission room built 1901; closed 1965, then used as village hall until 1990s. ***Schools and other institutions:*** school at Waverbridge, erected 1844; closed 1970 (now dwelling house).

DUNNERDALE WITH SEATHWAITE

Township in Kirkby Ireleth parish, Lonsdale hundred, Lancashire North of the Sands.

Acreage: 10,272 acres [4,157 ha], much of it unenclosed fell common. ***Population:*** rising from 143 in 1821 to 155 by 1851; then sharply to 289 in 1861. Peaked at 299 in 1881, then declined gradually to 119 in 2011. ***Landownership:*** part of section of Furness Fells allotted to William de Lancaster c.1160. Granted to Roger de Kirkby c. 1170, passing through Kirkby family until sold 1497. Held by earl of Derby after 1521; sold to Hesketh family 1610, but was sequestered by parliament c.1644; afterwards held by Penny family who sold it to Richard Towers of Duddon Hall after 1788. Towers' estates passed to Maj. W. S. Rawlinson 1860 and to G. H. Cheetham 1903. ***Economy:*** hill farming; medieval bloomeries; lead and copper mining, early 19th century; yarn spinning mill at Seathwaite, recorded 1820s; large-scale slate quarrying at Walna Scar and Stainton Ground in later 19th century. ***Places of worship:*** chapel of ease at Seathwiate recorded 1547; enlarged 1796; rebuilt as Holy Trinity Church 1874. Holy Innocents' Chapel, near Broughton Mills, built 1887. ***Schools and other institutions:*** school taught in chapel by curate 1720; school built 1782, and endowed 1841; closed; now village hall (Seathwaite Parish Room). Ulpha school stood in Dunnerdale with Seathwaite township (see Ulpha).

EAGLESFIELD

Township in Brigham parish, Allerdale above Derwent ward, Cumberland. Absorbed into Dean CP 1934.

Acreage: 1,998 acres [809 ha], including part of 1,000 acres [405 ha] of common land in Eaglesfield and Blindbothel, enclosed 1815. ***Population:*** rising from 310 in 1801 to 411 in 1831, then falling to stand at 233 in 1931 (last census year for which separate figure available). ***Landownership:*** one of the Five Towns in honour of Cockermouth (q.v.). *Manor of Eaglesfield* in possession of Eaglesfield family from 14th century until death of Richard Eaglesfield (d. 1557); divided between several owners by later 16th century. ***Economy:*** predominantly agricultural; lime-burning and limestone quarrying; extensive quarrying at Tendley Quarry since 19th century. ***Places of worship:*** John Dalton Memorial Church (dedicated to St Philip) built 1891 to serve enlarged ecclesiastical parish of Mosser, which included Eaglesfield. Quaker burial ground from 1693, at which meeting house built 1711; only used occasionally by 1901; sold 1980 and converted to dwelling. Wesleyan Methodist chapel built 1845; closed (used as mechanic's garage by 1980s). ***Schools and other institutions:*** served from early 19th century by Paddle School (see Blindbothel). Village hall opened 2009.

EAMONT BRIDGE *see* Yanwath and Eamont Bridge

EDENHALL

Ancient parish in Leath ward, Cumberland. Absorbed into Langwathby CP 1934.

Acreage: 3,429 acres [1,388 ha]. ***Population:*** estimated at 265 in 1688. Rose from 148 in 1801 to 315 in 1851, then declined to 216 in 1931 (last census year for which separate figure available). ***Landownership:*** *manor of Edenhall* was granted to Peter de Brus in 12th century; held by Turp family in 13th century, passing to Stapleton family by marriage in early 14th century and to Musgrave family by marriage of Joan Stapleton to Thomas Musgrave in 15th century. Descended through Musgraves until sold by Sir Richard George Musgrave (d. 1926). Eden Hall, rebuilt 1821, was demolished 1934. ***Economy:*** predominantly agricultural; fishery in River Eden noted in later 17th century. Quarrying (recorded in name Slatequarry Wood) and gravel digging. Brickworks in 20th century; closed and sold 2012 to become transport depot. ***Places of worship:*** medieval parish church of St Cuthbert (living held jointly with Langwathby since 1380s); enlarged and restored 1834. ***Schools and other institutions:*** school endowed 1771; new school built by Sir Philip Musgrave 1815; closed 1937. 'Female free school', supported by Lady Musgrave, recorded 1847.

EGREMONT

Borough and ancient parish in Allerdale above Derwent ward, Cumberland. Parish also included mining villages of Bigrigg and Moor Row. Became Egremont UD 1894-1934. Minor boundary changes in 1890s transferred land at south end of borough from Lowside Quarter and St John Beckermet CPs to Egremont.

Acreage. 2,750 acres [1,113 ha], excluding small detached portion (26 acres [11 ha]) near Linethwaite, transferred to Hensingham CP 1882. Parish included extensive, scattered areas of common land (perhaps one-third of parish acreage) enclosed 1783, along with commons in Lowside Quarter and St John Beckermet.

Landownership. Egremont castle was seat of *barony of Copeland or Egremont*, granted to William le Meschin early 12th century. Barony descended to his granddaughter Alice de Rumilly (d. 1186/7), coming to her grandson Richard de Lucy c.1200. On his death 1213 it passed to his daughter Mabel, wife of Lambert de Moulton and descended through Moulton family until death of John de Moulton 1334. Divided 1338 between his three sisters, who each obtained one-third of barony: Joan, wife of Robert Fitzwalter; Elizabeth, wife of Sir Robert Harrington; and Margaret, wife of Thomas de Lucy. Lucy share descended to earls of Northumberland (see Cockermouth), who obtained Fitzwalter share by exchange with other lands in early 16th century, and Harrington share in 1594. Barony then descended with honour of Cockermouth (q.v.).

Origins and growth of the town. Borough established, literally at foot of castle, by lords of Copeland barony c.1200; market charter granted 1267. Borough byelaws refer to dyers, weavers and fullers, suggesting economy based on textiles. Iron mining (at 'Thirnby', in north of parish) recorded from c.1150 and across 13th and 14th centuries. Borough suffered in late medieval period but was described as 'good market towne' by Sir Daniel Fleming 1671. Population estimated at 1,410 in 1688. 18th century saw beginnings of expansion of industry: there were four tanneries; two paper mills built mid-18th century; Greendyke flax mill built c.1760; High Mill (a flax mill by 1861) built as fulling mill 1760s. Population seems to have remained fairly stable, standing at 1,741 in 1821.

Population boom began in 1850s: by 1861 it stood at 3,481, increase being attributed to 'the opening of iron-ore pits and to the erection of flax mills', and two more decades of strong growth followed, taking population to 5,976 by 1881. As well as iron mining (there were eleven mining companies in parish

1901) and flax/linen industry, limestone and freestone quarrying and lime-burning were carried out extensively in 19th century. Town continued to serve as market centre in mid-19th century, with cattle fairs and hiring fairs for servants at Whitsuntide and Martinmas.

Most quarries and some iron mines had closed by 1900; most remaining mines closed in period 1914 to 1933. Egremont's population declined from 1890s, fluctuating around 6,000 across first half of 20th century. Light industries replacing older extractive and manufacturing industries in 20th century included chocolate factory established by Rowntrees 1950 and carpet, shoe, bed headboard and handbag factories. Sellafield nuclear plant also provided employment. Population rose again from 1950s to peak of 7,996 in 1981. It stood at 7,444 in 2001.

Places of worship. Church of St Mary (formerly chapel in St Bees parish) had gained parochial rights by 13th century; medieval church was rebuilt 1881-3. Site of medieval chapel, recorded 1338, preserved in field-name 'Chapel Close' in Brisco Mount area. Wesleyan Methodist chapel, Main Street, established 1821; replaced by new building 1876; still in use. Wesleyan Association Methodist chapel, Church Lane, built 1839; replaced by larger chapel, the United Methodist Free Church, Bookwell, built 1893; closed 1964; later used as Masonic hall. Primitive Methodist chapel, North Road, built c.1852; closed 1935; later used as rugby club house. St Bridget's Roman Catholic mission school and chapel built 1872; replaced early 20th century by St Mary's RC Church; in turn replaced by new building 1960. Cemetery to north of town, with mortuary chapels, established 1864. As Moor Row and Bigrigg grew in second half of 19th century, they acquired places of worship: church of St John the Evangelist, Bigrigg, built 1877-80. United Methodist chapel, Bigrigg, built 1862. Wesleyan Methodist chapel, Moor Row, built 1888; closed c.1957. Moor Row also had Primitive Methodist and United Methodist chapels.

Schools. Petty school recorded 1722. School built and endowed by Thomas Benn (d. 1743), vicar of Millom; rebuilt 1844. In 1847 there were also four other day schools. Infant school established 1855 in former poor house; closed c.1870. Board schools built at Bookwell (1872; now Bookwell Primary School), Moor Row (1880; now Moor Row Primary School) and Bigrigg (1884; closed and converted to housing). Central Infant School, Main Street, built 1904; became girls' junior and senior school 1920; then mixed senior school 1955; closed c.1964 when Wyndham School opened. Wyndham School opened 1964 as an early comprehensive community school; renamed West Lakes Academy 2008. Roman Catholic School built 1872-3; rebuilt on same site 1966; now St Bridget's RC Primary School. Schools at Orgill (now Orgill Primary School) established 1998 to serve late-20th century housing developments west of town centre.

Other institutions. Municipal buildings: Market Hall (used for concerts) built 1883; adjacent Town Hall built 1889-90. Mechanics' Institution, with library, established 1846. Recreation ground beside river opened 1881.

EGTON WITH NEWLAND

Chapelry in Ulverston parish, Lonsdale hundred, Lancashire North of the Sands. Included hamlets of Arrad Foot, Penny Bridge, Spark Bridge and Greenodd. Combined with Mansriggs and Osmotherley CPs 2003.

Acreage: 3,660 acres [1,481 ha], divided between hamlets thus: Egton: 1,524 acres [617 ha]; Newland: 2,136 acres [864 ha]. Common waste and mosses (c.2,000 acres [c.800 ha]) enclosed and drained 1823. ***Population:*** rising from 675 in 1801 to 934 in 1901; then dropping to 867 in 1951 and to 810 by 1971. Increase to 898 by 2001. ***Landownership:*** *manor of Egton with Newland* acquired by Furness Abbey by gifts from William de Lancaster 1276 and John Harrington 1320. After Dissolution descended to dukes of Buccleuch. ***Economy:*** Greenodd, highest point to which River Leven was navigable, formerly a port, shipping slate, iron bars, copper ore, gunpowder and other local products to Liverpool and Glasgow, and having small-scale boat building. Lakeside branch railway, built 1869, cut off access to river and sounded death-knell for port's trade. Blast furnaces at Newland forge (built 1747; closed 1891) and Penny Bridge (built 1748; demolished 1791; replaced by flax mill 1805). Iron foundry, Spark Bridge, converted to bobbin mill later 19th century; in use until 1950s. Cotton mill at Crake Mill, Spark Bridge, until 1860. ***Places of worship:*** said to have been chapel at Newland 1577. Chapel of ease of St Mary, Penny Bridge, built and endowed by William Penny (d. 1788); rebuilt 1831; chancel added 1855-6 and nave rebuilt 1864-5; tower taken down 1893; new tower built 1969. Wesleyan Methodist chapel, Spark Bridge, built 1864; closed 1999; converted to holiday accommodation 2013. ***Schools and other institutions:*** endowed school recorded 1818; replaced 1868 by public elementary school on different site at Penny Bridge, built with an endowment from Lindow's charity; now Penny Bridge CE Academy. Reading room and institute at Greenodd, built 1907; rebuilt 2003; now Greenodd and Penny Bridge Village Hall. Reading room at Spark Bridge recorded 1911. Village hall at Spark Bridge built 1928.

ELLENBOROUGH AND EWANRIGG

Township in Dearham parish, Allerdale below Derwent ward, Cumberland. Absorbed into Maryport (q.v.) 1929.

Acreage: 1,115 acres [451 ha], including Ellenborough common (103 acres [42 ha]), enclosed 1849. ***Population:*** 471 in 1801 growing steadily to 4,509 by 1911, the increase attributed 1871 to expansion of coal mining and opening of blast furnaces. ***Landownership:*** *manor of Ellenborough (or 'Alneburgh')*: Eaglesfield family from 13th century, passing by marriage to John Senhouse of Seascale 1528 and descending through Senhouses of Netherhall thereafter. *Manor of Ewanrigg (or 'Unerigg'):* Multon family in 14th century; had been acquired by Thwaites family of Thwaites, near Millom, by 1435. Sold 1627 to Richard Barwis; then c.1638 to Ewan Christian of Milntown, Isle of Man. Remained seat of Christians until death of William Bell Christian 1886, when estate was sold. ***Economy:*** coal mining on both Ellenborough (by 1730) and Ewanrigg estates by c.1750 (see Maryport). ***Places of worship:*** All Souls church, Netherton, built c.1886; south aisle added 1899-1906. Primitive Methodist chapel, Ellenborough, built 1860; used as Sunday school after new chapel in Old Road opened 1907; both closed. Wesleyan Methodist chapel, Netherton, built 1892; closed and demolished c.2012. ***Schools and other institutions:*** Ewan Christian (d. 1719) left building to be used as schoolhouse; taught by vicar in early 19th century. Replaced by Board schools at Ellenborough (closed and converted to housing) and Netherton (now Netherton Infant School) by 1900. Ellenborough and Ewanrigg Infant School built on new site 1970. Hospital for Infectious Diseases at Alneburgh House built 1894. Victoria Cottage Hospital built 1902-3; enlarged 1937-8.

EMBLETON

Chapelry and township in Brigham parish, Allerdale above Derwent ward, Cumberland.

Acreage: 3,948 acres [1,598 ha]. Commons totalling 3,150 acres [1,275 ha] in Embleton and Setmurthy enclosed 1824. ***Population:*** rising from 292 in 1801 to 442 in 1831, then slowly declining to stand in range 350-380 for much of later 19th and first half of 20th century; further decline to 297 by 2001. ***Landownership:*** *manor of Embleton* (which excluded surrounding wastes in forest of Derwentfells) granted to Orme de Ireby 1195. Descended through Ireby family for several generations; later purchased by Thomas

Brathwaite of Warcop (after 1589); his grandson Richard sold it to Philip Lord Wharton before 1688. ***Economy:*** predominantly agricultural. Drainage tile manufacture at Hundith Hill mid-19th century; short-lived lead-mining near Stanley Hall 1853 to 1860. Granite quarrying at Close Quarry (also known as Embleton Quarry), opened 1907; closed 1950. ***Places of worship:*** medieval chapel of ease, recorded c.1210; St Cuthbert's church, built 1806, probably on same site. Wesleyan Methodist chapel built 1863; rebuilt 1903; closed 1970. ***Schools and other institutions:*** school taught in chapel by curate in early 18th century. Embleton and Wythop Voluntary School (situated near Wythop Mill) built 1808; closed 1978. Village hall, rebuilt 2009.

ENDMOOR *see* Preston Richard

ENNERDALE AND KINNISIDE

Chapelry in St Bees parish, Allerdale above Derwent ward, Cumberland, containing townships of Ennerdale and Kinniside and extra-parochial waste of Copeland Forest or Stockdale Moor.

Acreage: 22,407 acres [9,068 ha], of which Kinniside township contained 6,722 acres [2,720 ha], Ennerdale 13,178 acres [5,333 ha] and Copeland Forest or Stockdale Moor 2,507 acres [1,015 ha]. Ennerdale Fell (7,637 acres [3,090 ha]) was enclosed 1872. Kinniside Common (5,189 acres [2,100 ha]) and Stockdale Moor (2,495 acres [1,010 ha]) remain unenclosed. ***Population:*** estimated at 395 in 1688. Across first half of 19th century, Ennerdale contained c.190 inhabitants and Kinniside in range 200-250. Combined population of townships declined from peak of 666 in 1871 to stand at 240 in 2001. ***Landownership:*** part of Copeland forest. At partition of barony of Egremont (q.v.) 1338, *Ennerdale* was assigned to Harrington share; descended to Grey family; escheated to crown 1554; leased to Sir James Lowther 1765 and bought outright by earl of Lonsdale 1822. *Kinniside* was assigned to Fitzwalter share of barony as part of 'Middleward' of Copeland forest and descended with barony of Egremont (q.v.) from 16th century. ***Economy:*** predominantly hill farming (sheep fair at Ennerdale Bridge in 19th century) and (from 1925) forestry. Medieval bloomery site at Smithy Beck; iron mines (let to Patrickson family) recorded 1688. Lead-mining at Kinniside from later 18th century, peaking in 1820s; closed c.1855; lead smelter in operation in 1820s. Water supply: Ennerdale Water supplied Whitehaven's water from

1885; new works opened 1995. Meadley Reservoir (in Kinniside) supplied Cleator Moor from 1880s; drained 2010. Youth hostel and field centre at Gillerthwaite at head of Ennerdale, later 20th century. ***Places of worship:*** chapel of ease at Ennerdale Bridge recorded 1534; rebuilt as St Mary's church 1856. Wesleyan Methodists met in farmhouse (possibly How Hall) before 1800. Supposed remains of Catholic chapel there mentioned 1847. ***Schools and other institutions:*** parochial school at Ennerdale Bridge by mid-19th century; rebuilt 1878; now Ennerdale and Kinniside CE Primary School.

ESKDALE AND WASDALE

Township in St Bees parish, Allerdale above Derwent ward, Cumberland. In 1934 Wasdale Head was transferred to Nether Wasdale CP and remainder of township absorbed Birker and Austhwaite to create Eskdale CP.

Acreage: 17,419 acres [7,049 ha], including extensive unenclosed fells on Eskdale Common, 7,590 acres [3,072 ha] of common land, and open summits of Sca Fell massif, which were not common land but lord's freehold. Fells around Wasdale Head were enclosed 1808. ***Population:*** rose from 201 in 1801 to c.350 for much of mid-19th century. Industrial activity resulted in peak of 535 in 1881, from which population fell back to mid-19th century levels, standing at 338 in 1931 (last census year before boundary changes). ***Landownership:*** southern section of Copeland forest, assigned to Margaret, wife of Thomas de Lucy 1338 and descending with Lucy share of barony of Egremont (q.v.) thereafter. Ownership of manorial waste passed to The National Trust 1979, who also owned several farms. Estate centred on Gatehouse, rebuilt 1896-1901 as villa for James Henry Rea, Liverpool ship-owner, included property in Miterdale, which passed to Forestry Commission 1957. ***Economy:*** predominantly hill farming; small fair called 'Dog-skin fair' held on St Catherine's day (25 November) in later 17th century. Industries in 19th century included bobbin mill at Longrigg in Miterdale; carding mill at Gill Bank; granite quarrying; iron mining. Iron mine at Nab Gill, opened 1871, resulted in construction of Ravenglass & Eskdale Railway (1873-5). Tourism developed from later 19th century: Stanley Ghyll built as hotel 1894; youth hostel built 1937; Ravenglass & Eskdale Railway ('the Ratty') became tourist honeypot from mid-20th century. Wasdale Head was centre for rock-climbing from later 19th century (Wastwater Hotel built by 1860) and Outward Bound centre at Gatehouse, Eskdale Green, became focus for outdoor pursuits in 20th century. ***Places of worship:*** St Catherine's church,

Eskdale, parochial chapel which gained burial rights 1445; chapelry also embraced Birker and Austhwaite (q.v.), south of River Esk. Chapel at Wasdale Head (now St Olaf's church), a tiny dales chapel, recorded 1552; without burial ground until c.1901. St Bega's mission church, Eskdale Green, built by Lord Rea, opened 1890. Wesleyan Methodist chapel at Key How built 1849; enlarged 1890; closed c.1964. ***Schools and other institutions:*** school in Eskdale endowed 1723. 'High School' near Boot built 1863; closed later 20th century (now bunk barn accommodation). School at Eskdale Green by 1860; 'Low School' there, held in St Bega's mission church from 1890; moved to new site in 1960s; now St Bega's CE Primary School. School at Wasdale Head by 1855; closed c.1900. Village hall, Eskdale Green, in former school on lower floor of St Bega's church.

ETTERBY *see* Stanwix

FARLAM

Ancient parish in Eskdale ward, Cumberland, comprising townships of East Farlam and West Farlam.

Acreage: 5,232 acres (2,117 ha), of which East Farlam contained 4,012 acres (1,624 ha) and West Farlam 1,220 acres (494 ha). After 1883 Farlam CP also included Carnetley (78 acres [32 ha]), formerly a detached portion of Lanercost parish. Farlam commons (2,300 acres [931 ha]) enclosed 1780. ***Population:*** estimated at 260 in 1688; rising from 592 in 1801 to 1,585 in 1881 as result of industrial activity; dropped back to 1,365 by 1901, then declined across 20th century to 590 in 2001. ***Landownership:*** *manor of Farlam* held by Windsor family in 12th century; direct lordship appears to have been with Farlam family (probably under-tenants of Windsors), through whom manor descended to 14th century, when John de Farlam, having no issue, granted it to Ranulph de Dacre, after which it descended with barony of Gilsland (see Brampton). ***Economy:*** coal mining; quarrying and lime-burning. Colliery at Windgates, near Templegarth, recorded 1688; wooden waggonway built 1799 to link earl of Carlisle's mines on Tindale Fell to Brampton coal staithe; replaced by new railway 1836; closed 1953. Main centres of industrial activity included Kirkhouse (coke ovens; iron foundry in 19th century, superseded by brick works in 20th century); Clowsgill (limestone quarries and lime works; had closed by 1900); and Roachburn (colliery from 1860s to 1912). ***Places of***

worship: medieval parish church at Kirkhouse; rebuilt as church of St Thomas à Becket on adjacent site 1859-60. Wesleyan Methodists met in private houses until chapels built at Hallbankgate (built 1856; enlarged 1883; closed and converted to dwelling) and Doleshole, near Tindale, (built 1870, rebuilt 1888; closed). ***Schools and other institutions:*** curate taught school in church 1703; three schools (two of them dame schools) recorded 1816. School at Hallbankgate (now Hallbankgate Village School) had been built by 1863. School at Doleshole, for Tindale spelter works (see Midgeholme), built 1860; now demolished. Mechanics' Institute and Library at Hallbankgate built 1857. Village hall (Lacy Thompson Memorial Hall) built 1898; extended 1995 and 2000. Reading room at Doleshole, Tindale, built 1868, moved into former Methodist chapel 1888. Church hall erected at Kirkhouse 1928.

FARLETON

Township in Beetham parish, Kendal ward, Westmorland. Absorbed into Beetham CP 1935.

Acreage: 1,204 acres [487 ha], including 301 acres [122 ha] of unenclosed common land on Farleton Knott. ***Population:*** rising from 92 in 1801 to peak of 118 in 1841, then falling below 100 by 1861 and to 63 by 1921 (last census year for which separate figure available). ***Landownership:*** *manor of Farleton* held under Beetham by Harrington family in 15th century, from whom it was confiscated c.1485 and granted to Stanley family. Sir Richard Hutton was lord of manor 1639 and it had descended by marriage to Col. Brierly of Grange (Yorks.) by 1692. ***Economy:*** predominantly agricultural; small-scale limestone quarrying and lime-burning in 19th century. ***Places of worship:*** 'Chapel hill' said (1692) to have been site of chapel. ***Schools and other institutions:*** none.

FAWCETT FOREST

Township straddling boundary between Kendal, Shap and Orton parishes, Kendal ward, Westmorland. In 1935 Fawcett Forest CP absorbed parts of Orton, Shap and Whinfell CPs.

Acreage: 6,470 acres [2,618 ha] before boundary changes. Walshaw Common and Hazel Bank (307 acres [124 ha]) enclosed 1870. ***Population:*** sparsely populated; falling from 81 in 1801 to 51 between 1851 and 1871, 37 in 1901, and 23 in 2001. ***Landownership:*** granted to Byland Abbey (Yorks.) in 12th century. After Dissolution acquired 1554 by Bellingham family of Levens (q.v.), with which it descended. ***Economy:*** hill farming; grouse shooting; small-scale quarrying. ***Places of worship:*** none. ***Schools and other institutions:*** none.

FINSTHWAITE *see* Colton

FIRBANK*

Township in Kirkby Lonsdale parish, Lonsdale ward, Westmorland.

Acreage: 2,985 acres [1,208 ha], including 1,085 acres [439 ha] of common land on Firbank Fell, enclosed 1851. ***Population:*** rising from 190 in 1801 to 233 in 1851; inflated to 345 in 1861 as result of influx of railway builders, then declining to 153 in 1901 and 97 in 2001. ***Landownership:*** granted c.1201 to William de Arundel; held with Killington (q.v.) from 13th century to later 16th century by Pickering family of Killington. Uptons of Ingmire Hall, Sedbergh, and Cavendish-Bentinck family of Underley, Kirkby Lonsdale, held large estates in township from 19th to mid-20th century. ***Economy:*** predominantly farming; small-scale quarrying; tourism from 20th century (caravan site at Waterside). ***Places of worship:*** chapel of ease recorded 1535; said to have been near Priestfield (on boundary with Killington) until rebuilt on Firbank Fell in 17th century. Chapel used by Separatist group ('Westmorland Seekers') in early 1650s; George Fox preached on hillside there 1652 (supposed site now 'Fox's Pulpit'). Chapel rebuilt 1742 (dedicated to St Martin); damaged by storms 1839-40 and demolished (graveyard remains). Rebuilt again as church of St John the Evangelist, on less remote site near Under Knotts, 1841. ***Schools and other institutions:*** school adjoining chapel on fell, recorded 1692; rebuilt on new site at Stocks 1858; closed 1938; building subsequently used as church hall.

FLIMBY

Parish (formerly chapelry) and township in Allerdale below Derwent Ward, Cumberland.

Acreage: 1,651 acres (668 ha). Flimby and Seaton Commons (1,000 acres [405 ha]) enclosed 1826. ***Population:*** estimated at 350 in 1688; rising steadily from 273 in 1801 to peak at 2,498 in 1921; then gradual decline to 1,763 in 1991. ***Landownership:*** granted to Holme Cultram Abbey in 12th century. After Dissolution granted to Thomas Dalston, from whom it was purchased by John Blennerhassett 1546. Remained in Blennerhassett family until 1772, when sold to Sir James Lowther. ***Economy:*** collieries, saltpans (said to be recently built) and sea fishing recorded 1688. Iron-making: blast furnace in 18th century (c.1715; short-lived). In 19th and 20th centuries major collieries were at Risehow (closed 1966), Gillhead (sunk 1867; closed 1959), and Flimby Colliery (Robin Hood and Watergate pits; closed 1909). Coke ovens at Risehow and brickworks at Gillhead. Industrial estate (including concrete works) on former Risehow colliery site. ***Places of worship:*** chapel of ease to Camerton ancient parish, recorded from 12th century; burials took place at Camerton until 1797. Church of St Nicholas rebuilt on site of earlier chapel 1794, restored and enlarged 1862. Wesleyan Methodist chapel built 1858; replaced 1875 by new chapel at Rye Hill which closed c.1970. Primitive Methodist chapel built 1862; replaced by new Westfield chapel, West Lane, 1929; still in use. ***Schools and other institutions:*** two unendowed schools recorded 1818. Board schools built at Rye Hill 1876; now Flimby Primary School. Ladies' boarding school, run by Miss Wilson, at Flimby Lodge by 1860; building bought by Cockermouth Poor Law Union 1886 and used as workhouse school and vagrancy ward; recorded as a boys' home 1938; derelict by 1977. Working Men's Reading Room and Public Hall built 1873.

FLOOKBURGH *see* Lower Holker

FRIZINGTON *see* Arlecdon

GAITSGILL *see* Dalston

GAMBLESBY*

Township in Addingham parish, Leath ward, Cumberland. Absorbed into Glassonby CP 1934.

Acreage: 9,810 acres [3,970 ha] excluding Glassonby Fell, a detached portion of rough fell land (706 acres [286 ha]) in Glassonby township, which was added to Gamblesby CP 1888. Gamblesby Low Fell and Viol Moor (1,480 acres [599 ha]) enclosed 1860; Gamblesby Fell (1,720 acres [696ha]) enclosed 1868. ***Population:*** rising from 222 in 1801 to peak of 301 in 1831, followed by downward trend to 197 in 1931 (last census year for which separate figure available). ***Landownership:*** descended with Glassonby (q.v.) until Edward I granted Gamblesby-with-Unthank to William, 2nd Lord Latimer (d. 1326), subsequently passing by marriage to Nevilles in later 14th century. Reverted to Crown 1471; then descended with honour of Penrith (q.v.). ***Economy:*** pastoral farming. Coal and lead mining on Hartside, recorded from 1688. Ironstone mining, quarrying and lime-burning in 19th century. Only Woolhead and Hartside quarries remained open by 1898. ***Places of worship:*** Wesleyan Methodist chapel built 1784; replaced by new chapel on same site 1864; still in use. Congregational chapel built 1824; replaced by larger chapel 1864; closed 1930s; converted to dwelling by 1955. St John's Church (Anglican chapel-of-ease to St Michael's, Addingham) built 1868; closed 2002 and converted to dwelling. ***Schools and other institutions:*** schoolhouse by 1751, when John Wesley preached in it. Board school built on new site, opened 1877; closed 1970; used as village hall until 2001 (when closed, as structurally unsound); hall rebuilt and reopened 2007.

GARRIGILL *see* Alston

GARSDALE

Chapelry and township in Sedbergh parish, Ewecross wapentake, West Riding of Yorkshire.

Acreage: 12,172 acres [4,926 ha], including extensive fell commons, parts of which were enclosed 1859. Baugh Fell (part of Bluecaster Common (7,556 acres [3,058 ha]), straddling boundary between Sedbergh and Garsdale) remains unenclosed. ***Population:*** rising from 571 in 1801 to peak of 911 in 1871 during building of Settle-Carlisle railway; then steady decline to 202 in 2001. ***Land ownership:*** *manor of Garsdale* belonged to St Agatha's Abbey, Easby, which leased it to Scropes of Castle Bolton. After Dissolution, sold to Sir William Garway 1621, who sold it on to five Garsdale yeomen. Braithwaite Otway of Ingmire Hall bought up the five parts and enfranchised tenants1740. *Grisedale* was part of manor of Ravenstonedale (q.v.). ***Economy:*** predominantly hill farming, with hand-knitting. Coal mining on Baugh Fell and quarrying (sandstone, limestone, and 'marble'), mid-19th century. ***Places of worship:*** church of St John the Baptist, chapel of ease to Dent, recorded mid-16th century; rebuilt 1861. Quaker meeting houses at Birkrigg in Garsdale (house bought 1703; closed 1900) and Stubstacks in Grisedale (built

1706; closed 1870, then used by Methodists; demolished c.1888). Protestant dissenters had chapel in mid-18th century. Wesleyan Methodist chapel, Low Smithy, built 1830 or 1831; still in use. Primitive Methodist chapel, Garsdale Street, built 1841; still in use. Mount Zion Chapel (Methodist), Garsdale Head (otherwise called Hawes Junction), opened 1876; still in occasional use. Further Wesleyan Methodist chapels at Garsdale Foot (built 1864; closed 1979) and Grisedale (built 1889; closed 1970). ***Schools and other institutions:*** school endowed by Thomas Dawson 1634; rebuilt as National school on new site, 1842, becoming Garsdale Primary School; closed 1985; converted to village hall.

GELTSDALE

Extra-parochial area of upland waste with status of hunting forest (King's Forest of Geltsdale) in Eskdale ward, Cumberland. Absorbed into Castle Carrock CP 2003.

Acreage: 5,245 acres [2123 ha]. ***Population:*** 18 in 1851 (first year for which figures available); 26 in 1891, then declining to a mere 6 in 2001. ***Landownership:*** granted to Lanercost Priory by Vaux family of Hayton, Geltsdale and Tryermaine; after Dissolution passed to earl of Carlisle. ***Economy:*** hill grazing; woodland (birch and alder) and red deer; since 1975 has been RSPB reserve. ***Places of worship:*** none. ***Schools and other institutions:*** none.

GILCRUX

Ancient parish in Allerdale below Derwent ward, Cumberland.

Acreage: 2,018 acres [817 ha]. Gilcrux Infield (open field) strips consolidated 1648; enclosure completed before 1749. Gilcrux Outfield (c.460 acres [186 ha]) enclosed 1754. Gilcrux Common and Green (c.110 acres [44 ha]) enclosed 1814. Grange Grassings (95 acres [38 ha]) enclosed 1816. ***Population:*** estimated at 220 in 1688; rising from 249 in 1801 to 653 in 1861, as result of expansion of coal mining; then falling to 413 by 1901 and to 303 in 2001. ***Landownership:*** *manor of Gilcrux* granted to Calder Abbey by Bonekil (or Buncle) family in mid-13th century. By 1632 had passed to Thomas Dykes of Wardhall in which family it remained until 20th century. ***Economy:*** farming; coal mining recorded by 17th century ('Gilcrouse Pit' working before 1636). Expansion in 19th century: Jane Pit (sunk 1815, closed 1863); Eliza Pit (opened 1854; closed before 1882) and Ellen Pit or Bullgill Colliery (sunk 1859; closed 1910). Small drift mine worked 1937/8 and barytes drift mine operated in Rose Gill 1920s. Some small-scale quarrying, later 19th century. ***Places of worship:*** medieval parish church of St Mary. Wesleyan Methodist chapel built 1875; closed 1985. ***Schools and other institutions:*** schoolmaster mentioned c.1663. Parish school built 1771; replaced by new school on different site 1864; closed 1971; building then used as village hall (refurbished 2012). Mission room (attached to vicarage) built 1892; demolished 1989. Reading room (privately owned); demolished c.1940.

GILSLAND (barony) *see* Brampton; (village) *see* Denton, Upper; Waterhead

GLASSONBY

Township in Addingham parish, Leath ward, Cumberland. CP enlarged by absorbing Gamblesby CP 1934.

Acreage: 1,581 acres [640 ha], excluding Glassonby Fell (a detached portion of 706 acres [286 ha] of rough fell land) added to Gamblesby CP 1888. Glassonby Fell, Viol Moor and Maughanby Common, totalling 1,720 acres [696 ha], enclosed 1867. ***Population:*** in range 145-165 across 19th and early 20th centuries; stood at 164 in 1931 (last census year before boundary change). Population of enlarged CP was 314 in 2001. ***Landownership:*** Glassonby and Gamblesby held jointly by descendants of Hildred of Carlisle from 12th century, passed to Sir Christopher Seton, on whose execution in 1306 they were forfeited to Crown. Glassonby then came to Dacres, whose successors sold it 1716 to Sir Christopher Musgrave of Edenhall, in whose family it remained in 19th century. ***Economy:*** predominantly agricultural. Quarrying in 19th century. Glassonby Beck Mill became paper mill (initially making pulp for Kirkoswald paper mill; later mill board and paper) from c.1860; closed by 1911. ***Places of worship:*** medieval parish church of Addingham thought to have been lost in River Eden; existing St Michael's church dates from 16th century (chancel built 1512); restored 1786 and 1898. Wesleyan Methodist chapel built 1839; rebuilt on new site 1869; closed 2011. ***Schools and other institutions:*** free school at Maughanby endowed by vicar, Edward Mayplett, 1634; rebuilt 1868 on new site with schoolhouse; closed and converted to dwellings. Village hall built c.1930; closed c.2000.

GLEASTON *see* Aldingham

GLENRIDDING *see* Patterdale

GOSFORTH

Ancient parish in Allerdale above Derwent ward, Cumberland.

Acreage: 8,160 acres [3,302 ha], including c.3,000 acres [c.1,215 ha] of common moor and fell, enclosed 1815. Parish was divided into four townships or constablewicks: Gosforth, Low Bolton, High Bolton and Boonwood & Seascale. Seascale was detached 1897 to form separate CP containing 1,063 acres [430 ha]. ***Population:*** estimated at 630 in 1688. Rose rapidly across mid-19th century, from 330 in 1801, reaching 1,327 in 1891 (last census year before Seascale was separated). Population of remainder of parish was c.930 in early and mid-20th century but rose from 1970s to stand at 1,230 in 2001. ***Landownership:*** parish contained three manors: Gosforth, Bolton and Newton or Seascale. *Manor of Gosforth* held by de Gosford family in 13th century; divided between five co-heiresses after death of Robert de Gosford 1316. *Manor of Bolton* in hands of de Waberthwaite family in 13th century; held by branch of Senhouse family by 18th century; had passed to Charles Lutwidge by 1777 and remained in Lutwidge family of Holmrook Hall for much of 19th century. *Manor of Newton and Seascale* was held by Senhouse family from c.1270 until lost by mortgage c.1707. Re-acquired by branch of Senhouses 1800 but sold after death of Sir Humphrey le Fleming Senhouse (d. 1841) to Anthony Benn Steward, from whom it passed to Falcon-Steward family. ***Economy:*** predominantly agricultural. Two cattle fairs annually at Boonwood (on former common land assigned for that purpose at enclosure); had discontinued by 1930s. Woollen textile industry implied by name Walk Mill (i.e. fulling mill). Sandstone quarrying and basket-making in mid-19th century. Afforestation (Blengdale Forest) mid-20th century.

SEASCALE developed as resort after arrival of Whitehaven & Furness Junction Railway 1849. Scawfell Hotel built beside station by 1860. Plans for seaside resort in 1879 not fully realised, though visitor accommodation and golf links created by 1900; expanded as dormitory settlement for munitions workers and, later, workers at Sellafield, from 1940s.

Places of worship: medieval parish church of St Mary (Gosforth cross and other Viking age sculpture suggest pre-Conquest foundation); rebuilt 1896-9. St Cuthbert's chapel of ease, Seascale, originated as iron church 1881; rebuilt

1890. Wesleyan Methodist chapels in Gosforth village (built 1874) and at Seascale (built 1887); both still in use. St Joseph's Roman Catholic Church, Seascale, built 1960. ***Schools and other institutions:*** parish school (unendowed) recorded 1818; rebuilt 1873; now Gosforth CE Primary School. School (near School Tarn) on Seascale Moor built c.1830; replaced by Seascale CE school, Halsenna Road, opened 1873; closed 1951 when new Seascale Council School built. Preparatory school at Seascale, recorded 1901. Harecroft Hall School (independent preparatory school) opened 1925; closed 2008. Library and reading room at Gosforth, established 1860. Parish hall built 1930.

GRANGE-OVER-SANDS

Resort town in Lonsdale hundred, Lancashire North of the Sands. For pre-urban history of area, see Broughton East, Allithwaite, Lower and Allithwaite, Upper.

Acreage. Grange CP and UD created 1894 from parts of Broughton East, Upper and Lower Allithwaite CPs; new CP covered 1,540 acres [623 ha]. Further 274 acres [111 ha] transferred from Lower Allithwaite to Grange 1914.

Origins and growth of the town. Grange developed as hydropathic centre and tourist town following opening of Ulverston & Lancaster Railway 1857; by 1870s it was described as 'in much local repute as a watering-place'. Resort development took off in 1860s, with building of Grange Hotel (completed 1866) and laying out of ornamental gardens on former marshland in front of railway station (1865). Kents Bank Hotel (at former departure point for cross-sands route to Hest Bank) opened 1876. Growth boosted in 1880s with arrival of hydropathic institutions: Grange Hydropathic and Hazelwood Hydro (now Grand Hotel), opened 1887. Former villas overlooking sea were converted into hotels, most notably Abbot Hall at Kents Bank (built 1840; became hotel 1916) and Netherwood (built 1893). From 1870s, town's economy centred on retail and service sectors, supporting tourism and caring businesses. Convalescent homes were built: North-East Counties Friendly Societies' home 1897 (present building dates from 1914-16) and Working Men's Club and Institute Union home (now Cartmel Grange Care Home) 1914. Promenade laid out 1902-04 and lido built 1932 (closed 1993). Other activities included coach-building, limestone quarrying, slipper manufacture and printing. By 1891 population was 1,733; by 1921 it stood at 2,920 (of whom almost 30% were visitors). By c.1900 Grange's character was as community of retired business people and manufacturers. Analytical and technical agricultural testing laboratories moved from Aynsome to Kentsford

House, Kents Bank, Grange 1954 and Nature Conservancy had offices in former villa at Merlewood from 1951 to 2002. Grange continued to grow across middle and later decades of 20th century, gaining significant retired population; by 2001 population had reached 4,042.

Places of worship. St Paul's Church (Anglican) built 1852; enlarged several times between 1861 and 1904 and partly rebuilt 1932 (adjacent clock tower, one of Grange's landmarks, built 1912). Grange Fell church (Anglican), built 1907, originated in church room on Grange Fell, recorded 1898. Wesleyan Methodist chapel built 1874; enlarged 1903 by addition of hall and Sunday school; still in use. St Charles' Roman Catholic Church built 1884; still in use. Congregational Church built 1894; now United Reformed Church. Cemetery on Grange Fell Road opened 1889.

Schools and other institutions. Sunday school (began 1811) became day school 1830. National school built 1864; enlarged 1884; closed (later used as Royal British Legion club) when new school built on new site in 1960s; now Grange CE Primary School. By later 19th century Grange possessed many of institutions of a small town: Working Men's Institute established 1866. Masonic Lodge built and opened 1895. Town hall and Victoria Hall built 1898 (opened 1901).

GRASMERE

Township in Grasmere parish, Kendal ward, Westmorland. Became Grasmere UD 1894 to 1935; then absorbed into Lakes UD.

Acreage: 7,319 acres [2,962 ha] the bulk of it consisting of unenclosed fell commons, totalling 5,072 acres [2,053 ha]. ***Population:*** rising from 270 in 1801 to 445 by 1851; increase to 604 in 1861 attributed to 'erection of Lake residences'; continued to rise, standing at 988 in 1931 (last census year for which separate figure available). ***Landownership:*** *manor of Grasmere* formed part of barony of Kendal (q.v.), with which it descended. ***Economy:*** hill farming; woollen textiles (fulling mill recorded 14th century). Slate quarrying at White Moss by late 17th century; expansion in 19th century; closed 20th century. Grasmere became major tourist centre from mid-19th century, noted for its association with William and Dorothy Wordsworth; Dove Cottage (and nearby Jerwood Centre, completed 2004) and their graves in churchyard becoming popular attractions. Tourist businesses included four hotels by 1885, lodging houses, Grasmere Gingerbread Shop (established 1854 in former school on edge of churchyard), Heaton Cooper art studio (established 1938), cafes, textile shops, and garden centre. ***Places of worship:*** parish church of

St Oswald (chapel of Kendal ancient parish until 14th century) contains late-medieval fabric; re-roofed c.1562; restored 1840. Wesleyan Methodist chapel built 1874; closed 2008. Our Lady of the Wayside Roman Catholic church built 1964-5. ***Schools and other institutions:*** school in corner of churchyard built by subscription c.1685. Rebuilt on new site 1854 (infant school built 1862) and enlarged 1879; now Grasmere CE Primary School. Working Men's Reading Rooms erected 1873; remain open. Grasmere New Hall opened 1903; still in use as village hall.

GRAYRIGG

Township in Kendal parish, Kendal ward, Westmorland. CP enlarged by absorbing Dillicar CP 1986.

Acreage: 3,753 acres [1,519 ha], including 1,022 acres [414 ha] of common land on Grayrigg Fell enclosed 1868. ***Population:*** rising from 199 in 1801 to peak of 264 by 1841, then gradual decline to low point of c.165 in mid-20th century; slight increase to stand at 223 in 2001. ***Landownership:*** granted by William de Lancaster, baron of Kendal, in marriage with his daughter Agnes to Alexander de Windsor c.1160-70, descending through Windsor family until death of Sir William Windsor 1385, when passed by marriage to Duckets, who held it until c.1685 when Anthony Ducket sold it to Sir John Lowther, who enfranchised the tenants 1695. ***Economy:*** largely agricultural; small-scale quarrying in 19th century. ***Places of worship:*** chapel of ease recorded 1469; rebuilt 1708; present church of St John built by subscription 1837-8; tower rebuilt 1869. Quaker meeting house at Beckhouses built 1696; closed 1841 but rebuilt 1871; closed again 1952. ***Schools and other institutions:*** school taught in Quaker meeting house by 1709; schoolhouse adjoining built late 18th century. Parochial school built by subscription 1818; rebuilt 1960s; now Grayrigg CE Primary School. Girls' school established 1845 and endowed by trustees of Quaker meeting house at Beckhouses. Six almshouses built and endowed 1869 by John, Richard and Mary Rowlandson of Gillbank. Grayrigg Institute or Coronation Hall built 1911.

GREAT CORBY *see* Wetheral

GREAT MUSGRAVE *see* Musgrave

GREAT ORTON *see* Orton, Great (Cumberland)

GREENODD *see* Egton with Newland

GREYSOUTHEN

Township in Brigham parish, Allerdale above Derwent ward, Cumberland.

Acreage: 1,646 acres [666 ha] including 500 acres [202 ha] of common moor, enclosed 1826. ***Population:*** rising from 455 in 1801 to peak of 758 by 1861; dropped back from 690 in 1881 to 490 in 1901 as result of closure of coal mines. Stood at 508 in 2001. ***Landownership:*** one of the Five Towns in honour of Cockermouth (q.v.); ownership divided between several free tenants by 16th century. ***Economy:*** farming and coal mining. Mining recorded by 1720s; five collieries by mid-19th century. In decline by 1880s: Melgramfitz Pit closed 1886. Linen thread manufacture at Oldfield Mill (steam-powered flax mill) in early/mid-19th century. ***Places of worship:*** Quaker meeting house built 1742; closed 1871. Wesleyan Methodist chapel built 1833; closed later 20th century and converted to dwelling. ***Schools and other institutions:*** voluntary school built c.1838; enlarged c.1900; closed c.1971. Village hall (Joseph Harris Memorial Hall) built later 19th century. Public garden (commemorating Queen Elizabeth's silver jubilee) opened 1977.

GREYSTOKE

Ancient parish in Leath ward, Cumberland, embracing 13 townships, of which 9 became separate CPs (for which see separate entries): Berrier & Murrah, Bowscale, Hutton John, Hutton Roof, Hutton Soil, Matterdale, Mungrisdale, Threlkeld and Watermillock. This entry treats Greystoke CP, which covered townships of Greystoke, Little Blencow, Johnby and Motherby & Gill.

Acreage: Greystoke CP contained 7,511 acres [3,040 ha], divided between constituent townships thus. Greystoke: 4,725 acres [1,912 ha]; Johnby: 1,971 acres [798 ha]; Little Blencow: 336 acres [135 ha]; Motherby and Gill: 479 acres [194 ha]. Johnby pasture, common and Craggs (638 acres [258 ha]) were enclosed 1795. Common field at Motherby survived into 19th century and was enclosed 1817. ***Population***: population of constituent townships in 1851 was: Greystoke: 345; Johnby: 133; Little Blencow: 56; Motherby and Gill: 95. Combined population for Greystoke CP rose from c.500 in early 19th century to peak of 608 in 1881; then fell to 473 by 1931. Rose again across later 20th century to stand at 733 in 2001. ***Landownership:*** Greystoke Castle (licence to crenellate 1353) was seat of *barony of Greystoke*, held by de Greystoke family, descended from Lyulph in early 12th century. Passed to Dacres on marriage c.1488 of Elizabeth, daughter of Robert de Greystoke, to

Thomas, Lord Dacre. On division of Dacre estates 1569, Greystoke was assigned to Anne (c.1557-1630), wife of Philip Howard, Lord Arundel, and descended with dukedom of Norfolk until 1815, when it passed to younger branch of Howard family, in whose hands it remains. Castle rebuilt late 17th century and again 1837-45; extended 1875-8. Greystoke and Motherby & Gill were held directly of barons of Greystoke but Johnby and Little Blencow were dependent manors. *Manor of Johnby,* first mentioned 1200, held successively by Musgrave and Wyvill families; sold to Roger Williams c.1650 and passed by marriage to Hasells of Dalemain, who sold it to duke of Norfolk 1783. *Manor of Little Blencow* granted to Adam de Blencowe in 1357 in recognition of his service at Crecy and Poitiers. Descended through Blencowe family until Henry Prescot Blencowe sold it to duke of Norfolk 1802. ***Economy:*** predominantly agricultural. Attempt to found borough at Greystoke (burgages recorded 1289; market granted 1350) was unsuccessful. Limestone quarrying and lime-burning in 19th century. Forestry in 20th century. ***Places of worship:*** medieval parish church of St Andrew (collegiate, with master and six chaplains, from 1382 to Reformation); restored 1818 and 1848. Ruins of supposed medieval chapel in grounds of Blencow Hall. Wesleyan Methodist chapel at Little Blencow built 1877. ***Schools and other institutions:*** school at Greystoke built 1838 by Henry Howard, who endowed it; now Greystoke Primary School. Girls' school at Little Blencow built by subscription 1856; now closed. Village hall at Greystoke built 1932.

GRINSDALE

Ancient parish in Cumberland ward, Cumberland. Part of Belle Vue CP was added to Grinsdale CP 1910 and Grinsdale CP was absorbed into Beaumont CP 1934.

Acreage: 838 acres [339 ha], increasing to 1,425 acres [577 ha] after boundary change. Grinsdale Common (350 acres [142 ha]) enclosed 1808. ***Population:*** in range 90-115 before boundary change, with peak of 138 in 1821. ***Landownership:*** held of barony of Burgh by de Grinsdale family until c.1200, when divided between two heiresses; by 1271 moieties were held by Thomas de Newton and William de Sore. Newton's moiety passed by marriage to Martindales in 14th century and later came to Dacres. Other moiety sold to Dentons of Cardew in 15th century; George Denton sold it to Sir John Lowther c.1686. ***Economy:*** farming; fishery in River Eden. Bone manure works at Grinsdale Bridge by 1860s; tar distillery there by 1900; both closed mid-20th century. ***Places of worship:*** medieval parish church of St Kentigern, ruinous before rebuilding in 1738-40; further restoration 1895. ***Schools and other institutions:*** parish served by school at Kirkandrews-on-Eden, as beneficiaries of Pattinson Charity. Village hall: built as recreation hall or reading room early 20th century; burnt down 1934 and rebuilt on smaller scale 1935.

GRIZEBECK *see* Kirkby Ireleth

HACKTHORPE *see* Lowther

HAILE

Ancient parish in Allerdale above Derwent ward, Cumberland, formerly divided into two townships, Wilton and Haile.

Acreage: 2,672 acres [1,081 ha], including common land on Haile Moor (900 acres [364 ha]), enclosed 1814. ***Population:*** in range 240-300 for most of 19th and early 20th centuries, reaching peak of 330 in 1851. Fell in later 20th century to low point of 155 in 1971; then rose to stand at 237 in 2001. ***Landownership:*** *manor of Haile* (one of group in Copeland barony of which lords of Gilsland held intermediate lordship) held 1295 by Agnes (who married William de Ponsonby) and Constance, daughters and heiresses of Alexander de Haile. Divided between their descendants in 14th century but reunited in hands of Ponsonby family by later 16th century; Haile Hall (present building rebuilt by Ponsonbys in late 16th century) remained their seat in early 21st century. Hamlet of *Wilton* remained in hands of lords of Copeland barony. ***Economy:*** farming; iron ore mining in mid-20th century (Haile Moor mine: shaft sunk 1939; mine active from 1941; closed 1973); afforestation in 20th century. ***Places of worship:*** medieval parish church (dedication unknown); rebuilt in 18th century; restored 1882-3. ***Schools and other institutions:*** Haile and Wilton National School built 1879; closed 1955; building became village hall 1974.

HARRABY *see* St Cuthbert Without

HARRINGTON

Ancient parish in Allerdale above Derwent ward, Cumberland. Northern part of parish was transferred to Workington MB 1934, when remainder was re-named Lowca CP.

Acreage: 2,360 acres [955 ha], of which parts which became Lowca CP contained 1,093 acres [442 ha]. Harrington and Lowca commons, totalling 534 acres [216 ha], were enclosed 1761. ***Population:*** estimated at 485 in 1688. Rose from 1,357 in 1801 to over 2,000 by 1851; almost doubled in late 19th/early 20th century, from 2,294 in 1871 to 4,371 in 1921 but had dropped slightly to 4,128 by 1931 (last census year before boundary changes). ***Landownership:*** *manor of Harrington* held of Workington by Harrington (or 'Haverington') family in 13th and 14th centuries. It descended by marriage to Henry Grey, duke of Suffolk (executed 1554) and in 1556/7 was granted by Crown to Henry Curwen of Workington; thereafter it descended with Workington. ***Economy:*** coal mines in Harrington Park: value put at £100 per year in 1688, when there was also salt pan at Lowca. Coal mining expanded in later 18th century. Ships were using small haven at Harrington Beckfoot by early 18th century and trade increased after Henry Curwen built new quay there c.1760. Port was known as 'Bella Port' (probably in honour of Curwen's daughter and heiress, Isabella). New town on grid of streets beside harbour appears to have been laid out by John Christian Curwen from 1780s (most of it was demolished 1966). Shipbuilding and associated industries (ropeyards and sailmaking) established by 1800. In 19th century Harrington was a port of some importance, exporting coal to Ireland and bringing in iron ore. In 1865 mines were taken over by Messrs Bain, Blair and Paterson of Harrington ironworks, who sank new pits from 1870s. By 1905 only one pit remained open; a further pit, sunk 1910-11, was worked until 1968. Port of Harrington closed 1928. Ironstone and fire-clay were mined at Harrington in later 18th century. Iron foundry established at Lowca by 1804; in 1840s it supplied engines for Maryport & Carlisle Railway; closed 1921. Arrival of railway 1843-5 encouraged further industrial expansion: blast furnaces built close to harbour 1857; demolished 1937. Harrington Brickworks established by mid-19th century; extensive brick works opened later at Micklam; closed 1976. Other industries included copperas works or 'chemical works' at Copperas Hill, built 1798; sold on bankruptcy of owner 1835; disused by 1860. Tanyard and brewery in village by mid-19th century. During Second World War 'Magnesite' works built on shore by Ministry of Aircraft Production 1940, to extract magnesium from seawater; closed 1953. Wind farm along coast at Lowca became operational 2000. ***Places of worship:*** medieval parish church (modern dedication to St Mary); rebuilt 1634; large eastern extension added 1807-8, producing profile which led to nickname 'the snail church'. Present church dates from 1884-5; tower rebuilt 1905-7. Roman Catholic chapel built for Catholic school 1874; rebuilt as St Mary's RC Church 1893. Presbyterian Church (now United Reformed Church) built 1881. Wesleyan Methodist chapel built 1828, with additional chapel in Gothic style in Victoria Square, built 1886 (closed; now Wesley House). Primitive Methodist chapel, Ellerbank, built 1891; still in use. Victoria Hall used by Salvation Army 1901 and by Plymouth Brethren 1938. Methodist chapel at Lowca built 1910; closed 2007 (converted to dwelling). ***Schools and other institutions:*** petty school recorded 1722; school near Hall Green Farm by late 18th century (graffiti from 1791). Board schools, Church Road, built 1875 and 1897; replaced 2005 by Beckstones Primary School on new site. Board school at Lowca; now Lowca Community Primary School. St Mary's Catholic schools. Victoria Hall built 19th century; housed literary association and library in 1901; demolished 1960s. Victoria Orange Hall, Primrose Terrace, built 1897. Reading room at High Harrington, converted from smithy. High Harrington Community Centre opened after Second World War in former nissen hut.

HARTLEY

Township in Kirkby Stephen parish, East ward, Westmorland.

Acreage: 3,229 acres [1,307 ha], including approximately 1,000 acres [c.400 ha] of common fell. ***Population:*** rising from 139 in 1801 to 177 by 1851; surged to 215 in 1861 during influx of railway construction workers; dropped back to range 115-145 across most of 20th century; stood at 133 in 2001. ***Landownership:*** *manor of Hartley* held by Harcla (i.e. Hartley) family until forfeited 1323 on attainder of Sir Andrew de Harcla, earl of Carlisle. Granted to Nevilles of Raby who sold it to Sir Thomas Musgrave prior to 1353; descended through Musgrave family to 19th century. Sir Christopher Musgrave moved to Eden Hall c.1680, demolishing most of Hartley Castle. ***Economy:*** farming; lead and copper mines with smelt mill in 18th century (disused and in ruins by 1850s); coal seam on fell mined mid-19th century; limestone quarrying and lime-burning. Quarrying expanded in 1920s: Hartley Quarry opened 1926, with rail connection until 1975 to transport quarry products; quarry mothballed 2008 but remains open. ***Places of worship:*** none. ***Schools and other institutions:*** none.

HARTSOP *see* Patterdale

HAVERBRACK

Township in Beetham parish, Kendal ward, Westmorland, which included part of village of Beetham, including parish church. Absorbed into Beetham CP 1935.

Acreage: 658 acres [266 ha]. ***Population:*** in range 90-120 across most of 19th and early 20th century; peak of 143 in 1891; stood at 82 in 1931 (last census year for which separate figure available). ***Landownership:*** *manor of Haverbrack* granted to Conishead Priory; after Dissolution, granted 1546 to William Thornburgh. Sold early 17th century to group intending to open lead mine on Kent estuary. Sold again mid-17th century to Edward Wilson of Nether Levens, ancestor of Wilsons of Dallam Tower, through whom it descended to 19th century. Dallam Tower rebuilt by Daniel Wilson 1720-5 and enlarged 1826. ***Economy:*** farming; quarrying and lime-burning. Large-scale limestone quarrying at Sandside Quarry, in operation by 1901; still worked. Heron corn mill (18th century), in operation until 1955; restored and opened to public as visitor attraction. ***Places of worship:*** medieval parish church of St Michael in Beetham village. ***Schools and other institutions***: none.

HAVERIGG *see* Millom

HAVERTHWAITE *see* Colton; Holker, Upper

HAWKSHEAD AND MONK CONISTON WITH SKELWITH

Township in Hawkshead parish, Lonsdale hundred, Lancashire North of the Sands. Divided 1894 between newly-created CPs of Coniston, Hawkshead and Skelwith.

Acreage: 10,429 acres [4,221 ha] in four divisions: Hawkshead (1,973 acres [798 ha]); Field Head (2,520 acres [1,020 ha]); Monk Coniston (3,109 acres [1,258 ha]); and Skelwith (2,828 acres [1,144 ha]). Common land at Hawkshead (1,852 acres [749 ha]) and Skelwith (725 acres [293 ha]) enclosed 1862. ***Population:*** rising from 920 in 1801 to peak of 1,362 in 1841, then in range 1,100-1,250 for remainder of 19th century; stood at 1,224 in 1891 (last census year before creation of new CPs). ***Landownership:*** *manor of Hawkshead* granted to Furness Abbey c. 1160 and descended with abbey's estates after Dissolution (see Barrow-in-Furness). ***Economy:*** livestock farming and woodland industries since medieval period. Coppice woodland and charcoal burning with associated activities (e.g. tanning: tanyard at Castle, 1840s). Slate quarrying, 17th-20th century; on large scale at Hodge Close Quarry from 19th century to 1960s; Brathay Quarry continues in use. Iron forge recorded 1671, perhaps the bloomery site near Blelham Tarn. Flax mill at Thursgill by 1789; converted to bobbin mill c.1800; had closed by 1910. Hawkshead gained charter for weekly market and fairs 1606; described by Sir Daniel Fleming 1671 as 'a little market for the sale of provisions, wooll and yarne'; wool market extinct by 1900. Tourism of growing importance since 19th century, Hawkshead becoming one of Lake District's main tourist destinations by late 20th century. ***Places of worship:*** chapel of ease in Dalton parish recorded c.1200; said to have gained parochial status 1578 (though registers begin 1568); parish church of St Michael dates largely from 16th and 17th century (date stones of 1578); restored 1875-6. Holy Trinity Church at Brathay, Skelwith, consecrated 1836. Baptist chapel at Hawkshead Hill, converted from cottage 1709 for congregation established 1678; congregation had dwindled to extinction by 1833 but revived c.1876 when chapel renovated; still in use. Hawkshead Methodist Church originated as Union Chapel, converted from cottage 1862; still in use. ***Schools and other institutions:*** Hawkshead Grammar School founded by Edwin Sandys, archbishop of York, 1585; rebuilt 1675; closed 1909 (building now a museum); library attached to school founded 1669, augmented 1717. National school for girls recorded 1836. New school built 1863 (and infant school built 1873); replaced by Hawkshead Esthwaite County Primary School on new site on edge of village by 1980s. Brathay District CE School, Skelwith Bridge, opened c.1863; closed c.1970; became village hall (Skelwith Community Hall). Town hall or market hall built 1790 on site of former shambles; enlarged 1887. Reading room at Outgate, built 1906; converted to housing early 21st century.

HAYTON (near Brampton)

Ancient parish in Eskdale ward, Cumberland, embracing townships of Hayton, Little Corby, Fenton & Faugh and Talkin, last of which was considered part of Brampton parish in 17th century and earlier.

Acreage: 7,845 acres [3,175 ha], divided between constituent townships thus: Hayton: 2,564 acres [1,038 ha]; Little Corby: 359 acres [145 ha]; Fenton & Faugh: 2,409 acres [975 ha]; Talkin: 2,513 acres [1,017 ha]. Common land at Hayton Low Common (1,053 acres [426 ha]) enclosed 1704; Hayton High Common (2,125 acres [860 ha]) stinted 1704 and subsequently divided; further common land at Hayton (200 acres [81 ha]) enclosed 1814; Talkin Fell (1,460 acres [591 ha]) enclosed 1854. ***Population:*** estimated at 370 in 1688; parish population rose from 1,015 in 1801 to peak of 1,604 in 1831, then declined to low point of 1,068 in 1911, rising again in later 20th century to stand at 2,180 in 2001. ***Landownership:*** *manors of Hayton and Talkin* were part of barony of Gilsland with which they descended (see Brampton). *Little Corby* was held by Howards of Corby Castle (see Wetheral). Graham family of Edmond Castle (built late 18th century; extended 1820s and 1840s) were resident landowners and dominant local family in 19th century. ***Economy:*** farming (noted in later 17th century for rye grown on its sandy soils); coal mining on Talkin Fell (Blacksike Colliery: first shaft sunk 1820-1; disused 1884); quarrying in 19th century; tile-making in Little Corby at Toppin Castle (from 1833 to 1901) and near Allan Grove (from 1843 to 1895). Woollen manufacture (tweed, rugs, blankets etc) at Hardbank mill, established 1823; closed c.1881. ***Places of worship:*** medieval parish church of St Mary Magdalene; rebuilt 1780; aisle and vault added 1793; chancel rebuilt 1842. Anglican chapel of ease at Talkin built by Thomas Henry Graham of Edmund Castle 1842. Faugh school used as mission room in 1901. Wesleyan Methodist chapels at Corby Hill (built 1852; rebuilt 1894; still in use), Fenton (built 1850; closed) and Talkin (built 1870; closed 1959; converted to dwelling). ***Schools and other institutions:*** school near Talkin endowed by John Milburn 1798; rebuilt 1858 as National school; closed and converted to dwelling. School in Hayton, funded by Graham family, built 1818; enlarged 1853; rebuilt by T. H. Graham 1871; now Hayton CE Primary School. School at Faugh by 1861; closed c.1933. Reading rooms at Hayton (established by T. H. Graham 1856; replaced by new building 1925; extended 2011), Talkin (built 1898) and Fenton. Parish workhouse at Acres, near Hayton village, built 1826.

HAYTON AND MEALO

Township in Aspatria parish, Allerdale below Derwent ward, Cumberland.

Acreage: 1,868 acres [753 ha]. ***Population:*** rising from 174 in 1801 to 461 in 1851, then gradual decline to 229 in 2001. ***Landownership:*** *manor of Hayton* passed from Mulcaster family by marriage of daughter of Robert Mulcaster to Sir Piers Tilliol (d. 1435), passing again by marriage of Margaret daughter of William Tilliol to Nicholas Musgrave in later 15th century. It remained in Musgrave family, whose seat was at Hayton Castle (medieval tower house remodelled by Sir Edward Musgrave 1609), until marriage in 1769 of Eleanor, daughter of Sir Richard Hylton (formerly Musgrave) to William Jolliffe (d. 1802), whose descendants remained lords of manor until 20th century. ***Economy:*** largely agricultural. Market and fair at Hayton granted 1278; did not survive. Brick and tile manufacture in 19th century. ***Places of worship:*** domestic chapel, Hayton Castle, repaired 1719 and survived until early 19th century. St James' Church (Anglican) built 1868. Congregational chapel built 1844; closed. ***Schools and other institutions:*** school on Hayton village green by 1860s; replaced by Board school outside village, erected by 1890s; closed. Public hall or reading room built 1909.

HELBECK (formerly Hillbeck)

Township in Brough parish, East ward, Westmorland.

Acreage: 2,760 acres [1,117 ha], including extensive rough grazings on Hillbeck Fell (1,000 acres [c.400 ha]), enclosed 1859, and Hillbeck Intack (418 acres [169 ha]), enclosed 1867. ***Population:*** fluctuating in range 40-100 across 19th and early 20th century (peak: 101 in 1821); stood at just 26 in 1961 (last census year for which separate figure available). ***Landownership:*** *manor of Helbeck* held by Helbeck family, passing by marriage to Blenkinsops c.1300. They held it for many generations, until Thomas Blenkinsop sold part of manor to Richard Burton 1638 and remainder to his son Thomas Burton 1657. Held by Baker family in early 18th century and Carleton family later 18th century. Bought by William Wilkinson for his nephew J. M. Breeks 1852 and remained in Breeks family until death of Olive Breeks (d. 1951), when it descended to A. J. Blackett-Ord. ***Economy:*** hill farming; coal mining in 17th and 18th centuries; lead-mining in 17th and 18th centuries (smelt mill recorded 1669); limestone quarrying (continuing) and lime-burning from later 18th century. Cotton mill on Yosgill Beck built by John Carleton, then lord of manor; opened 1784; closed 1794. Moorland used as training ground by Ministry of Defence and as grouse moor in 21st century. ***Places of worship:*** none. ***Schools and other institutions:*** none.

HELSINGTON*

Township in Kendal parish, Kendal ward, Westmorland, which included Sizergh and part of Brigsteer. CP lost territory to Kendal borough 1935 and gained remainder of Brigsteer 1970.

Acreage: 3,327 acres [1,346 ha] before boundary changes (126 acres [51 ha] transferred to Kendal 1935). Commons (c.1,000 acres [405 ha]) enclosed 1855. ***Population:*** rising from 230 in 1801 to 360 in 1891; stood at 319 in 1931 (last census year before boundary changes). Modern CP contained 288 inhabitants in 2001. ***Landownership:*** *manor of Helsington* held by barons of Kendal (q.v.) and descended as part of Lumley fee. Manor house at Helsington Laithes in hands of Bellingham family of Levens by 1517, with whom it descended until 20th century. *'Manor' of Sizergh* descended from Deincourts to Stricklands on marriage of Elizabeth Deincourt to Sir William de Strickland 1239; it remained with Strickland (now Hornyold-Strickland) family in 21st century. Sizergh Castle largely rebuilt by Walter Strickland in 16th century. ***Economy:*** farming; peat-cutting (for Kendal market until 19th century). Marble works (polishing mill for 'marble' from local limestone) in mid-19th century (had closed by 1895). Saw mill converted to snuff mill by Gawith, Hoggarth & Co. in 1880s; closed 1991. New Sedgwick Gunpowder works established 1857; closed 1935. Tourism increasingly important from later 20th century. ***Places of worship:*** church of St John erected by subscription 1726; restored 1858. Domestic chapel at Sizergh Castle from 1431-2; Roman Catholic chapel there by 1773. ***Schools and other institutions:*** school recorded 1778; schoolhouse built 1790s. Replaced by new endowed school, built 1854; closed 1965; used as residential centre for children from Leeds area from 1987; redeveloped as bunkhouse 2013. Howard Orphan Home built 1865 to educate and train girls for domestic service; closed 1954. Village hall (former nissen hut) opened 1958; rebuilt 2010.

HELTON *see* Askham

HENSINGHAM

Township in St Bees parish, Allerdale above Derwent ward, Cumberland. CP enlarged by inclusion of Low Keekle CP 1881 but abolished 1934, when divided between Whitehaven, St Bees and Weddicar CPs.

Acreage: 2,184 acres [884 ha] before boundary changes; enlarged CP of Hensingham and Low Keekle contained 2,249 acres [910 ha]. Common land on Hensingham Moor (562 acres [227 ha]) enclosed 1768. ***Population:*** steady rise from 590 in 1801 to 1,538 in 1861; then rapid increase across 1860s to 2,060 by 1871; continued to rise to 2,312 in 1911; stood at 2,116 in 1931 (last census year for which figure available). ***Landownership:*** moiety of *manor of Hensingham* held by Robert de Branthwaite 1276 descended by marriage c.1320 to Whitrigg family (lords of Little Bampton), and from them by marriage in mid-14th century to Skelton family. Sold by Thomas Skelton of Branthwaite 1608 to Thomas Salkeld of Brayton, who sold it (except manor house and demesne lands) to Anthony Benn 1650. Mining royalties conveyed to Sir John Lowther 1683. By mid-19th century there were several villas in township. ***Economy:*** coal mining by 1688. Linen thread and check factory recorded 1829, closed by 1901. Limestone quarry and lime works at Overend by mid-19th century; closed early 20th century. By 19th century much employment was in mines and other industries of Whitehaven. Economic development in 20th century included Seker's Silk Mill, opened 1938, becoming a major employer; closed 2006. Smith Brothers printing and packaging works, Ivy Mill, opened early 1950s; site later used as workwear factory. Other small factory units opened from 1960s at Sneckyeat. ***Places of worship:*** St John's Church built 1791; endowed, restored and consecrated 1808; extended 1843 and 1859; rebuilt on new site (old church demolished) 1911. Keekle Mission opened 1875; closed 2013. Wesleyan Methodist chapel at West View built 1856; replaced by new chapel on new site 1901; still in use. ***Schools and other institutions:*** two schools recorded 1818. National school enlarged 1873 with attached infant school (opened 1889); closed 2003, when Junior School, opened 1959, became Hensingham Primary School. School at Goosebutts in 1868; closed 1932 and became reading room. Keekle Terrace Voluntary School opened 1875 in same building as Keekle Mission (enlarged 1885, 1899 and 1903); closed 1958. Overend Secondary Modern School opened 1960 and Whitehaven Grammar School relocated on adjacent site 1968, thus moving all Whitehaven secondary education to Hensingham. In 1990s they merged to form Whitehaven School. St Benedict's RC High School opened 1971. Galemire Infectious Diseases Hospital opened 1879; closed 1966. Homewood (former private villa) served as chest unit from 1955 until opening of West Cumberland Hospital, built 1959-64 to replace Whitehaven Infirmary. Public halls in Hensingham for Conservative Association (opened 1885) and Liberals (erected 1886); both closed. Cinema, Hensingham Square; closed. Former Keekle Mission, Keekle Terrace, reopened as village hall 2015.

HESKET-IN-THE-FOREST

Ancient parish (thought to have been chapelry in Carlisle St Mary parish until 14th century) in Leath ward, Cumberland, embracing townships of High and Low Hesket (which also included village of Armathwaite), Calthwaite, Plumpton Street, Itonfield and Petteril Crooks. CP enlarged by absorbing Plumpton Wall CP 1934.

Acreage: 16,733 acres [6,890 ha], including c. 4,000 acres [c. 1,600 ha] of common land enclosed under Inglewood Forest enclosure award 1819. Parish acreage divided between constituent townships thus: High and Low Hesket: 3,658 acres [1,480 ha]; Calthwaite: 1,888 acres [764 ha]; Plumpton Street: 2,714 acres [1,098 ha]; Itonfield: 2,980 acres [1,206 ha]; and Petteril Crooks: 5,159 acres [2,088 ha]. ***Population:*** 1,545 in 1811, rising to 2,150 in 1871; then falling back to 1,822 in 1931 (last census year before boundary changes). Population of enlarged Hesket CP stood at 2,363 in 2001. ***Landownership:*** Hesket parish covered core of *royal forest of Inglewood* which, as defined 1301, comprised tract of lowland between Carlisle and Penrith, lying west of River Eden as far as Thursby, Caldbeck and Blencow. Held by Crown with honour of Penrith from reign of Richard III until 1696, when William III granted forest and honour to William Bentinck, duke of Portland, whose descendant, 3rd duke, sold it to his brother-in-law William Cavendish, 5th duke of Devonshire, 1787. Inglewood forest court or 'swainmote' met at Court Thorn, Low Hesket. *Manor of Armathwaite* held by Skelton family from 14th century until 1712 when Richard Skelton sold it to William Sanderson; then descended to William Milbourne 1741; sold to earl of Lonsdale 1846, later passing to Edward Ecroyd (1833-1914). *Manor of Nunclose* (land in Hesket belonging to nunnery at Ainstable) was granted to William Graham 1552, passing to Lowthers who exchanged it for other land 1695 with Dalstons of Acornbank, who sold it 1762 to William Milbourne, thus uniting it with Armathwaite. Large sales of land in forest in years 1789-93: Calthwaite sold to Dixon family; Itonfield to Oliphants. ***Economy:*** predominantly agricultural; fishery in River Eden at Armathwaite, said 1688 to be worth £50 per annum. Quarrying for freestone on Great Barrock and for grindstones at Ivegill in 19th century. Tile works at Calthwaite and Plumpton Street in mid-19th century. ***Places of worship:*** medieval parish church of St Mary, High Hesket; rebuilt 1720. Chapel at Armathwaite built 1401-2; rebuilt 1668; now chapel of Christ and St Mary. Supposed site of early chapel at Chapel Hill, Calthwaite. All Saints Church, Calthwaite, built 1913. Mission room at Broadfield recorded 1901. United Presbyterian chapel, Plumpton Street (date stone 1709); sold to

Wesleyan Methodists 1899; rebuilt 1905; still in use (known as Plumpton Backstreet Chapel; now Cottage Wood Centre). Methodist chapel at Low Hesket built 1869; closed late 20th century. ***Schools and other institutions:*** school at High Hesket endowed by John Brown 1763; rebuilt 1853; now High Hesket CE Primary School. Endowed school at Armathwaite recorded 1818; rebuilt 1854; replaced by new school on different site 1884; now Armathwaite Primary school. School at Calthwaite built by subscription 1852; rebuilt 1875; now Calthwaite CE Primary School. Reading room at Calthwaite, built 1888; converted to dwelling. Village hall at Low Hesket opened 1954.

HESKET NEWMARKET *see* Caldbeck

HETHERSGILL

Township in Kirklinton parish, Eskdale ward, Cumberland.

Acreage: 5,464 acres [2,211 ha]. Commons in Hethersgill and Kirklinton Middle Quarter, totalling 4,264 acres [1,726 ha], enclosed 1735. ***Population:*** 665 in 1801, rising to peak of 799 in 1841. Decline thereafter, falling to 447 by 1931 and to 382 in 2001. ***Landownership:*** part of barony of Kirklinton (see Kirklinton Middle Quarter). ***Economy:*** predominantly agricultural. Quarrying in Hether Burn valley and clay digging at Prior Rigg for brick-making later 19th century; commercial peat extraction on Bolton Fell in later 20th century. ***Places of worship:*** church of St Mary, built 1876 as chapel of ease. Wesleyan Methodist chapel at Ullermire built 1833; replaced by chapel in Hethersgill village, built 1901; closed 1979. Quaker meeting house at Sykeside built 1688; rebuilt 1736; closed 1913. ***Schools and other institutions:*** endowed school at Shaw Foot, near Ullermire, rebuilt 1871 and enlarged 1895; closed 1980s. Reading room by church at Hethersgill, built early 20th century; now demolished. Village hall built 1950s.

HEVERSHAM

Township in Heversham ancient parish, Kendal ward, Westmorland. Included Milnthorpe (q.v.) which became separate CP 1896.

Acreage: 2,297 acres [930 ha] before loss of Milnthorpe in 1896; 1,582 acres [640 ha] thereafter. Commons in Heversham and Milnthorpe (547 acres [221 ha]) enclosed 1815. ***Population:*** joint population of Heversham and Milnthorpe rose from 968 in 1801 to 1,545 in 1881, majority living in Milnthorpe. In 1891 separate figure for Heversham was 356, rising to 741 in 1981, before falling slightly to 647 by 2001. ***Landownership:*** one-third of *manor of Heversham* granted to St Mary's Abbey, York, early 12th century; granted after Dissolution to Richard Bouskill, whose grandson sold it to Edward Wilson of Levens 1614. Remaining two-thirds given to Agnes, daughter of William de Lancaster on her marriage to Alexander de Wyndesor c.1260; sold successively to Neville family 1398; to Thomas Bradley of Arnside c.1577, and to James Bellingham of Levens 1597. Since 17th century lordship held jointly by owners of Levens Hall and Dallam estates. ***Economy:*** predominantly agricultural. ***Places of worship:*** medieval parish church of St Peter; roof and much of interior destroyed by fire 1601; restored (and new tower built) 1868-70. ***Schools and other institutions:*** Heversham Grammar School founded before 1619 by Edward Wilson. Originally located on Heversham Head above church; relocated to present building in village c.1870; now part of Dallam School, Milnthorpe, established 1984. School established in churchyard 1839 (in building now church hall); moved to new premises in Leasgill 1891; now Heversham St Peter's CE Primary School. Girls' boarding school in Leasgill Cottage, later 19th century. The Athenaeum in Leasgill established as community centre and reading room by Argles family, 1870.

HINCASTER

Township in Heversham parish, Kendal ward, Westmorland.

Acreage: 667 acres [270 ha]. ***Population:*** rising from 102 in 1801 to 156 in 1831; falling to 117 in 1881 and rising again to 195 in 2001. ***Landownership:*** part of manor of Burton-in-Kendal (q.v.) until 14th century. Passed to Strickland family by 1426 and seems to have descended with Sizergh thereafter (see Helsington). ***Economy:*** predominantly agricultural. ***Places of worship:*** none. ***Schools and other institutions:*** none.

HOFF

CP created 1894 from part of Appleby St Lawrence ancient parish, East ward, Westmorland. Hoff CP consisted of townships of Burrells and Drybeck and most of township of Hoff and Hoff Row.

Acreage: former townships contained following acreages: Burrells: 692 acres [280 ha]; Drybeck: 1,371 acres [555 ha]; Hoff and Hoff Row: 1,818 acres [736 ha], totalling 3,881 acres [1,571 ha]. Hoff CP as constituted 1894 covered 3,660 acres [1,481 ha]. Common land at Burrells Moor enclosed 1774; Hoff and Drybeck commons (975 acres [395 ha]) enclosed 1830. ***Population:*** each of constituent townships had c.90 inhabitants in early 19th century. Population of Hoff CP was in range 180-200 across 20th century and stood at 189 in 2001. ***Landownership:*** *manor of Hoff* held by Sir Hugh de Morville late 12th century; descending to Moultons, lords of barony of Gilsland (see Brampton) with which it descended until purchased by William Williams (d. 1680) of Johnby Hall, from whom it descended to Winder and Milward families. By 19th century was in hands of Lord Hothfield, lord of barony of Appleby (q.v.), who also owned Drybeck and Burrells. ***Economy:*** predominantly agricultural. Quarrying and lime-burning in 19th century; tile kiln on Burrells Moor late 19th century. Saw mill and bobbin manufacture at mill at Rutter Falls, which generated hydro-electricity to supply Asby in first half of 20th century. ***Places of worship:*** Wesleyan Methodist chapel in Hoff built 1895; closed and converted to dwelling. Anglican services held monthly in school room 1938. ***Schools and other institutions:*** CE elementary school (Oakbank School) built midway between Hoff and Drybeck 1865; closed; converted to dwelling c.2006.

HOLKER, LOWER

Township in Cartmel parish, Lonsdale hundred, Lancashire North of the Sands, containing villages of Cark, Flookburgh and Holker.

Acreage: 2,387 acres [966 ha]. Commons enclosed 1809 under Cartmel enclosure award. ***Population***: in range 1,000-1,200 across 19th century, with peak of 1,225 in 1851. Rose to 1,835 in 1921, with development of Ravenstown in 1917 as proposed airship station (which was never built); fell back to 1,425 in 1931, remaining at that level across mid-20th century; rose again to stand at 1,808 in 2001. ***Landownership:*** part of manor of Cartmel (see Lower Alllithwaite). Preston family acquired estate at *Holker* after Dissolution, building Holker Hall as their seat. Passed by marriage successively to Lowther and Cavendish families; Holker Hall was rebuilt by 7th duke of Devonshire 1838-41 and again (after a fire) 1871-3. ***Economy:*** farming (Holker estate providing employment especially from 19th century); fishing since medieval period: inter-tidal fish traps; cockle fishing, shrimping. Fishing industry benefitted from wider access to markets after arrival of railway and opening of Cark-in-Cartmel station 1857. Fish processing plant (Youngs) in Cark from 1950; closed 1970s. Flookburgh granted market charter 1412; attempts to found borough there failed, though village provided accommodation and other services as mid-point between two sections of over-sands route across Morecambe Bay. Cotton mill in Cark village established c.1782; had closed by mid-19th century but Cark remained commercial centre. Tourism increasingly important across later 20th century; several caravan parks. ***Places of worship:*** chapel of ease at Flookburgh, recorded 1520; rebuilt 1778; demolished 1900 and replaced by church of St John the Baptist at Flookburgh, built 1897-1900. Wesleyan Methodist chapel at Cark built 1904; closed 2005. St Cuthbert's Roman Catholic Church, Flookburgh, built 1935; closed c.2015. ***Schools and other institutions:*** curate served as schoolmaster at Flookburgh in 1670s and 1680s, but no school recorded in 18th century. Holker School opened 1865 and Flookburgh School 1875; both closed when new school (Flookburgh CE Primary School) was built 1971. Village halls at Flookburgh, built 1887, and Cark (Lower Holker Village Hall), built 1908.

HOLKER, UPPER

Township in Cartmel parish, Lonsdale hundred, Lancashire North of the Sands, including villages of Backbarrow and Low Wood and part of Cartmel village. Part of Upper Holker CP was separated to help create Haverthwaite CP 1927 and remainder was absorbed into Lower Allithwaite CP 1949.

Acreage: 7,140 acres [2,890 ha] before boundary changes. Commons enclosed under Cartmel enclosure award 1809. ***Population:*** rising from 835 in 1811 to 1,134 in 1851 but falling back to 850 by 1871 as result of closure of cotton mills. Stood at 825 in 1901. After creation of Haverthwaite CP population of remaining rural areas was 223 in 1931. ***Landownership:*** part of manor of Cartmel (see Lower Allithwaite). ***Economy:*** farming; fishing (recorded in name Fish House); woodland industries (charcoal-burning; bark stripping; basket making). Part of Cartmel village, west of River Eea, including market place and race course, lay in Upper Holker township. Leven valley from Backbarrow to Low Wood saw concentration of industrial sites from 17th century, served from 1868 by Furness Railway (see Colton for sites on west side of valley). Iron forge at Burnbarrow, near Low Wood, 1603-20. Blast furnace built at Low Wood 1747; closed 1785; served by quay called Bigland Dock on River

Leven. Paper mill at Backbarrow burned down 1782 and rebuilt as cotton mill by cotton spinners Birch, Robinson and Walmelsey. Rebuilt after fire 1856; closed 1867; rebuilt again as ultramarine ('Dolly Blue') works; closed 1982. Site now occupied by Whitewater Hotel and Lakeland Motor Museum. Second cotton mill (north mill) had become woollen mill by later 19th century. Gunpowder works at Low Wood, established by Daye Barker & Co. 1798; closed 1935. Hydro-electricity generation using weir at Backbarrow from c.2000. ***Places of worship:*** none. ***Schools and other institutions:*** school at Brow Edge endowed by George Bigland 1685; rebuilt 1873; closed 1929. Cartmel Grammar School (see Lower Allithwaite) moved to new site north of race course 1790; extended 1862; closed 1960s; later used as hotel and care home.

HOLME (Westmorland)

Township in Burton-in-Kendal parish, Lonsdale ward, Westmorland.

Acreage: 1,648 acres [667 ha]. Commons enclosed under Dalton, Burton and Holme enclosure award 1815. ***Population:*** steadily rising across earlier 19th century from 226 in 1801 to 1,154 in 1851, including considerable numbers of Irish migrants. Dropped back, then stable in range 700-800 across later 19th century, before declining to 627 by 1961. Renewed growth in later 20th century, to stand at 1,167 by 2001. ***Landownership:*** *manor of Holme* held by Thomas son of Gospatrick, ancestor of Curwens of Workington, c.1200. Descended in Curwen family to late 16th century. Acquired by Prestons of Preston Patrick, who sold it to Francis Charteris of Hornby Castle 1717. Later sold to Wilsons of Dallam Tower. ***Economy:*** farming; industry important from late 18th century. Flax mill at Holme Mills, south of village, built by Charles Parker and John Waithman by 1790, resulting in creation of industrial colony there. No longer working by 1860; became coconut and jute matting factory under Edward Shepherd and Sons (1864-1916) and William Goodacre (1916-1974); closed 1975; let as industrial units. Lancaster canal, with wharf at Holme, opened 1819. Second flax mill on canal side in village, recorded 1857; had closed by 1896. Westmorland Cooperage and Saw Mills, making boxes, packing cases and small casks; relocated to Glasson Dock (Lancs.) 1915. Small-scale limestone quarries on Holme Park Fell, 19th century. Large limestone quarry, Holme Park Quarry, open by 1920s; still in operation. ***Places of worship:*** Holy Trinity Church built 1839. Wesleyan Methodist chapel built 1833; replaced by new Methodist chapel 1924; closed c.2010 and converted to housing. Roman Catholic chapel at Sheernest 1848; sold c.1950 and became housing. ***Schools and other institutions***: National school built by subscription 1840; closed 1911 (now Holme Parish Hall); replaced by new school on different site 1911; now Holme Community School. Working Men's Institute (now Holme Social Club) established 1875.

HOLME ABBEY or ABBEYTOWN

Township in Holme Cultram parish, Allerdale below Derwent ward, Cumberland, including village of Abbeytown.

Acreage: 5,616 acres [2,273 ha], including 472 acres [191 ha] in small detached portions, intermixed with land of other townships in parish. Holme Abbey was centre of ancient parish and lordship of Holme Cultram (covering 24,828 acres [10,048 ha]), which embraced townships of Holme Abbey, Holme East Waver, Holme Low and Holme St Cuthbert. Together, these formed Holme Cultram UD from 1894 to 1934. Commons in Holme Cultram parish, totalling c.6,000 acres [c.2,430 ha] and including Colt Park common field and stinted pastures on marshes, enclosed 1814. Benwray, a common field of 110 acres [45 ha] in Holme Abbey was enclosed 1843. ***Population:*** 590 in 1801, rising to 962 in 1861, then declining to 720 in 1951 before rising again (peak 891 in 1961) to stand at 776 in 2001. ***Landownership:*** *lordship of Holme Cultram* (coterminous with ancient parish) was granted to Cistercian abbey of Holme Cultram (located at Abbeytown) on its foundation 1150. After Dissolution it remained in Crown hands until 1693, when sold to Frances Villiers, who sold it to Bartholomew Burton of Suffolk 1695. Passed from Burtons to Edward Stephenson 1732, in whose family (Rowland Stephenson having changed family name to Standish 1834) it remained in 20th century. ***Economy:*** predominantly agricultural. Weekly market and fairs at Easter and in October held at Abbeytown in 17th century; village was focal point for whole parish and contained retailers and other services by 19th century. Carlisle & Silloth Railway arrived, with station at Abbeytown, 1856; beside it was oilcake mill by 1860s. Agricultural engineering works, Abbey Implements Works, established before 1900. ***Places of worship:*** nave of medieval abbey of St Mary served as parish church from Dissolution. Tower fell down 1600 and church was burnt 1604 and again 2006; restored and reopened 2015. Wesleyan Methodist chapel built 1859; in use early 21st century. ***Schools and other institutions:*** school at Aldoth built on site of older one 1851; enlarged 1909; closed 1957. School at

Abbeytown built 1853; replaced by modern building on new site (now Holm Cultram Abbey CE Primary School). Library and reading room built 1856. Assembly room built 1890; adopted as village hall 1932.

HOLME EAST WAVER

Township in Holme Cultram parish, Allerdale below Derwent ward, Cumberland, including village of Newton Arlosh (or Long Newton).

Acreage: 5,572 acres [2,255 ha], including 135 acres [55 ha] in detached portions. Commons enclosed under Holme Cultram enclosure award 1814; Ellercar and Feather Close common meadow (300 acres [121 ha]) enclosed 1851. Mosslands of Wedholme Flow, part of which lies in township, remain registered common land. ***Population:*** rising from 371 in 1801 to peak at 526 in 1861; then steady decline to 358 in 1921; post-War increase (404 in 1951), then settling back to c.300 at end of 20th century. ***Landownership:*** part of Holme Cultram lordship (see Holme Abbey). ***Economy:*** predominantly agricultural. Salt-making in medieval period (name Salt Cotes recorded 1305). Newton Arlosh obtained market charter 1304 to replace Skinburness, after it was washed away by sea: little evidence of urban development. Brick- and tile-making in 19th century. ***Places of worship:*** medieval chapel of St John at Newton Arlosh; building authorised 1304; fell into ruin after Dissolution; reconstructed 1843. ***Schools and other institutions:*** Parkhill School (one mile south of village) rebuilt 1868; closed and converted to housing. Mission hall built early 20th century; now used as village hall.

HOLME LOW

Township in Holme Cultram parish, Allerdale below Derwent ward, Cumberland, including Silloth (q.v.), which became separate CP 1934.

Acreage: 6,160 acres [2,493 ha], including 156 acres [63 ha] in small detached portions, before boundary changes. Adjacent to township's lands, which were protected by medieval sea dyke, lay Skinburness Marsh (1,178 acres [477 ha]), common grazing to townships of Holme Low, Holme Abbey and Holme St Cuthbert, which remains common land. Other commons enclosed under Holme Cultram enclosure award 1814. ***Population:*** 637 in 1801, rising to c.950 in mid-19th century. Growth of Silloth led to rapid population increase, to reach 3,590 in 1921. After separation of Silloth CP 1934, population of remaining rural areas of Holme Low stood at c. 350 through to 2001. ***Landownership:*** part of Holme Cultram lordship (see Holme Abbey). ***Economy:*** predominantly agricultural. Holme Cultram abbey attempted to found borough at Skinburness in 1301 but it had been washed away by sea by 1305. It remained fishing village and minor harbour with ferry to Annan from 1787 (daily boat service still in operation 1829). Tourism (caravan parks) and light industry in later 20th century, based on hangars at disused Silloth Airfield. ***Places of worship:*** supposed site of medieval chapel near Grune Point. St Paul's church (Anglican) at Causewayhead, built 1845. ***Schools and other institutions:*** Greenrow Academy, a large boarding academy, established 1780; closed 1871. School on crossroads at Causewayhead by 1866; replaced by Board school (Colt Park School) nearby, erected 1875; closed and converted to dwelling. Sunday school with attached reading room and library built 1890s; closed and converted to housing. Silloth Airfield, base for training bomber crews and for aircraft maintenance, opened 1939; closed 1960; later used for private aircraft.

HOLME ST CUTHBERT

Township in Holme Cultram parish, Allerdale below Derwent ward, Cumberland.

Acreage: 6,301 acres [2,550 ha], including 141 acres [57 ha] in small detached portions. Commons enclosed under Holme Cultram enclosure award 1814. ***Population:*** 589 in 1801 rising to 822 in 1851, then falling back to 750 until 1901, since when gradual decline to c.400 by late 20th century. ***Landownership:*** part of Holme Cultram lordship (see Holme Abbey). ***Economy:*** predominantly agricultural; windmills at Mawbray (working 1858) and Hards Mill, Aikshaw (disused by 1860s). Sand extraction near Tarns 20th century. Tourism (caravan parks) from later 20th century. ***Places of worship:*** possible medieval chapel of St Roche, dedicated 1327, near Tarns. Chapel of St Cuthbert, recorded 1538. (One of these perhaps recorded in name Chapel Hill near New Cowper). New church of St Cuthbert at Rowks erected 1845; tower built 1924. Quaker meeting house at Beckfoot built c.1690; rebuilt 1745; closed 1940 (used by Methodists until 1965). Wesleyan Methodist chapel at Mawbray built 1838 or 1843; closed and converted to housing c.2000. Wesleyan chapel at Foulsyke built 1898-9; closed 1992 and converted to dwelling. Congregational chapel at New Cowper built late 19th century; closed 1948; demolished 1970. ***Schools and other institutions:*** National school near St Cuthbert's Church built 1845; became infant school when new Board school built nearby 1873; now Holme St Cuthbert School. School on sea front at Old Mawbray in 1860s. Reading room and library at Overby, established 1897. Village hall (Culterham Hall), Mawbray, opened 1951.

HOLMROOK *see* Drigg

HOUGHTON *see* Stanwix

HUGILL

Township in Kendal parish, Kendal ward, Westmorland, including village of Ings. Merged with Nether Staveley and Over Staveley CPs 2004.

Acreage: 2,901 acres [1,173 ha]. Common land enclosed, along with commons in Applethwaite and Troutbeck, 1842. ***Population:*** rising from 237 in 1801 to peak of 443 in 1851; in range 350-400 across later 19th and early 20th century (temporarily inflated to 559 in 1891 by influx of workmen constructing Thirlmere aqueduct); had dropped to 341 by 1961; increased across later 20th century to stand at 416 in 2001. ***Landownership:*** part of manor of Staveley (see Over Staveley). ***Economy:*** farming; lead-mining with smelt mill near Scroggs Bridge late 18th to early 19th century; bobbin manufacture at Ings, Fellfoot, Rawes and Newgate mills in mid-19th century. Motor mechanic's garage at Ings opened 1931 (building used as bullet factory during Second World War). ***Places of worship:*** St Anne's chapel of ease at Grassgarth recorded 1546; ruins still visible 1692. Replaced by chapel at Ings, built and endowed by Rowland Wilson 1635. Rebuilt as present church of St Anne 1743; north transept added 1877. ***Schools and other institutions:*** school attached to chapel at Ings endowed 1654; further endowment 1854; rebuilt 1869 as CE primary school; closed 1972; building now parish hall. Almshouses at Grassgarth, endowed c. 1743; closed 1989.

HUNSONBY AND WINSKILL

Township in Addingham parish, Leath ward, Cumberland. Combined with Little Salkeld CP to create Hunsonby CP 1934.

Acreage: 1,687 acres [683 ha]; including 702 acres [284 ha] of the 1,400 acres [567 ha] of Little Salkeld Common enclosed 1849. ***Population:*** rising from

117 in 1801 to 284 in 1881 (numbers swelled to 362 in 1871 as result of nearby railway navvy camp); stood at 300 in 1931 (last census year for which separate figure available). ***Landownership:*** part of manor of Little Salkeld (q.v.). ***Economy:*** farming; sandstone quarrying at Briggle Quarry 19th century. Industrial development in vicinity of Little Salkeld station (opened 1876) on Settle-Carlisle railway: coal merchants and haulage contractors by 1900. Agricultural manure and seed merchants nearby; premises became paper and board mill; closed 2003. ***Places of worship:*** Wesleyan Methodist chapel in Hunsonby c.1823; rebuilt on new site 1862; extended 1905 and 1923; closed 2001 and converted to dwelling. Anglican mission room (the 'Institute') in Hunsonby built 1905; extended 1928; closed c.1975. ***Schools and other institutions***: school founded by Joseph Hutchinson (d. 1726) but endowment only received after his mother's death c.1765. Existing building pre-1849; extended 1897; closed 1974. Sons of Temperance hut at Winskill built before 1925; demolished c.2010. Open air swimming pool built 1937.

HUTTON-IN-THE-FOREST

Ancient parish in Leath ward, Cumberland. Absorbed into Skelton CP 1934.

Acreage: 2,462 acres [996 ha], divided between two townships: Hutton-in-the-Forest (1,753 acres [709 ha]) and Thomas Close (709 acres [287 ha]). Common land enclosed under Inglewood Forest enclosure award 1819. ***Population:*** estimated at 285 in 1688; rising from 200 in 1801 to peak of 285 in 1871, then falling to 204 in 1931 (last census year for which separate figure available). ***Landownership:*** manor of Hutton-in-the-Forest held by Hutton family, hereditary foresters of Plumpton Hay, from 12th century until 1606, when sold by Lancelot Hutton to Sir Richard Fletcher of Cockermouth, whose descendants, Fletcher-Vane family (latterly Barons Inglewood) continue to be main landowners. ***Economy:*** predominantly agricultural. ***Places of worship:*** medieval parish church of St James; rebuilt 1714; restored 1867-8. Wesleyan Methodist chapel at Hutton End built 1841; enlarged 1885; closed by 1980. ***Schools and other institutions:*** Hutton Marr school founded by Thomas Fletcher 1715; moved to Hutton End Wesleyan chapel 1909; closed 1985. School at Thomas Close recorded 1860. Former Methodist chapel at Hutton End now used as village hall.

HUTTON JOHN

Township in Greystoke parish, Leath ward, Cumberland. Combined with Hutton Soil to form Hutton CP 1934.

Acreage: 690 acres [279 ha], including detached portion of 133 acres [54 ha] at Lacet. ***Population:*** in range 20-50 across 19th and early 20th century (peak 50 in 1891); stood at 22 in 1931 (last census year for which separate figure available). ***Landownership:*** *manor of Hutton John* held of Greystoke barony by Hutton family from 13th century until death of Thomas Hutton c.1628, when it passed to his sister Mary, who married 1564 Andrew Hudleston of Farington (Lancs.), a younger son of Hudlestons of Millom. Their descendants sold manor to duke of Norfolk (lord of Greystoke barony), 1787 but retained demesnes. ***Economy:*** predominantly agricultural; some lime-burning in 19th century. ***Places of worship:*** none. ***Schools and other institutions:*** none.

HUTTON ROOF (Cumberland)

Township in Greystoke parish, Leath ward, Cumberland. Absorbed into Mungrisdale CP 1934.

Acreage: 2,643 acres [1,070 ha]; commons enclosed 1796. ***Population:*** in range 160-200 for most of 19th century, with peaks of 214 in 1821 and 209 in 1851; gentle decline to 108 by 1931 (last census year for which separate figure available). ***Landownership:*** part of barony of Greystoke (q.v.). ***Economy:*** predominantly agricultural; some lime-burning and freestone quarrying early 19th century. ***Places of worship:*** school licensed for worship 1847. ***Schools and other institutions:*** school funded by charitable endowment built 1830; burnt down 1905; rebuilt following year; replaced by new school 1968, which closed 1984.

HUTTON ROOF (Westmorland)

Township in Kirkby Lonsdale parish, Lonsdale ward, Westmorland.

Acreage: 2,715 acres [1,099 ha], including 271 acres [110 ha] of common land enclosed 1822. Newbiggin Crags, Holme Stinted Pasture and Hutton Roof Crags [274 ha] remain common land. ***Population:*** rising from 179 in 1801 to peaks of 351 in 1831 and 366 in 1891 (result of quarrying industry), then falling to low point of 131 in 1971 before rising again to 193 in 2001.

Landownership: *manor of Hutton Roof* held by Huttonroof family from 12th century to mid-14th century, probably passing by marriage to Hugh de Moresby who held it 1375. John Moresby held it 1411; later descent of manor not known. ***Economy:*** farming; quarrying for freestone, flags and grindstones mentioned 1692; quarrying expanded later 19th century; had declined by 1910. Timber yard, early 21st century. ***Places of worship:*** chapel of ease (now St John's Church), possibly medieval in origin; built 1610 (date stone); rebuilt 1757 and again 1881. Methodist chapel built 1850, effectively closed 1912; sold 1936. ***Schools and other institutions:*** school built 1773 at edge of township, shared with Lupton (q.v.). Rebuilt next to church 1852; closed 1979. Reading room built 1890s; replaced by village hall built 1981-7.

HUTTON SOIL

Township in Greystoke parish, Leath ward, Cumberland, which included villages of Penruddock and Troutbeck. Combined with Hutton John to form Hutton CP 1934.

Acreage: 4,349 acres [1,760 ha] before acquiring 133-acre [54-ha] detached portion of Hutton John CP at Lacet, 1884. Common land totalling c.3,500 acres [1,416 ha], and open field in Penruddock, enclosed 1817. ***Population:*** rising from 233 in 1801 to peak of 415 in 1871; then falling back to stand at 346 in 1931 (last census year for which separate figure available). ***Landownership:*** part of barony of Greystoke (q.v.). ***Economy:*** predominantly agricultural. Several inns along Penrith-Keswick road, turnpiked 1762. Livestock auction mart established in 1880s at Troutbeck Station on Cockermouth, Keswick & Penrith Railway (opened 1865). Stock sales for cattle, sheep and pigs held fortnightly; closed 1991. ***Places of worship:*** Presbyterian chapel at Penruddock erected 1712; rebuilt 1789; closed 2011. Anglican mission church of All Saints, Penruddock, built 1902. Wesleyan Methodist chapel at Beckses built 1840. Troutbeck station waiting room used for religious purposes by St Kentigern's church, Mungrisdale, by agreement with railway company. ***Schools and other institutions:*** school established 1756 by Andrew Hudleston of Hutton John; rebuilt 1872; now Penruddock Primary School. Penruddock and District Public Room built as reading room 1905; now used as village hall.

INGLEWOOD (royal forest) *see* Hesket-in-the-Forest

INGS *see* Hugill

IREBY, HIGH

Township in Ireby parish, Allerdale below Derwent ward, Cumberland. CP abolished on creation of Ireby (later Ireby & Uldale) CP 1934.

Acreage: 2,733 acres [1,106 ha], including 1,596 acres [646 ha] of common fell, enclosed 1816. Scawthwaite Myre, Gilltongue and Armathwaite Head had been enclosed by private agreement 1711. ***Population:*** fluctuating in range 100-150 for most of 19th and early 20th centuries, with peak of 185 in 1831; stood at 125 in 1931 (last census year for which separate figure available). Combined population in Ireby (later Ireby & Uldale) CP fell from 541 in 1951 to 428 in 2001. ***Landownership:*** *manor of High Ireby* granted in 12th century to Gospatric son of Orm, lord of Workington and Seaton, who gave it to his second son Orm, known as Orm de Ireby. It descended through his family for several generations but was in hands of Richard Orfeur of Plumbland by 1546, when he sold it to Anthony Barwis. Both High Ireby and Ruthwaite were sold by Richard Barwis to Sir Richard Fletcher of Cockermouth in 1636 and subsequently descended to Fletcher-Vane family of Hutton-in-the-Forest. ***Economy:*** predominantly agricultural; some quarrying and lime-burning. Barytes mine at Ruthwaite from 1870s; closed by 1920. ***Places of worship:*** none, unless name Chapel House records a former chapel. ***Institutions:*** none.

IREBY, LOW (otherwise Market Ireby)

Township in Ireby parish, Allerdale below Derwent ward, Cumberland. CP abolished on creation of Ireby (later Ireby & Uldale) CP 1934.

Acreage: 1,200 acres [486 ha]. ***Population:*** fluctuating around 300 for most of 19th and early 20th centuries (peak 337 in 1861); had fallen to 248 by 1931 (last census year for which separate figure available). ***Landownership:*** *manor of Low Ireby* held by William de Ireby in 13th century, whose daughter Christian sold it to Adam de Middleton 1302-3. By 1367 it had passed to Tilliols, from whom it descended through marriage of Felicia, daughter of William Tilliol or Colville, to William Musgrave (d. 1487) of Crookdake. On death of William Musgrave (d. 1664), it descended with Crookdake to

Ballantines and from them to Ballantine-Dykes family of Dovenby, who continued to hold it until early 20th century. ***Economy:*** predominantly agricultural; limestone quarries and lime-burning in 19th century; brick works late 19th century. Market charter granted 1237. Ireby's market was considered threat to Cockermouth's in 16th century and Thomas Denton wrote 1688 that 'In this market at Ireby all sorts of corne & graine (big [barley] & oates especially), and also salt, are sold at farr cheaper rates, and by a larger measure, then in any market in the north'. By 1900 market had long since ceased but two livestock fairs were still held each year. Market cross and former moot hall survive in centre of village. ***Places of worship:*** medieval parish church (surviving as 'Old Chancel') stood in fields on western side of township; replaced by new church of St James at northern end of village, built 1845-6. Wesleyan Methodist chapel built 1870; had closed by 1988 (subsequently converted to dwelling). ***Schools and other institutions:*** endowed school founded 1726 or 1729; rebuilt on new site at Parkin Croft 1880; now Ireby CE Primary School. Former school building became Oddfellows' Hall and subsequently (from 1952) Women's Institute Hall; second village hall, The Globe Hall, built 1908.

IRELETH *see* Dalton-in-Furness

IRTHINGTON

Ancient parish in Eskdale ward, Cumberland, comprising townships of Irthington, Laversdale, Newby and Newtown.

Acreage: 7,390 acres [2,991 ha], distributed between constituent townships thus: Irthington: 1,038 acres [420 ha]; Laversdale: 3,754 acres [1,519 ha]; Newby: 860 acres [348 ha]; Newtown: 1,738 acres [703 ha]. Commons (3,600 acres [1,457 ha]) enclosed 1781. ***Population:*** rising from 870 in 1801 to 1,049 by 1841; then declining to 898 in 1871 and 646 in 1931; increased to 841 in 1951 after building of Carlisle airport, then fell back to stand at 712 in 2001. ***Landownership:*** Irthington was caput of Gilsland until 1335 and descended with barony (see Brampton). ***Economy:*** predominantly agricultural; brick and tile works at Laversdale Lane End 1830s to late 1860s. Carlisle Airport built for RAF use 1940-1; purchased by Carlisle Corporation 1960 and civil airport opened 1961: limited services and flying training, with extensions and upgrades 1970s-1980s. ***Places of worship:*** medieval parish church of St Kentigern in Irthington village; restored 1849-53; clock tower added 1896-7. Independent chapel at Newtown recorded 1860; perhaps forerunner of Gospel Hall (so-named 1899) termed Methodist chapel 1924; now converted to private house. ***Schools and other institutions:*** Irthington village school established 1792; rebuilt 1850; replaced by Board school 1875; now Irthington Village School. Another school taught by curate 1816. School at Laversdale endowed 1790s; rebuilt 1856; closed 1952 and sold to community 1970 for use as village hall. Public hall in Newtown built 1906.

IRTON

Ancient parish in Allerdale above Derwent ward, Cumberland, containing townships of Irton and Santon with Melthwaite.

Acreage: parish contained 6,066 acres [2,455 ha]; the CP (known as Irton, Santon and Melthwaite) was slightly larger than parish (it included Easthwaite in St Bees parish) and contained 6,180 acres [2,501 ha]. Common land extending to 1,877 acres [760 ha] on Irton Fell, Huerigg Moor, Cragg Moor and Irton Moor enclosed 1813. ***Population:*** estimated at 515 in 1688; rising from 490 in 1811 to peak of 614 in 1881 before falling back to early 19th-century levels. Rose to second peak of 677 in 1951, before dropping back to 373 by 2001. ***Landownership:*** *manor of Irton* held by Irton family from 13th century until death of Samuel Irton 1866; passed to Burns-Lindow family and was sold to Thomas Brocklebank 1896. *Manor of Santon* held by de Copeland family in 13th and 14th centuries; had passed to Winder and Irton families by 1777; by 1842 was in hands of Skeffington Lutwidge of Holmrook Hall and remained in Lutwidge family in 1901. ***Economy:*** predominantly agricultural; seed and plant nurseries at Hall Santon recorded 1842-1901. ***Places of worship:*** medieval parish church of St Paul, rebuilt 1856-7; extended 1872; 9th-century cross in churchyard suggests pre-Viking origin. Wesleyan Methodist chapel at Hall Santon built 1828; closed; converted to holiday accommodation. ***Schools and other institutions:*** National school on Irton Moor endowed by Henry Caddy 1716; built 1717; rebuilt c.1856; closed c.1972; converted to guest house. Two other schools, supported by 'quarterage', recorded 1818. Village hall at Santon Bridge; building dated 1864.

ISEL OLD PARK

Township in Isel parish, Allerdale below Derwent ward, Cumberland. Absorbed into Blindcrake CP 1934.

Acreage: 1,844 acres [746 ha]. Isel Common, covering northern part of township, enclosed 1813. ***Population:*** rose from 88 in 1801 to a peak of 108 in 1831; then declined erratically to stand at 68 in 1931 (last census year for which separate figure available). ***Landownership:*** part of manor of Isel (see Blindcrake, Isel and Redmain). ***Economy:*** largely agricultural; salmon and trout fishery at Isel bridge in 17th century; managed timber production in park woods by Sir Wilfrid Lawson in 17th century; some modern afforestation. ***Places of worship:*** medieval parish church of St Michael, Isel, (restored 1878) stands within township. ***Schools and other institutions:*** none.

IVEGILL *see* Dalston

JOHNBY *see* Greystoke

KABER

Township in Kirkby Stephen parish, East ward, Westmorland.

Acreage: 4,533 acres [1,835 ha], including extensive moorland commons regulated under Winton and Kaber Common Regulation Award 1915. ***Population:*** rising from 135 in 1801 to 268 in 1861, during later stages of building of South Durham & Lancashire Union Railway. Fell back to fluctuate between 136 and 200 from 1871 to 1961 but declined thereafter, falling to 89 in 2001. ***Landownership:*** *manor of Kaber* held by Kabergh family until mid-14th century when it was acquired by Fulthorpe family, in which it descended until passing by marriage to Wandesfords of Kirklington (Yorks.) before 1558. George Wandesford sold it 1604 to Robert Wadeson of Yafforth (Yorks.), whose son sold it to the tenants 1618. ***Economy:*** hill farming; cattle fair held in February since 1883. Coal pits on Kaber Fell. ***Places of worship:*** Primitive Methodist chapel built 1859; enlarged 1875; rebuilt 1891; closed 2004 and converted to holiday accommodation. ***Schools and other institutions:*** school endowed by Thomas Waller 1689; rebuilt 1862; enlarged 1880; closed c.1988; now used as village hall.

KENDAL (formerly Kirkby Kendal)

Borough, market and manufacturing town in Kendal ward, Westmorland. This entry covers medieval borough of Kendal and townships of Kirkland and Nether Graveship.

Acreage. Kendal CP (which included Kirkland) contained 2,242 acres [907 ha] in later 19th century; Nether Graveship CP covered 358 acres [145 ha]. Kirkland and Nether Graveship CPs were absorbed into Kendal in 1908 and municipal boundary was further extended by transfer of 78 acres [32 ha] from Scalthwaiterigg in 1919 and 1,005 acres [407 ha] from surrounding CPs in 1935. As well as built-up area, Kendal CP included open land in large medieval park attached to Kendal Castle on east of River Kent and common land (158 acres [64 ha]) on Kendal Fell, to west of town, enclosed 1767.

Landownership. Kendal was seat of *barony of Kendal*, covering much of southern Westmorland. Granted to William de Lancaster c.1150, it descended to William III de Lancaster (d. 1246) and was divided between his nephews, Peter de Brus (d. 1272) and Walter de Lindsay (d. 1271). Lindsay share later became known as Richmond fee; Brus share was divided again, on death of Peter de Brus, between his sisters Margaret, wife of Robert de Ros, and Lucy, wife of Marmaduke de Thweng. Ros share became known as Marquis fee; Thweng share as Lumley fee. *Richmond fee* descended through Lindsay family, passing by marriage of Christian de Lindsay to Ingelram de Guisnes (d. 1324) and through their descendants to Philippa (d. 1411), wife of Robert de Vere, earl of Oxford. On her death it reverted to Crown, and was subsequently leased until granted by Charles II to Queen Catherine of Braganza; then leased to Lowther family 1694, who purchased it 1810. *Marquis fee* passed by marriage of Elizabeth de Ros (d. 1395) to Sir William Parr and descended in Parr family to William Parr, marquis of Northampton (d. 1571), brother of Queen Katherine Parr. It then passed to Crown and descended with Richmond fee. *Lumley fee* descended in Lumley family until exchanged for other lands with Henry VIII 1531-2. Lands in the fee were purchased piecemeal by Bellingham family in mid-16th century and Lumley fee descended thereafter with manor of Levens (q.v.).

Origins and growth of the town.

Medieval Kendal. Presence of Roman fort of ALAVANA at Watercrook, pre-10th century religious site at Kendal (suggested by place-name element Kirkby, 'church settlement') and pre-Conquest estate centre just north of town at 'Strickland' (recorded in Domesday Book) all suggest that vicinity of later town had become focal point for surrounding area long before foundation of borough. Gilbert fitz Roger fitz Reinfred acquired market charter for Kendal 1189; this, and borough charter granted to inhabitants by Gilbert's son, William de Lancaster III, in first half of 13th century, probably confirmed an existing trading centre. Borough, focused on market place at junction of route from north over Stramongate Bridge and main route up Kent valley, grew up north of church and close to seats of baronial power, the Norman motte and bailey castle (now Castle Howe) and its 13th-century successor, Kendal Castle, to east of river. Medieval town served as market, religious and administrative centre for whole of barony, though influence of barons of Kendal would have been weakened by partition of barony after 1246. References to fulling mill 1274 and dyeworks 1310 suggest woollen textile industry well established by 1300.

Kendal c.1500-1820. By 1500 Kendal's textile trade had national market, its coarse woollen cloth known as 'Kendal green' being fabric worn by the common man. Kendal was finishing centre for cloth manufactured across rural southern Cumbria, its 'shearman-dyers' dominating trade. Town was incorporated by charter 1576 (confirmed and amplified by further charters 1636 and 1684), its government being vested in mayor and 12 aldermen. By 1600 population had probably risen to 2,200-2,600. It suffered during mortality crises of 1590s and 1623; by end of 17th century it had not fully recovered: in 1695 borough (excluding Kirkland) had 2,159 inhabitants. After slump in textile trade in early 17th century, Kendal specialised and diversified: Sir Daniel Fleming, writing 1671, reported that Kendalians 'seem to be shaped out for trade; improving themselves not only in their old manufacture of cottons, but of late in the makeing of Druggets, Serges, Hats, worsted-stockings &c, whereby many poor are dayly set on work and the town much enriched'. Thomas Denton, writing 1688, said: 'The people are generally all tradesmen, very industrious & provident, which by the help of their cloathing & cotten trade (which they disperse all over England) makes them much richer then their neighbours. They are a sober, orderly sort of people, and it's a corporation as regularly govern'd as any, where ever I came; being constituted by the name of maior, aldermen & comminalty. Their revenue is but small, but well managed.' Cloth industry dominated: tenter grounds (with frames on which to stretch finished cloth) spread out onto surrounding open ground, on Fellside and Gooseholme. Town's motto, *Pannus mihi panis* ('cloth is my bread') had been coined by early 17th century. Other industries in 18th-century Kendal included making of wool cards; tanning and leather working; snuff making (from 1792); and manufacture of horn combs (established 1794). Population grew rapidly, particularly from the 1730s: by 1801 it had reached c.8,000 (borough containing c.7,000 inhabitants and Kirkland a further 1,000). Its growing importance as regional centre was reflected in newspaper publishing: short-lived titles, *The Kendal Weekly Courant* and *Kendal Weekly Mercury* had appeared in 1730s; by 1820 it boasted two papers, the *Westmorland Advertiser and Kendal Chronicle* (founded 1811; replaced 1834 by the *Kendal Mercury*) and *The Westmorland Gazette* (founded 1818).

Kendal since c.1820. Kendal's manufacturing base was boosted by Lancaster canal, which reached it 1819 (closed 1944) and Lancaster & Carlisle railway, which opened 1846, with spur through Kendal to Windermere opening 1847. Town gained a significant industrial base across the 19th century, not only in textiles but also in iron founding and shoemaking:

Textiles: introduction of steam power from 1843 accompanied an increase in scale of woollen industry. Dockray Hall Mill (or Gandy Mill) operated as woollen mill from 1788; diversified into carpet-making from c.1830; ceased to make woollen cloth after First World War; demolished 1940. Castle Mills (woollen), on site of medieval corn mill, later fulling mill; then carding and spinning mills. Rebuilt 1806 and further expanded 1855 for power loom weaving by J. J. and W. Wilson; closed 1933. Low Mills, rebuilt as woollen fustian mill 1806. Carpet making, established 1822, became an important specialism, first at Dockray mills, then at Castle Mills, where it continued into late 20th century. Hosiery knitting established by E. W. Thompson & Sons 1878 at Sandes Avenue and Bridge Mills, Stramongate ('the sock mill'); closed 1980s.

Iron foundries developed after canal and railway allowed coal and iron to be transported cheaply. Foundry established at Low Mills c.1840. Canal Iron Works (on site of former warehouses at Canal Head, redundant after arrival of railway) began as agricultural implement manufacturer 1853; developed turbine-making; sold to Gilkes 1887. Isaac Braithwaite & Son established laundry machine engineering works (Ibis Works) on Shap Road by 1927.

Shoemaking grew out of long-established leather trade: in 1829 there were 7 tanners and c. 30 shoemakers. Somervell Brothers, who established leather business in Kendal from 1842, started shoe manufacture from 1860s, registering trade mark 'K Shoes' 1875. By 1900 it was a major employer, producing 170,000 pairs of shoes annually. Survived until 1990s.

Other specialist industries included snuff manufacture, established outside town at Meal Bank 1792 (see Scalthwaiterigg). Three main firms of snuff manufacturers in Kendal by later 19th century: Samuel Gawith; Gawith &

Hoggarth and Illingworths; the first two merged 2015; Illingworths closed 1983. Limestone quarrying, lime-burning and polishing stone into 'marble', took place on Kendal Fell from c.1800.

Kendal's population more than trebled across 19th and 20th centuries. Total number of inhabitants (Kirkland and Nether Graveship included) grew from c.8,000 in 1801 to 14,183 by 1901. Economic base was transformed across 20th century, with loss of both textile industry and shoemaking. Some manufacturing remained on new industrial estates to north of town (such as Farley's milk food works, established at Mintsfeet in 1970s). Service industries and administration provided employment, notably Provincial Insurance Company, based in Kendal from 1919. As manufacturing industries declined, Kendal evolved across second half of 20th century into administrative, cultural and tourist centre, its position on edge of Lake District attracting significant retired population. As well as being seat of Westmorland County Council (1889-1974), it continued to serve important local government function after 1974, containing offices of South Lakeland District Council and Lake District National Park Authority and several departments of Cumbria County Council. Population grew rapidly in second half of 20th century, rising from 18,541 in 1951 to 27,505 in 2001.

Places of worship. Medieval parish church of Holy Trinity, almost certainly pre-Viking foundation, remained mother church to much of south Westmorland, its parish covering 25 townships. Other places of worship in medieval Kendal included Allhallows chapel on Fellside, St Anne's chapel at Dockray Hall and St Leonard's hospital. St George's church (Anglican) in Market Place, built 1755; demolished 1855. As population expanded, new Anglican district churches were established: St Thomas', Stricklandgate, built 1835-7, and new church of St George, Castle Street, built 1838-41. Mission church, Allhallows chapel, on Fellside, built 1864-6. Town's cemetery opened 1855.

Kendal had strong and influential nonconformist element. Quaker presence by 1656, when land on Sepulchre Lane was bought for burial ground. Quaker meeting house, Stramongate, built 1688; replaced by present building 1816. Presbyterian chapel in corner of Market Place built by 1702; present Market Place chapel built 1720. Market Place congregation became Unitarian and Orthodox Presbyterian seceders built chapel on Beast Banks 1763, which was sold 1811 ('Scotch' burial ground there remained in use until 1855). Subsequent splintering resulted in building of Independent chapel in Lowther Street 1782 (closed 1929), from which further secession 1823 led to building of another Independent chapel in former theatre in Woolpack Yard. Much of that congregation, in turn, left to build Zion chapel (now United Reformed church), in yard off Highgate, 1844; rebuilt 1896. Remainder of Woolpack Yard congregation moved to new chapel, St John's Church, Sandes Avenue, 1897; demolished in 1980s. Inghamite chapel, in barn on edge of High Beast Banks, established c.1756; rebuilt 1844; converted to housing 1985.

First Methodist church was established in former playhouse (later Working Men's Institute) in Market Place 1787 and remained there until purpose-built chapel was built at head of Stricklandgate 1828; rebuilt 1882. Wesleyan chapel in Chapel Lane built 1872. Primitive Methodists had built chapel in Castle Street 1823 and moved to new chapel 1870. Job Pennington Memorial Primitive Methodist Mission Hall built on Low Fellside 1899. Sandylands Methodist Cornerstone church built on Sandylands housing estate 1937. Roman Catholic church, rebuilt 1793, was replaced by church of Holy Trinity & St George, New Road, built 1835-7. Evangelical mission hall, Gillingate, built by Quakers 1894; transferred to Methodists 1939; later used by Salvation Army. Other denominations included Christian Brethren, who had chapel near Stramongate in later 19th century; Salvation Army, who had citadel behind Highgate in mid-20th century; Jehovah's Witnesses, who used former Oddfellows' Hall as Kingdom Hall; Evangelical church, Parr Street, built 1970; and New Life Community Church, which took over Shakespeare Centre 1994.

Schools. Free Grammar School founded 1525; rebuilt on Aynam Road, 1889. Bluecoat school, part of Sandes Hospital charity, established c.1670; closed and amalgamated with Grammar School in 1880s. Friends School, Stramongate, first established 1698; moved to larger premises 1792; closed 1932. Expansion of elementary education during 19th century saw National schools at top of Allhallows Lane (founded 1818; girls' school opened 1824 and infant school 1874; became Central Primary CE School; closed 1985); at St Thomas' church (1842); at St George's church (1854) and in Kirkland (girls' school, 1861). British schools founded in Castle Street (1835) and Kendal Green (1873). Roman Catholic community served by St Mary's & St Joseph's schools, New Road, in former Catholic church building. In 1885, as well as these schools, there were 10 private academies. New RC primary school, Dean Gibson School, Gillingate, built 1898. In 20th century new county primary schools were established at Stramongate and at Ghyllside. Secondary education expanded with opening of Kendal High School for Girls 1890; closed 1980 when it merged with boys' Grammar School. Stramongate Boys' School opened 1937. Post-war growth of secondary schooling saw new secondary modern schools at Longlands (opened 1959 and 1960) and St Mary's RC, Hawesmead Avenue (opened 1971; site taken over by Dean Gibson RC Primary School 1977). Re-organisation 1983 resulted in two comprehensive schools, for northern and southern areas of town respectively: at Longlands (now Queen Katharine School) and former Grammar School site (now Kirkbie Kendal School). Vocational education provided by James Allen Technical Institute, Beezon Road, built 1912, forerunner of Kendal Technical College (now Kendal College), Milnthorpe Road, built 1971.

Other institutions. New Biggin, in middle of Highgate, contained 'shops' built c.1500; demolished 1803. Market house built 1855 (used as public library from 1891; demolished 1909); replaced by new Market Hall, opened 1889. White Hall, built as cloth exchange, rebuilt as assembly rooms 1824-5. Moot Hall in Stricklandgate, replaced by Town Hall, converted 1858-9 from former Whitehall assembly rooms. House of Correction (Kendal's town prison) built at northern end of town, probably pre-1700, to replace dungeon under Moot Hall. Rebuilt 1785; became county gaol after closure of gaol at Appleby 1888; closed as civilian prison 1894; site sold for redevelopment 1907. On establishment of Westmorland County Council 1889, Kendal (rather than ancient county town of Appleby) became seat of county government; County Hall was built 1937-9.

Poor house on Fellside built in 17th century and another for township of Kirkland, probably c.1740. From 1725-1768 Kendal's workhouse was at Brownsword House, Stricklandgate. Income received by corporation from enclosure of Kendal Fell 1767 used to build new workhouse, opposite House of Correction. This subsequently became Kendal Poor Law Union Workhouse from 1836; workhouse buildings used as Kendal Green Hospital for elderly from 1948 to 1970. Almshouses: Sandes Hospital built 1659; endowed for 8 poor widows 1670. Dowker's Hospital, endowed 1831; cottage almshouses built in Highgate 1833; demolished c.1960. Sleddal Jubilee Almshouses, Aynam Road, endowed by John Sleddal 1887. Howard Orphan Home for girls, at Stonecross, opened 1865; later used as local authority old people's home; closed 1990. Cropper Memorial Hospital, Captain French Lane, built 1869-70, became Westmorland County Hospital; closed 1991 when new Westmorland General Hospital opened on Burton Road. Helme Chase maternity home closed 1993. Public bath house, Allhallows Lane, 1863, with swimming baths added 1883.

'Economic Library' founded 1797; merged with Mechanics' Institute (established 1824). Oddfellows' Hall, Highgate (now converted to housing), established 1833 on site of Unicorn Inn and taken over by Mechanics' Institute 1857. Working Men's Reading Association opened Workmen's Newsroom in Market Place 1843. Christian and Literary Institute, Kent Street, (similar in objectives to Mechanics' Institute) established 1852. Museum established by William Todhunter 1796 taken over 1835 by Kendal Literary and Scientific Society (founded 1817), which established museum and library in Stricklandgate House, which they leased 1854. Museum moved to new premises (formerly wool warehouse) adjacent to Allen Technical Institute in 1918; now Kendal Borough Museum. Abbot Hall Art Gallery opened 1962, with adjacent Museum of Lakeland Life and Industry. Subscription library, founded 1794, taken over by Kendal Literary and Scientific Society; earliest public library was that opened in Workmen's Newsroom 1847. Free public library established in former Market Hall 1891 and replaced by Carnegie Library, Stricklandgate, opened 1909. Playhouse (in what later became Working Men's Institute, Market Place) founded 1758; moved c.1780 to The Fold, Stricklandgate. Purpose-built theatre in Woolpack Yard, known as New Theatre until 1807 and Theatre Royal thereafter; taken over by Independent chapel 1823. Further theatres built in Old Shambles 1828 and behind Shakespeare Inn 1829; latter closed under pressure from nonconformists 1834; subsequently used as ballroom (latterly known as Shakespeare Centre). St George's Hall, Stramongate, built 1878, used for theatrical productions and as cinema in early 20th century, becoming bingo hall in 1960s; demolished 1993. 'Roxy' cinema opened 1913 (closed 1950s); Palladium, Sandes Avenue, opened 1931 (closed 1989). New theatre built 1993 in Brewery Arts Centre (opened 1972), converted from former brewery.

KENTMERE

Chapelry and township in Kendal parish, Kendal ward, Westmorland.

Acreage: 6,613 acres [2,676 ha], divided into four quarters – Green Quarter, Hollowbank Quarter, Crag Quarter and West Wreay Quarter – for local administrative purposes. Common land in Kentmere (131 acres [53 ha]) enclosed 1850; Green Quarter stinted pasture (753 acres [305 ha]) enclosed 1854. ***Population:*** rising from 166 in 1801 to peak of 212 in 1811 and 1821, then declining to stand at 164 in 1901, 91 in 1961 and 95 in 2001. ***Landownership:*** *manor of Kentmere* was assigned to Laderina, sister and co-heir of Peter de Brus (d. 1272), lord of moiety of Kendal barony, and descended through her daughter Sibil, who married Miles de Stapleton, to

Stapleton family, in whose hands it remained until sold by Gilbert Stapleton to Nicholas Fisher of Stainbank Green 1626. Sold on death of his descendant Henry Fisher to Thomas Wilson of Kendal 1745. Kentmere Hall was held by Gilpin family by later 14th century; sold early 17th century to Philipsons of Calgarth. ***Economy:*** hill farming; slate quarrying at head of valley from 18th century. Quarrying at its peak from 1847 to 1914; reduced activity continued at Steel Rigg until 1950s. Mining of diatomite (silicaceous earth used in industrial products) from bed of Kentmere Tarn: factory opened 1929; production ceased 1995. Kentmere Reservoir built 1845-8 to feed mills further down Kent valley. ***Places of worship:*** chapel of ease of St Cuthbert recorded 1453; restored and tower rebuilt 1866. ***Schools and other institutions***: curate licensed to teach 1696; school recorded 1818; rebuilt later 19th century; closed 1957. Kentmere Institute (village hall) built 1925-6.

KESWICK

Market town and tourist centre in Crosthwaite parish, Allerdale below Derwent ward, Cumberland. This entry covers area which became Keswick UD 1894, namely township of Keswick (including small extra-parochial area known as Briery Cottages) and part of Underskiddaw township, including Great Crosthwaite, south of railway line. Further boundary changes 1934 resulted in net gains from Castlerigg, St John's & Wythburn CP.

Acreage. Keswick township covered 728 acres [295 ha] and extra-parochial area of Briery Cottages (also known as Greta Mills) 46 acres [19 ha]. From 1894, having gained parts of Underskiddaw township, Keswick UD contained 1,166 acres [472 ha]. Common land at Town Cass and Chestnut Hill (c.70 acres [28 ha]) enclosed 1849.

Landownership. Mostly in *manor of Castlerigg and Derwentwater*, held by Adam de Derwentwater 1216, subsequently passing to Radcliffes of Dilston (Northumb.) by marriage of Elizabeth de Derwentwater to Sir Nicholas Radcliffe c.1417. It descended in Radcliffe family until execution of Sir James Radcliffe, third earl of Derwentwater, 1716, for his part in 1715 Jacobite rebellion, when it was forfeited to Crown. Rents and profits from estate supported Royal Greenwich Hospital for Seamen from 1735. Manorial rights and Greenwich Hospital's Keswick estate sold to Marshall family 1832.

Origins and growth of the town. Market charter granted 1276; Keswick became small seigniorial borough. Thomas Denton described character of market in later 17th century thus: 'the number of tanners and shoomakers, which dwell in and about the town, causeth the markets to abound with raw hides and leather'; little corn was sold but butter and cheese were plentiful; cattle fairs were held fortnightly across summer months. Town's prosperity grew with expansion of non-ferrous mining in surrounding areas under Mines Royal from 1564 to Civil War. German miners made headquartered on Vicar's Island, with water-powered ore processing and smelting works at Brigham on River Greta. Writing in 1671, Sir Daniel Fleming noted, 'Heretofore this towne was rich getting much by the minerall-men, but those being dispersed upon the giveing over of the workes, it is become much poorer'.

'Discovery' of Lake District by tourists from 1750s led to Keswick becoming focal point for picturesque tourism from 1770s, facilitated by new turnpike roads. In 1780s attractions included annual regattas on Derwentwater and two competing museums run by Peter Crosthwaite and Thomas Hutton. Town's population, calculated at 1,078 in 1787, grew rapidly across 19th century, doubling from 1,350 in 1801 to 2,618 by 1851. By 1852 tourist numbers during summer season estimated at 12,000 to 15,000.

Keswick's industrial base also expanded: leather production declined but woollen textile industry (carding, spinning, dyeing and weaving) grew in late 18th century, with water-powered mills along River Greta, both in town and at Greta Mills at Briery: two woollen mills, bobbin mill, brewery and pencil mill along river side by mid-19th century. Pencil manufacture expanded from late 18th century, becoming town's largest industry with founding of Joseph Banks' pencil factory at Greta Mill 1835.

Arrival of Cockermouth, Keswick & Penrith Railway 1864 (with Keswick Hotel of 1869) stimulated further tourism. Population had grown to 3,760 in 1891 (population of Keswick UD stood at 4,451 in 1901) and town expanded north of River Greta and to east of old urban core. Across 20th century tourism came to dominate, as industrial base declined. Woollen mill at Briery Cottages had closed by 1900; bobbin mill there closed 1958. Pencil manufacturing remained important: Hogarth & Hayes, at Southey Hill, and Robert Wilson & Co. (Cumberland Pencil Co. from 1899) at Greta Bridge, had merged on Southey Hill site by 1916; works rebuilt 1937; closed 2006. Pencil Museum remains on site. Population fairly stable across 20th century, standing at 4,984 in 2001. By early 21st century, 60% of town's employment was in tourism.

Places of worship. Medieval parish church of St Kentigern in Great Crosthwaite (restored 1844-6 and 1889) served extensive ancient parish. Church of St John the Evangelist, Keswick, built 1836-8 (extended 1862, 1882 and 1889); assigned parochial district 1839. Independent nonconformist congregation established 1705 or before; chapel built in Lake Road 1803; rebuilt 1858-9; now Keswick Congregational Church. Quaker meeting house at High Hill, Crosthwaite, built 1715; closed c.1775. Wesleyan Methodist chapel built 1814; replaced by chapel in Southey Street, built 1863; extended 1909; still in use. High Street chapel (Bethesda Free Chapel) of Christian Brethren built 1851; closed. Annual Keswick Convention (evangelical Christian) started 1875 and continues. Gospel Hall, St John's Street, still in use. Quaker meeting re-established when meeting house, Church Street, opened 1921; replaced by new meeting house at Greta Grove, Tithebarn Street, 1995. Roman Catholic church of Our Lady of the Lakes and St Charles built 1927-8; completed (including tower) 1962-4.

Schools. Free grammar school beside parish church at Crosthwaite founded by 1571. National School of Industry established beside River Greta 1820. Keswick township had twelve day schools 1833. St John's school (infants and girls) opened 1840; boys' school, located at Brigham, built 1851. Grammar school moved from Crosthwaite to new site on Keswick side of Greta Bridge, opening in 1898 as co-educational school; 1898 building became Rawnsley Hall in 1940, when new premises built behind it. Secondary modern school built at Lairthwaite 1951. Keswick School and Lairthwaite combined to form comprehensive secondary school 1980, initially on split site but merging during 1990s at Lairthwaite; now Keswick School. Primary schools at two St John's sites replaced by Trinity School for juniors, on new site near Windebrowe housing estate, 1974, with adjacent St Kentigern's School for infants and nursery built 1993. Trinity and St Kentigern combined to become St Herbert's School 2005. Keswick School of Industrial Arts, founded by H. D. Rawnsley in parish hall 1883-4; moved to purpose-built premises 1894; closed 1984 (building became restaurant).

Other institutions. Moot Hall, former manor court house, built 1695; rebuilt by Greenwich Hospital 1813, lower floor being used as market hall on Saturdays. Police station from 1857; replaced by police station, council offices and magistrates court in Bank Street 1902. Poor house, on site later occupied by Post Office, built following bequest from Sir John Banks (d. 1644); closed after 1862. Mary Hewitson Hospital opened 1891 and remains Keswick's cottage hospital.

Lending library in Sunday school attached to Crosthwaite church from 1827. Library with lecture hall opened in town 1849; Mechanics' Institute, in Market Place, formed same year. Oddfellows' Hall built 1850. Thomas Hutton's museum opened c.1775; closed c.1840. Peter Crosthwaite's museum established 1780; collections sold 1870. New Keswick museum, Station Road, built 1897, adjacent to Fitz Park, which had opened 1887. Theatre recorded c.1810; tradition of travelling theatre companies, notably seasonal Century Theatre (mid-20th century), accommodated from 1976 in fixed auditorium, the Blue Box, at Lakeside. Replaced by new purpose-built premises, the Theatre By The Lake, 1999. Alhambra cinema built c.1914; still in operation.

KILLINGTON*

Township in Kirkby Lonsdale parish, Lonsdale ward, Westmorland.

Acreage: 4,938 acres [1,998 ha], including 2,110 acres [854 ha] of common land enclosed 1821. ***Population:*** in range 250-300 across most of 19th century (peak 335 in 1821); declining across 20th century to stand at 152 in 2001. ***Landownership:*** *manor of Killington and Firbank* held from 13th to 16th century by Pickering family of Killington. On death of Sir Christopher Pickering (d. 1548) it passed to his daughter Anne; then to her son by her third marriage, Francis Vaughan, who sold it to the tenants 1585. ***Economy:*** predominantly agricultural. Killington Lake, reservoir supplying Lancaster-Kendal canal, created 1819; service station on M6 motorway opened 1972. ***Places of worship:*** medieval chapel of ease, now church of All Saints, made parochial 1585; repaired 1824; re-roofed 1863; restored 1895. ***Schools and other institutions:*** school endowed 1677; taught in chapel in 18th century; separate school built by 1858; enlarged 1873; closed 1948; building converted to village hall; refurbished 1999.

KINGMOOR

Extra-parochial place in Eskdale ward, Cumberland. Kingmoor CP was enlarged to include Cargo township (see Stanwix) but lost Kingstown area to Carlisle in 1951.

Acreage: 1,169 acres [473 ha] before boundary changes. ***Population:*** upward trend across 19th and early 20th century, from 103 in 1801 to 1,026 in 1931 (last census year before major boundary changes), as Kingstown grew from small hamlet as result of ribbon development along A7 north of Carlisle. ***Landownership:*** as name indicates, Kingmoor was moorland vested in Crown, over which citizens of Carlisle had common rights. From 1682 city corporation leased out sections of moor, creating new holdings. ***Economy:*** Carlisle burgesses held guild races on moorland near Kingstown until 19th century. By mid-19th century several beer sellers and inns on main road north from Carlisle at Kingstown and Greymoor Hill. Brick-making well established by 19th century: in 1901 main companies were Kingmoor Brick & Tile Co. and Kingstown Brickworks. Arrival of railway and expansion of Carlisle transformed the area: marshalling yards covered extensive acreages alongside railway; Kingmoor Depot for engine maintenance opened 1874; closed 1968. Extensive light industry at Kingstown Industrial Estate late 20th century. ***Places of worship:*** allocated to Houghton for ecclesiastical purposes 1841. Services held in school at Kingstown 1901. St Peter's church, Kingmoor, built early 20th century. ***Schools and other institutions:*** infant school at Greymoor Hill by 1860s. National school at Kingstown enlarged 1898; its successors, Kingmoor Infant and Junior Schools, opened on new site 1969. James Rennie Special School had been established at Kingstown by 1980s. Stanwix Cemetery, which opened 1887, lay in Kingmoor.

KING'S MEABURN

Township in Morland parish, West ward, Westmorland.

Acreage: 2,388 acres [966 ha], including 411 acres [166 ha] of common, enclosed 1779. ***Population:*** rising from 178 in 1801 to 216 in 1851, then gradually falling to 105 in 2001. ***Landownership:*** part of *manor of Meaburn*, divided between Sir Hugh de Morville (d. 1173/4) and his sister Maud, who was allocated Maulds Meaburn (see Crosby Ravensworth). King's Meaburn was forfeited by Morville for his role in murder of Thomas à Becket; then granted to Robert de Veteripont c.1200, descending thereafter with barony of Westmorland (see Appleby). ***Economy:*** predominantly agricultural; some quarrying. ***Places of worship:*** Wesleyan Methodist chapel built 1866; rebuilt 1932; still in use. ***Schools and other institutions:*** village school built 1831; enlarged 1907; closed 1983. Village hall opened 1914.

KINGWATER

Township in Lanercost parish, Eskdale ward, Cumberland.

Acreage: 18,770 acres [7,596 ha], consisting largely of moorland and peat mosses of Spadeadam Waste and Butterburn Flow. ***Population:*** in range 350-400 until 1871 (peak 407 in 1851), then falling to c.220-250 in early 20th century; further decline to 152 by 1951; stood at 164 in 2001. ***Landownership:*** part of Gilsland barony and thus Naworth Castle estate (see Brampton). ***Economy:*** livestock farming. Spadeadam wastes said 1688 to have been 'of late much improved'. Large-scale afforestation on Spadeadam from 1950s, surrounding RAF Spadeadam, site of testing ground for Blue Streak missile, constructed from 1956. Subsequently used as Electronic Warfare Tactics Range from 1976. ***Places of worship:*** Wesleyan Methodist chapels at West Hall (built 1835; in use 1991; closed) and Nickies Hill (built 1838; closed 1973). ***Schools and other institutions:*** unendowed school at Crowsgate, near Lees Hill, recorded 1847; closed 1876.

KIRKANDREWS MIDDLE QUARTER

Township in Kirkandrews-on-Esk parish, Eskdale ward, Cumberland. Combined with Kirkandrews Moat and Kirkandrews Nether to form Kirkandrews CP 1934.

Acreage: 4,312 acres [1,745 ha]. ***Population:*** rising from 573 in 1801 to peak of 624 in 1821; then falling steadily to stand at 255 in 1931 (last census year for which separate figure available). ***Landownership:*** see Kirkandrews Moat. ***Economy:*** predominantly agricultural; free stone quarry at Glinger Burn. ***Places of worship:*** medieval church of St Andrew (perhaps originally chapel of ease to Arthuret); in ruins by 1606. Modern parish founded 1632; new church built 1637; repaired 1685; then rebuilt 1775 by Dr Robert Graham of Netherby; restored 1892-3 and 1993. ***Schools and other institutions:*** schoolmaster at Bush Head by 1847; Rhodds school built 1871; closed 1931. Small almshouse for aged families erected by Sir James Graham by 1847.

KIRKANDREWS MOAT QUARTER

Township in Kirkandrews-on-Esk parish, Eskdale ward, Cumberland. Combined with Kirkandrews Middle and Kirkandrews Nether to form Kirkandrews CP 1934.

Acreage: 1,701 acres [688 ha]. ***Population:*** rising from 291 in 1801 to 300 in 1821, then declining trend (apart from surge in 1850s from 199 in 1851 to 376 in 1861) to stand at 136 by 1931 (last census year for which separate figure available). ***Land ownership:*** Liddel Strength was caput of *barony of Liddel* (later termed *manor of Esk*) granted to Turgis Brundos c.1100; granted to Nicholas de Stuteville (d. c.1206) and passed by marriage of Joan de Stuteville (d. 1276) and Hugh Wake to Wake family. On death of Thomas Wake (d. 1349), passed to his sister Margaret, wife of Edmund Plantagenet; their daughter Joan married the Black Prince, bringing barony to Crown. Sold by James I to earl of Cumberland 1603 and 1610; purchased 1628 by Sir Richard Graham and descended through Grahams of Netherby to 21st century. They extended Netherby Hall in mid-18th century and again in 1830s. ***Economy:*** predominantly agricultural; red freestone quarry at High Moat worked by Herbertsons Ltd of Galashiels from 1868; closed mid-20th century. Tile-making in 19th century. ***Places of worship:*** none. ***Schools and other institutions:*** schoolmaster by 1829; new school built 1869; closed 1946 and became Moat village hall.

KIRKANDREWS NETHER QUARTER

Township in Kirkandrews-on-Esk parish, Eskdale ward, Cumberland. Combined with Kirkandrews Moat and Kirkandrews Middle to form Kirkandrews CP 1934.

Acreage: 4,878 acres [1,974 ha], much of it mossland and marsh, including Solway Moss (site of battle 1542). 4,000 acres [c.1,600 ha] of 'Roland Marsh' destroyed by sea in 17th century and 600 acres [240 ha] devastated by eruption of Solway Moss 1771. ***Population:*** doubled from 245 in 1801 to 516 by 1821; then declined to 275 by 1901. Increased during First World War, as result of employment in Gretna cordite factory, to stand at 487 in 1931 (last census year for which separate figure available). ***Landownership:*** see Kirkandrews Moat. ***Economy:*** farming; commercial peat extraction by Richardson's Moss Litter Co. Ltd from 1898; still operating as Solway Moss Works. Slaughter house and bone fertilizer manufacturers at Mill Hill 1930s. German prisoner of war camp at Mossband during Second World War. ***Places of worship:*** services held at Blackbank school in 1901. ***Schools and other institutions:*** schoolmaster at Blackbank by 1829; Board school built there 1873; enlarged 1899 and 1914; moved to Guards 1915; then moved to Gretna 1916. Blackbank school buildings sold to Ministry of Munitions 1917 and occupied by Gretna munitions factory.

KIRKANDREWS-ON-EDEN

Ancient parish in Cumberland ward, Cumberland. Absorbed into Beaumont CP 1934.

Acreage: 1,010 acres [409 ha]. Kirkandrews Moor divided 1686-1708; remaining common land (140 acres [57 ha]) enclosed 1857. ***Population:*** estimated at 100 in 1688; in range 100-140 across 19th and early 20th century (peak 149 in 1901); stood at 145 in 1931 (last census year for which separate figure available). ***Landownership:*** part of barony of Burgh by Sands (q.v.). ***Economy:*** predominantly agricultural; some gravel pits in 18th and 19th centuries. Bone manure works beside Carlisle and Silloth railway later 19th century. ***Places of worship:*** medieval parish church of St Andrew; seems to have survived in 1688 but long since demolished (site marked by St Andrew's well and graveyard); parish served by Beaumont church since 1692. Kirksteads (one mile south of village) said to mark site of ancient chapel. ***Schools and other institutions:*** school built 1817 to serve villages of Kirkandrews, Grinsdale and Beaumont, benefitting from Thomas Pattinson's charity. Replaced by Board school on new site, erected 1876; closed 1984. Old school used as reading room since 1876. Carlisle's smallpox isolation hospital was built in south end of parish c.1906; closed c.1981. Village hall (Beaumont Parish Hall) built mid-20th century.

KIRKBAMPTON

Ancient parish in Cumberland ward, Cumberland, comprising townships of Great Bampton, Little Bampton and Oughterby (which included hamlet of Studholme).

Acreage: 3,733 acres [1,511 ha], divided between constituent townships thus: Great Bampton: 1,367 acres [553 ha]; Little Bampton: 1,357 acres [549 ha]; Oughterby: 1,008 acres [408 ha], of which Studholme formed detached portion of 262 acres [106 ha]. Commons at Borrow Moor, Midmoss Close, Westmoss Close and High Moor, totalling 200 acres [81 ha], enclosed 1775. ***Population:*** estimated at 255 in 1688; rising from 456 in 1801 to peak of 546 in 1851, then declining steadily to 348 in 1961; rising again to stand at 471 in 2001. ***Landownership:*** part of barony of Burgh. Held by Hildred de Carlisle in 12th century; partitioned between heirs after his death and remained in divided lordship. *Manor of Little Bampton* sold 1582 by Thomas Brisby to John Southaik and Richard Tolson who sold it to John Dalston 1586;

he sold it to the tenants. ***Economy:*** agricultural: windmill recorded in name Windmill House; market gardening in 20th century. ***Places of worship:*** medieval parish church of St Peter; restored 1871-2 and 1882. ***Schools and other institutions:*** school taught in church 1703. Three schools by 1847: by church, at Ploughlands and at Oughterby. Replaced by St Peter's National School at Longrigg Bridge, in centre of parish between the villages, built 1876 with addition 1897. In turn replaced by modern building on edge of village, now Kirkbampton CE Primary School. Reading room in Laurel Cottage from c.1920; replaced by village hall (ex-army hut) c.1930; condemned as dangerous 1990s; rebuilt on new site 2001.

KIRKBRIDE

Ancient parish in Cumberland ward, Cumberland.

Acreage: 1,667acres [674 ha], including Kirkbride common (c. 600 acres [c.240 ha]), enclosed 1816. ***Population:*** rising from 249 in 1801 to 383 in 1831; then falling slightly before rising again from 1891 to reach 519 by 2001. ***Landownership:*** *manor of Kirkbride* granted by Adam son of Odard to his second son c.1200. His descendants took name Kirkbride and held manor for several generations until it passed by marriage of Elizabeth, daughter of George Kirkbride (d. 1511), and John Dalston to Dalston family of Dalston. Sold 1764 to Joseph Wilson of Pontefract; then to William Matthews and later to earl of Lonsdale. ***Economy:*** predominantly agricultural. Kirkbride airfield, opened 1939 as maintenance depot; after Second World War became storage depot, closing 1960; later light aviation centre. ***Places of worship:*** medieval parish church of St Bridget; restored 1895. Quaker meeting house built 1698; closed 1854. Primitive Methodist chapel built 1865; rebuilt 1906; closed 1963 and converted to dwelling. Wesleyan Methodist chapel built 1869. ***Schools and other institutions:*** school taught by curate 1827; village school described as 'small and poor' 1860. Board school erected 1875; replaced by modern building; now Kirkbride Primary School. Reading room built 1881 with library added later; now village hall.

KIRKBY IRELETH (Kirkby-in-Furness)

CP in Kirkby Ireleth parish, Lonsdale hundred, Lancashire North of the Sands, including villages of Grizebeck, Sandside and Soutergate and townships of Woodland and Heathwaite.

Acreage: 9,702 acres [3,926 ha], of which 3,376 acres [1,366 ha] consists of unenclosed common land on Kirkby Moor. Common land at Sandside (300 acres [121 ha]) enclosed 1867. Settled areas consisted of four townships: Woodland: 1,315 acres [532 ha]; Heathwaite: 1,393 acres [564 ha]; Middle Quarter: 1,047 acres [424 ha] and Low Quarter: 2,571 acres [1,040 ha]. ***Population:*** rising from 1,041 in 1801 to peak of 1,809 in 1841; had fallen back to 1,477 by 1901 and continued to decline to 1,247 by 2001. ***Landownership:*** *manor of Kirkby Ireleth* held by Kirkby family from 12th century until mortgaged by Roger Kirkby 1689, after which it passed to duchess of Buckingham, who bequeathed it to her grandson Constantine Phipps, later Lord Musgrave, who sold it 1771 to Lord John Cavendish, son of duke of Devonshire. On his death 1796, manor was united with Cavendish family's Furness Abbey and Holker estates. ***Economy:*** farming, estuarine fishing, woodland industries. Medieval bloomeries (forge at Kirkby had 'gone to decay' by 1671); woollen textiles (recorded in name Tenter Bank, in Heathwaite). Lime-burning in Low Quarter; bobbin mill in Woodland; sickle smithy at Grizebeck in 19th century. Major industry from mid-19th century was Burlington slate quarries on Kirkby Moor: quarrying recorded from 18th century; major expansion from 1843 (peak output 1850-1880), with slate exported by tramway to flat boats on quay at Sandside until opening of Furness Railway 1847. Quarrying continues today. ***Places of worship:*** name 'Kirkby' (Scandinavian for 'settlement with a church') suggests church founded by 10th century. Medieval parish church of St Cuthbert at Beckside; tower rebuilt 1829 and again 1904; rest of church restored 1881, 1884. Chapel of ease at Woodland pre-Reformation; rebuilt 1689; rebuilt again 1822 and 1864-5; now church of St John the Evangelist. Church of Good Shepherd (originally mission church) at Grizebeck built 1898; closed 2012. Baptist chapel by St Mary's Well opened 1829; closed 1876; replaced by new chapel at Wall End, built 1876, which became Christian Brethren meeting house; now Church of Christ. Methodist chapel at Marshside, built 1870; now Kirkby-in-Furness Methodist Church. Gospel hall, Sandside, built 19th century; closed 2010. ***Schools and other institutions:*** endowed schools at Beckside and in Woodland, recorded 1818. School at Grizebeck built c.1829; rebuilt as Board School 1878; closed 1984. Girls' school recorded at Grizebeck 1849. Burlington School built 1877, replacing school at Beckside; now Burlington CE Primary School. Former Beckside school now village hall (Beckside Rooms). Village hall at Grizebeck built 1928; rebuilt 2010-11. Parish room at Woodland. Community centre at Beckside, built 1970s.

KIRKBY LONSDALE*

Market town and township in Kirkby Lonsdale parish, Lonsdale ward, Westmorland, including hamlets of Biggins, Kearstwick, Tearnside and Underley. Became Kirkby Lonsdale UD from 1894 to 1935.

Acreage. 3,254 acres [1,317 ha], including common land at Kirkby Moor (1,124 acres [455 ha]), enclosed 1810.

Landownership. *Manor of Kirkby Lonsdale* granted to St Mary's Abbey, York, by 1100; after Dissolution granted to Thomas Carus of Halton 1558, passing by marriage to Curwens of Workington. By 1640 was held by Prestons of Holker who sold it before 1672 to Sir John Lowther. *Manor of Deansbiggin* held by Copeland family in 13th and 14th centuries; passed through several families (Tunstall, Middleton, Booth, Borrett, Batty) before being sold on death of Henry Bickersteth (d. 1821) to Alexander Nowell of Underley. *Underley estate* in hands of Wilson family from 17th century; sold 1732 to Hugh Ashton, whose descendant sold it 1808 to Alexander Nowell, who rebuilt house 1825-8. Nowell sold Underley to William Thompson 1840, from whom it descended to Cavendish-Bentinck family. Largely broken up in 1940s.

Origins and growth of the town. Market grew up adjacent to parish church, probably in 12th century: market charter granted to parson, John de Kirkby, 1227, with weekly market on Thursday and livestock fairs (held fortnightly from Maundy Thursday to Michaelmas in late 17th century). In 1692 market and fairs said to 'afford a great store of cattle and all sorts of grain in great plenty'. By later 17th century Sir Daniel Fleming described Kirkby as 'a Towne of note, whither all the people round about repair to church and mercate' and said it was second only to Kendal in size. Market places in narrow streets close to church replaced by new Market Square 1820, where trading was governed by precise and elaborate rules, laid down 1822. As market centre close to county boundary and on main Kendal-Skipton road, Kirkby Lonsdale contained numerous inns. By late 18th century tanning and weaving had been established: carpet and blanket factories, two tanneries and water-powered textile printing works recorded in mid-19th century but most industrial activity had died out by 1900. Town remained modest in size, probably in part because railway did not reach it (nearest station 2 miles away at Cowan Bridge). Population rose from 1,283 in 1801 to only 1,802 in 1891, then fell, to reach a low point of 1,240 in 1951, before rising again to 1,771 in 2001. Economy increasingly depended on tourism by later 20th century. Livestock market closed by 1990s. Rural parts of township remained largely agricultural, with limestone quarrying and lime-burning in 19th century. Small business park beside A65 to west of town from 2007.

Places of worship. Name 'Kirkby' (Scandinavian, meaning 'settlement with a church') implies presence of church by 10th century, probably on site of later parish church of St Mary (contains fabric from c.1100; extended 15th/16th century; north aisle added 1574; restored 1866-8). Other medieval religious sites included hospital of St Leonard at Spittal and chapel at Tearnside (recorded by 1250; demolished 17th century). Methodist meeting house by 1778; replaced by new chapel 1834; still in use. Independent (Congregational) chapel established 1815; closed 1965; sold to Roman Catholics. Sandemanian chapel built 1828; had closed by 1873. St Joseph's RC church established in former Congregational chapel 1966.

Schools. School recorded 1550. Queen Elizabeth Grammar School established 1591 near church; rebuilt on Biggins Lane 1848; became comprehensive school 1985. Subscription school opened 1856 in former workhouse; replaced by National school 1858; replaced 1986 by new school on new site, now St Mary's CE Primary School. School in Independent chapel and several dame schools in mid-19th century. School of handicrafts established 1891 in former Sandemanian chapel; continued until 1940s. Underley Hall became residential school from 1940s, first as preparatory school; then as seminary (from 1960); latterly (from 1976 to 2012) school for children with special needs.

Other institutions. Workhouse at Mill Brow (newly built 1735) extended 1812 to serve Gilbert Union (Kirkby Lonsdale and sixteen surrounding townships), formed 1811; closed after formation of Kendal Poor Law Union 1836 (may have continued in use as casual ward in mid-19th century). Library (above vestry in parish church) in 17th century; book club, with library of 500 volumes, recorded 1794. Mechanics' Institute founded 1854. Institute on New Road built 1895. Public library in UD Council offices, Beckhead, until moved 1980 to former Sandemanian chapel. Reading room and institute at Kearstwick built 1902; closed 1990s and converted to dwelling.

KIRKBY STEPHEN

Market town in Kirkby Stephen parish, East Ward, Westmorland.

Acreage. 3,135 acres [1,269 ha], including Kirkby Stephen Common (1,491 acres [603 ha]), enclosed 1855. Shared pastures called Intake (215 acres [87 ha]) and High Intake enclosed 1850 and 1852.

Landownership. Part of barony of Westmorland (see Appleby), with which manorial lordship descended. By 17th century Wharton and Musgrave families had considerable landholdings in town. Manor said 1885 to be held in three portions by Lord Hothfield, earl of Lonsdale and trustees of Sir Richard Musgrave.

Origins and growth of the town. Market had probably grown up around parish church by 12th or 13th century: market place located at entrance to churchyard (divided from it by The Cloisters, serving as market house and piazza, built 1810). Charter granting market on Fridays and fairs in April and October granted 1351, almost certainly ratifying existing market. New market charter obtained 1606, moving market to Mondays with fairs at Whitsuntide and in October. By 19th century six livestock fairs held (between March and October) but by 1906 only those on 29 September (for horses) and 27 October (St Luke's day 'tup fair' for sheep) were of significance. Hiring fairs for servants were held at Whitsuntide and Martinmas and on two consecutive Mondays in June/July.

By later 17th century Kirkby Stephen had become established as centre of stocking knitting. In 1671 Sir Daniel Fleming described its market as 'much improved by the trade of stockings, lately taken up' and Thomas Denton, writing 1688, listed town's staple commodity as 'wooll & the manufacture of wool, vizt. yarn, stockings, caps, cloath &c'. This specialism maintained across 18th century but attempts to move into larger-scale textile production failed: unsuccessful attempts to establish cotton mill and manufacture of coarse woollen hats c.1800. Tannery, founded early 19th century, also short-lived. Other economic activity, growing in importance in 19th century, included brewing, brockram and limestone quarrying and lime burning. Coming of railways (South Durham & Lancashire Union Railway with Eden Valley branch 1861-2, and Midland Railway's Settle-Carlisle line 1875-6) had only limited effect, both stations lying at a distance, south of town. Population increase was modest, rising from 1,141 in 1801 to 1,339 in 1851 but (except for slight increases associated with railway building) remaining steady at around 1,500-1,600 from later 19th and across 20th century, standing at 1,832 in 2001. In 20th century, town's manufacturing base extended with new light industry, including Heredities cold-cast bronzed ware (1968-2003); C. & G. Wilkinson's bobbin factory (on site of disused East Station, 1969-1992); cable assembly (Brockhill Enterprises Ltd, from 1986); and mechanised stocking manufacture.

Places of worship. Name 'Kirkby' (Scandinavian for 'settlement with a church') suggests church founded by 10th century. Medieval parish church (early dedication unknown; now St Stephen's) partly rebuilt 1847-51 and 1871-4. Sandemanian chapel, New Inn Yard, later used by Baptists and, from 1810, by Independents; disused. Wesleyan Methodist chapel (Centenary Chapel) built 1839; still in use. Primitive Methodist chapel built 1864 (later used as Masonic Hall); replaced by new chapel 1902; closed 1979; now Youth Hostel. Congregational chapel built 1864 (replacing earlier chapel in New Inn Yard); closed c.1945; became Roman Catholic church from 1946 until Catholic congregation moved to share parish church from 1990s; now a shop. Brethren established Melbecks meeting room 1878; Gospel Hall (evangelical Christians) from mid-20th century; still in use. Baptist church built 1891; now Upper Eden Baptist Church. Quaker meeting house, Town Head, opened 1930. Cemetery, at north end of town, opened 1861.

Schools. Grammar School founded by Thomas Lord Wharton 1566; rebuilt and enlarged after 1878; and occupied latterly by girls' school 1908-1955. Several private academies in 19th century. National boys' school built 1868. Board school built 1877; now Kirkby Stephen Primary School. Christian Head co-educational secondary school built 1955; combined with Grammar School to become comprehensive school on Christian Head site, retaining name Kirkby Stephen Grammar School, 1959.

Other institutions. East Ward Union Workhouse (in former cotton mill building – therefore known as 'The Factory') opened 1818 as Gilbert Union workhouse; became public assistance institution known as Eden House 1930; closed 1961; demolished 1974. Temperance Hall and Literary Institution, containing hall, library and reading room, built 1856. Another reading room in Market Place recorded 1860. Oddfellows' Hall built 1855; replaced by new building 1890. Jubilee Park opened 1887.

KIRKBY THORE

Township in Kirkby Thore parish, East ward, Westmorland.

Acreage: 2,502 acres [1,013 ha.]; commons (1,175 acres [476 ha]) enclosed 1820. ***Population:*** rising from 247 in 1801 to lie in range 450-500 from mid-19th to mid-20th century; rapid increase across 1960s (from 486 in 1961 to 708 in 1971); stood at 731 in 2001. ***Landownership:*** *manor of Kirkby Thore* held by Kirkby Thore family (descended from Whelp who held it mid-12th century) in 13th-14th centuries; passed to Wharton family by 1450s, through whom it descended until death of William Wharton mid-18th century; then passed to Tufton family, lords of barony of Appleby (q.v.). ***Economy:*** farming. Gypsum quarrying by 19th century; major expansion early 20th century with plaster and plasterboard plant from 1910; still in operation. ***Places of worship:*** name 'Kirkby' (Scandinavian for 'settlement with a church') suggests church founded by 10th century, forerunner of medieval parish church of St Michael, close to site of Roman fort of BRAVONIACUM. Wesleyan Methodist chapel built c.1800; enlarged or rebuilt 1828; still in use. Bethel Chapel (United Methodist) built 1851; converted to dwelling by 1970s. Independent (Congregational) chapel built 1849; used as parish room by 1897; replaced by Memorial Hall on same site. ***Schools and other institutions*:** school endowed 1823 and built 1832; replaced by new Board school outside village 1875; enlarged 1883 and 1890; now Kirkby Thore Primary School. Old school became reading room and meeting room with library; converted to dwelling 21st century. Memorial Hall built c.1920 as village hall and First World War memorial.

KIRKLAND AND BLENCARN

Township in Kirkland parish, Leath ward, Cumberland. Absorbed into Culgaith CP 1934.

Acreage: 4,066 acres [1,646 ha], including extensive commons on Pennine edge: Kirkland Fell pasture (264 acres [107 ha]) enclosed 1857; Kirkland Fell (936 acres [379 ha]) enclosed 1866. Kirkland Fell (1,085 acres [439 ha]), Blencarn Fell (693 acres [280 ha]) and The Rigg, Blencarn (72 acres [29 ha]) remain registered common land. ***Population:*** rising from 188 in 1801 to 256 in 1851; then falling to low of 116 in 1921; stood at 129 in 1931 (last census year for which separate figure available). ***Landownership:*** part of barony granted to Adam son of Swein c.1100. *Manor of Blencarn* was granted to William L'Engleys after attainder of Andrew Harcla, earl of Carlisle, passing by marriage to Restwold family with whom it descended until sold to Thomas Lough 1540; on marriage of Mary Lough to John Carleton of Skirwith Hall 1724 it descended to Lough Carleton, who enfranchised the tenants 1792. *Kirkland* was part of manor of Skirwith (q.v.), held by Flemings of Rydal from 1606. ***Economy:*** farming; extensive mining for lead on slopes of Cross Fell and for coal and iron at Ardale Head. Smelt mill at Bank, Kirkland, in mid-19th century. ***Places of worship:*** medieval parish church of St Lawrence in Kirkland; rebuilt 1768 and again 1880. Wesleyan Methodist chapel, Blencarn, built 1840; closed c.1987. ***Schools and other institutions:*** free school for children of Culgaith and Blencarn founded 1775 (see Culgaith); superseded in Blencarn by Board school, built 1878; closed 1933; became Kirkland and Blencarn recreation room and village hall.

KIRKLINTON MIDDLE QUARTER

Township in Kirklinton parish, Eskdale ward, Cumberland.

Acreage: 3,043 acres [1,231 ha]. Commons in Kirklinton Middle Quarter and Hethersgill, totalling 4,264 acres [1,726 ha], enclosed 1735. ***Population:*** rising from 389 in 1801 to peak of 536 in 1841, then declining to low of 260 in 1971, recovering to stand at 369 in 2001. ***Landownership:*** part of barony of Levington, given in early 12th century to Richard Boyvill, whose posterity took name de Levington. After death of Levington heiress, Helewise (d. 1272), wife of Eustace de Balliol, manor passed to six co-heiresses, eventually descending to Musgraves of Hayton, who sold it to Edmund Appleby in 1661; he built core of Kirklinton Hall. His son Joseph Appleby married co-heiress of Dacres of Lanercost and their descendants took surname Dacre. Kirklinton passed to Rev. William Dacre (1827-1903), who sold it to George Graham Kirklinton before 1900. Kirklinton-Saul family greatly enlarged Kirklinton Hall 1875. ***Economy:*** predominantly agricultural. Charter for market and fair at 'Levington' granted 1263 but market seems not to have survived. Smithfield said to have acquired its name from informal market on Longtown-Brampton road, where carriers sold their goods. ***Places of worship:*** medieval parish church of St Cuthbert; rebuilt 1845. Quaker meeting house, used for funerals at burial ground at Meg's Hill, built 1749; sold 1964. ***Schools and other institutions:*** several schools in parish in early 19th century, including National school by c.1830. Fir Ends School, founded by Joseph Dacre (d. 1868); rebuilt on new site at Smithfield 1968; now Fir Ends Primary School.

KIRKOSWALD*

Township in Kirkoswald parish, Leath Ward, Cumberland. CP enlarged by absorbing Renwick and Staffield CPs 1934.

Acreage: 5,737 acres [2,322 ha], divided for local administrative purposes into High and Low Quarters. Common land at How Moor and Viol Moor (280 acres [115 ha]) enclosed 1699-1712; Haresceugh Fell (2,470 acres [1,076 ha]) and other small areas of moorland enclosed 1876. ***Population:*** rising from 634 in 1801 to peak at 768 in 1831; then declining to 466 in 1931 (last census year before boundary change). ***Landownership:*** *manor of Kirkoswald:* part of barony granted to Adam son of Sweyn c.1120; said to have been given in marriage to Robert Trivers, lord of barony of Burgh by Sands (q.v.) with which it descended to Morvilles. Divided between heiresses of Hugh de Morville 1202; reunited in hands of Thomas de Moulton of Gilsland 1272; then descended with barony of Gilsland (see Brampton) to 17th century. Awarded to Francis Lennard 1652; sold to Sir Christopher Musgrave 1716, descending in Musgrave family until sold to Sir Francis Ley 1913. *Manor of Haresceugh* granted to Lanercost Priory 1167; after Dissolution granted to Sir Thomas Dacre of Lanercost c.1550; sold 1663 to Dr Peter Barwick and gifted to chapel of Witherslack (q.v.) 1666; sold 1920. *The College* restored by John Maye late 16th century; sold to Henry Fetherstonhaugh 1590 and descended in same family to present; they remodelled it 1696. ***Economy:*** farming; charter for market and fair granted 1201: in 17th century weekly market held on Thursdays and fair around Whitsuntide; market had ceased by mid-19th century. Mineral extraction on Haresceugh Fell: quarrying and lime-burning; coal mining from 1746 to mid-19th century; lead mining from later 18th century; mine reopened for barytes extraction 1915; closed 1940s. Textile industry recorded from medieval period (fulling mill recorded 1399). Paper-making by 1715; former fulling mill converted to paper mill c.1789; paper-making ceased c.1882. Former paper mill in turn converted to woollen mill 1847; closed 1906. ***Places of worship:*** medieval parish church of St Oswald, which was collegiate 1523-1547; restored 1845, 1878-9 and 1893. Independent chapel at Parkhead built 1711; closed 1927; sold 1975 and converted to dwelling. Independent chapel at Sandhill, in Kirkoswald village, built before 1851; rebuilt 1865, sold 1938. Wesleyan Methodist chapel built 1821; replaced by new chapel 1871; still in use. ***Schools and other institutions:*** schoolmaster recorded 1613. School at Highbankhill probably founded by curate by 1684; endowed by John Lowthian of Staffield Hall 1742. Replaced by new National school at Sandhill 1857; now Kirkoswald CE Primary School. Library and reading room established by 1882; used by boy scouts from 1977; closed and converted to dwelling 1988. Church Institute built 1910.

KIRKSANTON *see* Millom Rural

KNOCK *see* Long Marton

LAMBRIGG

Township in Kendal parish, Kendal ward, Westmorland.

Acreage: 1,803 acres [730 ha], including common land on Lambrigg Fell, enclosed with Docker Fell 1886 (totalling 641 acres [259 ha]). ***Population:*** rising from 124 in 1801 to 176 in 1831; then falling back to 133 in 1901 and declining to 80 by 1961; stood at 90 in 2001. ***Landownership:*** granted to Gilbert of Burneside 1273, descending by marriage in mid-14th century to Bellinghams. Later acquired by Ducket family; Anthony Ducket sold it to Sir John Lowther 1680. ***Economy:*** predominantly agricultural; small-scale quarrying; copper vein worked in 18th century. Wind farm on Lambrigg Fell in operation since 2000. ***Places of worship:*** none. ***Schools and other institutions:*** none.

LAMONBY *see* Skelton

LAMPLUGH

Ancient parish in Allerdale above Derwent ward, Cumberland, embracing townships of Lamplugh, Kelton, Murton and Winder. CP enlarged by absorbing Salter and Eskett CP 1934.

Acreage: 6,343 acres [2,567 ha], divided between constituent townships thus: Lamplugh: 2,562 acres [1,037 ha]; Kelton: 1,897 acres [768 ha]; Murton: 1,066 acres [431 ha]; and Winder: 816 acres [330 ha]. Lamplugh commons (1,455 acres [589 ha]) enclosed 1862. ***Population:*** rising from 535 in 1801 to peak at 1,261 in 1881; then declined to 1,033 in 1921 before marked drop across 1920s, attributed to closure of mines and quarries; stood at 731 in 1931 (last census year before boundary change). ***Landownership:*** *manor of Lamplugh* held of lords of Workington by Lamplugh family from 12th century until sold 1919. *Manor of Murton* held directly of lords of Egremont barony by Murton family before passing to Lamplugh family in 14th century. *Manor of Kelton* seems to have descended with Harrington (q.v.) in later medieval centuries, being granted c.1550 to Christopher Morys or Moorhouse. Substantial estates in parish included Stockhow, held by Patrickson family in 16th and 17th centuries, and Red How and Streetgate, an estate accumulated by Dickinson family from 17th century. **Economy*:*** farming; quarrying, lime-burning and iron mining. Extensive iron mines in Murton and on Kelton

Fell from 1850s with major expansion from 1869; served by Rowrah and Kelton Fell mineral railway, opened 1877. Kelton Fell mine closed 1913 and most other mines in 1920s. Iron forge (making spades, shovels and edge tools) at Lanefoot, mid-19th century, on site of Walk Mill (i.e. fulling mill), recording earlier woollen textile industry. ***Places of worship:*** medieval parish church of St Michael; restored 1658 and 1771; rebuilt on same site 1870. Mission church at Kirkland built 1886; still in use. Primitive Methodist chapel at Kirkland by 1898; closed. ***Schools and other institutions:*** school, taught by curate and endowed by Richard Briscoe 1730s or 1747. Schoolhouse near High Mill Gill Head replaced by National school (Lamplugh Parochial School) at Murton, built 1868; closed 1951; converted to dwelling. Lamplugh & Kelton Board School at Kirkland built 1879; now Lamplugh CE Primary School. Reading room at Lamplugh Cross hosted meetings of Lamplugh Friendly Society established 1788 (second oldest in country). Women's Institute Hall built 1928; became Lamplugh Village Hall 2009.

LANERCOST *see* Burtholme

LANGDALE (East ward, Westmorland) *see* Orton (Westmorland)

LANGDALES

Chapelry in Grasmere parish, Kendal ward, Westmorland, covering valleys of Great Langdale and Little Langdale.

Acreage: 9,510 acres [3,849 ha], including extensive unenclosed fell commons totalling 4,994 acres [2,022 ha]. Common field named Langdale High and Low Common Field (77 acres [31 ha]) enclosed 1853. ***Population:*** around 300 in early 19th century, then rising from 442 in 1841 to peak of 846 in 1901; in range 650-800 across earlier 20th century; stood at 563 in 1971 (last census year for which separate figure available). ***Landownership:*** *Great Langdale* was hunting forest in barony of Kendal and descended with Windermere (q.v.). *Manor of Baysbrown* in Great Langdale granted to Conishead Priory by William de Lancaster; divided after Dissolution. *Manor of Little Langdale* held by Penningtons of Muncaster from 13th century; enfranchised to tenants c.1692. ***Economy:*** hill farming. Woollen textiles: two fulling mills recorded 1454; woollen mill at Colwith recorded late 17th century. Iron making at Hackett forge, Little Langdale, established 1620s by conversion of two fulling mills to bloom forge; converted to finery forge c.1711; ceased operation 1760s. Forge at Baysbrown late 17th century. Extensive slate quarrying: by mid-19th century major quarries at Thrang Quarry, near Chapel Stile, and Elterwater (which remains active). Elterwater Gunpowder works established 1824; closed 1928. Langdale Linen Industry ('arts and crafts' revival of cottage industry) established 1883; had declined by c.1920. Tourism increasingly important from 19th century; former gunpowder works at Elterwater converted to holiday resort later 20th century. ***Places of worship:*** chapel of ease in Great Langdale recorded 1571; rebuilt 1753; rebuilt again 1857 as Holy Trinity church; extended 1878. Tradition of early chapel at Chapel Mire in Little Langdale. ***Schools and other institutions:*** school taught in chapel by curate in early 18th century; two schools recorded 1818. School at Chapel Stile, Great Langdale, built 1824 by Elterwater Gunpowder Company; now Langdale CE Primary School. Little Langdale school built 1865; closed 1965. Langdale Institute, founded 1891, included billiard and reading rooms and library. Mission room at Chapel Stile by 1897; now Langdale village hall.

LANGRIGG AND MEALRIGG

Township in Bromfield parish, Allerdale below Derwent ward, Cumberland. Absorbed into Bromfield CP 1934.

Acreage: 2,138 acres [865 ha], including part of Bromfield Common (enclosed 1825). ***Population***: rising from c.200 in early 19th century to peak of 324 in 1861, then falling back to 215 in 1931 (last census year for which separate figure available). ***Landownership:*** *manor of Langrigg* came to Osmotherley or Osmunderley family on marriage of Agnes de Langrigg to Ranulph de Osmunderlaw c.1290. Descended through male line until Cuthbert Osmotherley sold it 1735 to his son-in-law Thomas Barwis, who rebuilt Langrigg Hall. His descendants sold manor and Hall to Joseph Bowerbank of Cockermouth 1876. ***Economy:*** predominantly agricultural; tile-making in 19th century; windmill at Langrigg (early 19th century). ***Places of worship:*** none. ***Schools and other institutions:*** infant school (by 1867), built by Barwis family of Langrigg Hall (now private house).

LANGWATHBY

Ancient parish in Leath ward, Cumberland. CP enlarged by absorbing Edenhall CP 1934.

Acreage: 2,012 acres [814 ha], including Langwathby Moor (494 acres [200 ha]), enclosed 1850. ***Population:*** estimated at 210 in 1688; rising from 226 in 1801 to 346 in 1861; temporary surge to 417 by 1871 when railway under construction; then falling to 287 by 1931 (last census year before boundary change). Population of enlarged CP stood at 575 in 1951 and increased to 748 by 2001. ***Landownership:*** *manor of Langwathby* was granted to Henry son of Swein early 12th century but became part of honour of Penrith (q.v.) from 13th century. ***Economy:*** predominantly agricultural. Race course on Langwathby Moor 17th century. Station of Settle-Carlisle line of Midland Railway opened 1876. Feed mill/biofuel depot and chicken processing plant near railway from late 20th century. ***Places of worship:*** parish church of St Peter (united with Edenhall from 1380) rebuilt on site of medieval predecessor in 1718; extended 1836. Wesleyan Methodist chapel built 1860; enlarged 1900; still in use. ***Schools and other institutions:*** school (largely for girls) by c.1830. Parish school built 1867; replaced by new building on new site 1966; now Langwathby CE Primary School. Lending library established 1841. Reading room/public hall built 1904 (closed and converted to private house). Village hall opened 1931.

LAZONBY

Township in Lazonby parish, Leath ward, Cumberland.

Acreage: 8,378 acres [3,360 ha]. Commons in Lazonby parish enclosed under Inglewood enclosure award 1819. ***Population:*** rising from 320 in 1801 to 570 by 1861; temporary surge to 1,123 in 1871 when railway under construction. In range 650-720 from 1881 to 1961; then rising to stand at 980 in 2001. ***Landownership:*** *manor of Lazonby* held by Stutevilles in 12th century, descending with barony of Burgh (q.v.) to Dacres; forfeited on attainder of Leonard Dacre 1570; restored to Sir Francis Leonard 1652; sold to Sir Christopher Musgrave of Edenhall 1716, in whose family it remained until early 20th century. ***Economy:*** farming; quarrying (millstone quarry recorded late 17th century and mid-19th). Auction mart established by railway station late 1870s. Later 20th century developments included establishment of bakery, toy-making workshop and pottery. ***Places of worship:*** medieval parish church of St Nicholas; replaced by present building 1864-6. Primitive Methodist chapel built 1847; replaced 1898, now closed. Wesleyan Methodist chapel built 1850; still in use. ***Schools and other institutions:*** vestry school from 1748; two day schools by c.1830. Church school rebuilt on new site 1863; enlarged 1907; now Lazonby CE School. Reading room and library established 1859; enlarged as village hall 1928. Swimming pool opened 1964.

LEECE *see* Aldingham

LEVENS

Township in Heversham parish, Kendal ward, Westmorland.

Acreage: 3,517 acres [1,423 ha] including 103 acres [42 ha] in small detached portions; 1,751 acres [709 ha] of commons, largely mossland and marsh in Lyth Valley, enclosed 1815. ***Population:*** rising from 442 in 1801 to 993 in 1841; then standing at around 950 for remainder of 19th century; declined to 708 in 1911; recovered to 844 by 1951; then rising from 1970s to stand at 1,007 in 2001. ***Landownership:*** *manor of Over Levens*, including Levens Hall, held from 12th century by de Redman family, who sold it 1562 to Alan Bellingham; purchased 1688 by Colonel James Grahme from trustees of Alan Bellingham (d. 1690). Passed by marriage to Howard family and on marriage of Hon. Frances Howard to Richard Bagot 1783 to Bagots, in which family it descended to present. Levens Hall rebuilt by Bellinghams in late 16th century; formal gardens with famed topiary laid out for James Grahme 1692-7. *Manor of Nether Levens* was held by de Levens family from 12th century; sold to Preston family before 1452, and to Edward Wilson of Dallam Tower 1694. ***Economy:*** predominantly agricultural; peat cutting on Lyth Valley mosses; tourism (notably Levens Hall) from 20th century. ***Places of worship:*** chapel of St John the Evangelist built 1826-8; spire added 1831. Wesleyan Methodist chapel at Beathwaite Green, converted from house 1795; replaced by new chapel 1892; still in use. ***Schools and other institutions:*** girls' school recorded 1810 and boys' school c.1820. New boys' school built 1824; closed 1907 when boys, infants and girls united on site of current Levens CE Primary School, part of which was original girls' school. Village Institute (former barn) bought by public subscription 1902 in celebration of coronation of Edward VII.

LIDDEL (barony) *see* Kirkandrews Moat Quarter

LINDAL & MARTON *see* Dalton-in-Furness

LINDALE *see* Allithwaite, Upper

LINSTOCK *see* Stanwix

LITTLE MUSGRAVE *see* Musgrave

LONG MARTON

Ancient parish in East ward, Westmorland, including townships of Long Marton, Brampton and Knock.

Acreage: 6,947 acres [2,811 ha], distributed between constituent townships thus: Long Marton: 1,757 acres [711 ha]; Brampton: 1,628 acres [659 ha]; Knock: 3,562 acres [1,442 ha]. Brampton Common enclosed 1772; commons in Long Marton (c.500 acres [c.200 ha]) and stinted pasture called Marton Park (c.500 acres [c.200 ha]) enclosed 1809; Knock Common (3,181 acres [1,287 ha]) enclosed 1823. ***Population:*** rising from under 600 in early 19th century to lie in range 750-820 from 1830s to 1871; fell back to 585 by 1901; in range 600-700 across 20th century; stood at 704 in 2001. ***Landownership:*** *manor of Marton* held by Veteriponts in 13th century, passing to Cliffords, then to Holland family. Held by Gray family in 15th century but in Crown hands by 1526-7; subsequently passed to Lowthers. *Manor of Brampton:* held by barons of Greystoke (q.v.) from early 14th century until 1422, when it passed to Lancaster family. Divided between three co-heiresses in 16th century; one third subsequently acquired by earl of Thanet; remainder sold to tenants. *Manor of Knock Shalcock:* held by Boyville family in 14th century, later passing to Rookby, Soulby and Lancaster families before coming to lords of barony of Appleby (q.v.) by 19th century. ***Economy:*** Long Marton and Brampton townships predominantly agricultural. Lead mining in Knock township from 17th century; smelt mill by Knock Ore Gill from before 1739 to c.1785. Quarrying and lime-burning on Knock Fell 19th century. Tourism from later 20th century. ***Places of worship:*** medieval parish church of St Margaret and St James; restored 1880. Anglican services held at Knock school in 1880s. Mission room built 1906. Wesleyan Methodist chapels at Long Marton (built 1818; no longer in use), Espland Hill, Brampton (built 1834; rebuilt 1866; closed by 1998) and Knock (built 1840; rebuilt 1873). ***Schools and other institutions:*** parish school near church endowed 1824; built by subscription 1833; enlarged 1900; replaced by modern building in village; now Long

Marton Community School. School in Knock recorded 1885. Former railway station used by North Salford High School, Manchester, as centre for school holiday outings in 1980s; sold 1993. Library established 1858. Long Marton Parish Institute built 1893-4; still in use as village hall. Brampton Institute and Parish Reading Room built 1922.

LONGSLEDDALE

Chapelry and township in Kendal parish, Kendal ward, Westmorland.

Acreage: 6,734 acres [2,725 ha], including fell commons of Sleddale Forest (1,012 acres [410 ha]), enclosed 1849, and Sadgill Common (1,990 acres [805 ha]), enclosed 1868. ***Population:*** downward trend since 1800, falling from 187 in 1801, to 144 in 1901, 89 in 1951, and 73 in 2001. Temporarily boosted by construction of Haweswater aqueduct in 1920s and 1930s (stood at 271 in 1931). ***Landownership:*** Leyburn and Thornburgh families prominent landowners from 13th century. James Leyburn sold extensive property in Longsleddale to Thomas Bellingham 1579 and William Thornburgh conveyed his property there to Sir James Bellingham 1629. Descended thereafter with Levens (q.v.). ***Economy:*** predominantly hill farming. Slate quarrying: Wren Gill quarry open by 1728; closed c. 1847; reopened briefly in late 19th century; last worked by Italian prisoners of war during Second World War. Smaller quarry at Stockdalebank. Woollen textiles: two fulling mills in 1627. Bobbin mill at Sleddalefoot Mill (on boundary with Strickland Roger) in mid-19th century. ***Places of worship:*** chapel of ease recorded 1571; rebuilt and made parochial 1712; present church of St Mary built 1863; renovated 1903. ***Schools and other institutions:*** school taught in chapel by curates in early 18th century. School attached to chapel built 1717. New school built opposite church 1863; closed 1946 and became village hall.

LONGTOWN *see* Arthuret

LORTON

Township in Brigham parish, Allerdale above Derwent ward, Cumberland.

Acreage: 5,318 acres [2,152 ha], excluding 170 acres [69 ha] at Swinside, a detached portion of Buttermere transferred to Lorton 1887. Commons (3,867 acres [1,566 ha]) enclosed 1835. ***Population:*** rising from 298 in 1801 to peak of 456 in 1861; thereafter declining to 233 in 1971; stood at 250 in 2001. ***Landownership:*** part of forest of Derwentfells in honour of Cockermouth. *Manor of High Lorton* granted to Carlisle Priory before 1158; transferred to Dean and Chapter of Carlisle after Dissolution. *Low Lorton* held by Thomas Mariscal by 1230; divided into thirds by 1305, one of which was held by Winder family from before 1397 to 1699, becoming Lorton Hall estate, which was acquired by Joshua Lucock (Bragg) 1800 and greatly developed; sold to Dixons of Rheda 1881, who substantially rebuilt Lorton Hall 1889-90; estate sold 1947. ***Economy:*** predominantly agricultural until 19th century: vale of Lorton noted for its fertility in 17th century. Woollen textile industry from medieval period: fulling mill at Tenters built c.1480. Brewery established at High Lorton in 1820s by Jennings family (moved to Cockermouth 1887). Flax thread spinning on former fulling mill site at Tenters from 1830s to early 20th century. Forestry and tourism from mid-20th century. ***Places of worship:*** medieval chapel of ease (chaplain recorded 1198) dedicated to St Cuthbert, had become parochial by 16th century; rebuilt 1807-9. Wesleyan Methodist chapel, High Lorton, built 1840; closed and sold late 20th century. ***Schools and other institutions:*** curate serving as schoolmaster recorded 1598; school taught in chapel in early 18th century; endowed c.1710. School built in High Lorton 1809; enlarged 1859 and 1895; now Lorton Primary School. Former malthouse attached to brewery in High Lorton became village hall ('Yew Tree Hall') 1909.

LOWESWATER

Chapelry and township in St Bees parish, Allerdale above Derwent ward, Cumberland, including hamlet of Mockerkin.

Acreage: 9,415 acres [3,810 ha] including extensive commons on Loweswater fells (5,871 acres [2,376 ha]), enclosed 1865. ***Population:*** 294 in 1801, rising to 454 in 1831, when lead mines were active; then declining slowly to 174 in 1961; recovered slightly to 209 in 2001. ***Landownership:*** *manor of Loweswater* separated from barony of Egremont 1230 and known in 14th century as manor of 'Balnes' (i.e. Bowness, from manor house and park beside Crummock Water). Granted to Alice de Lucy wife of Alan de Multon in 1230 partition and descended with Lucy share of honour of Cockermouth (q.v.) to earls of Northumberland. *Mockerkin and Sosgill* descended with honour of Cockermouth but manor of Loweswater (including Thackthwaite) sold by Crown to Richard Robinson 1546. Passed via Thomas Stanley to Anthony

Patrickson between 1583 and 1593, who sold to Sir Wilfrid Lawson of Isel 1624, in whose family it remained until purchased by John Marshall of Leeds 1814; sold to The National Trust and Balliol College 1937. ***Economy:*** predominantly agricultural; sheep fair held in September by mid-19th century. Woollen industry (fulling mills recorded from 14th century). Iron forge recorded 1305, perhaps near Scale Force where small-scale iron mining in 19th century. Lead mining at several locations in 19th century, including Nether Close (opened 1829; closed 1841) and Whiteoak (from 1860s to c.1891). Tourism of growing importance since late 18th century. ***Places of worship:*** chapel of ease recorded c.1125; raised to parochial status 1281; dedicated to St Batholemew; rebuilt 1827-9; altered 1884. ***Schools and other institutions:*** school taught by curate in mid-18th century. School on Mockerkin common built and endowed 1782; closed c.1890. School at Loweswater built 1780; replaced 1839 by larger school on new site given by lord of manor; closed 1948, becoming village hall.

LOWICK

Chapelry in Ulverston parish, Lonsdale hundred, Lancashire North of the Sands.

Acreage: 2,261 acres [915 ha], including Lowick Common (305 acres [123 ha]). Settled area divided between hamlets of Lower Lowick (544 acres [220 ha]) and Upper Lowick (1,412 acres [571 ha]). ***Population:*** rising from 278 in 1801 to 468 in 1861, then falling back to 279 in 1901 and declining to 224 in 2001. ***Landownership:*** granted to Robert Towers in 12th century and descended in Towers family until 15th century when Isabel daughter of William Towers married John Ambrose. Descended in Ambrose family until 1684 when bequeathed to John Latus from whom it descended via Blencowe, Everard and Gaskarth families to Montagu family who held it in later 19th century. ***Economy:*** farming; small-scale slate quarrying; basket-making in 19th century (continued into 20th). Tanyards at Moss and Lowick Green mid-19th century, latter becoming Lowick Tannery which continued into 20th century. Spade forge at Lowick Green built 1776; closed 1906, then used as saw mill and basket workshop. ***Places of worship:*** chapel of ease recorded 1292; rebuilt 1817; replaced by St Luke's Church, built on same site 1885. ***Schools and other institutions***: school taught by curate from 1696. Parish school near church replaced by Lowick School built on edge of common 1857; closed 2004; now outdoor activities centre. Old school by church now Lowick Community Hall.

LOWSIDE QUARTER

Township in St Bees parish in Allerdale above Derwent ward, Cumberland, containing villages of Braystones, Coulderton, Middletown and Nethertown. Part of Lowside Quarter CP was transferred to St John Beckermet CP 1934.

Acreage: 1,962 acres [794 ha] before boundary changes. Commons enclosed under Egremont enclosure award 1783; other small patches of waste enclosed 1859. Middletown outfield was enclosed 1794. ***Population:*** fluctuated in range 250-350 across 19th century, then rose to 624 in 1911; stood at 533 in 1931 (last census year before boundary change). ***Landownership:*** 'Coulderton' and park attached to Egremont castle retained in demesne by lords of barony of Egremont (q.v.) until divided between heiresses 1338, leading to creation of three villages of 'Over Coulderton' (now Coulderton); 'Middle Coulderton' (now Middletown) and 'Nether Coulderton' (now Nethertown). ***Economy:*** farming; water-powered industries on River Ehen: paper mill by 1755 (2 paper mills by 1774); iron foundry at Low Mill established 1749; flax mill there mid-19th century. Furness Railway built 1850, with stations at Braystones and Nethertown. Coastal tourism: Braystones described as 'health resort' by 1911, with summer bungalows on shore there and at Nethertown by 1910; caravan sites developed later at Nethertown and Braystones. ***Places of worship:*** none. ***Schools and other institutions:*** Board school at Nethertown by late-19th century; closed. Tower at Braystones, erected 1897 for Queen Victoria's Diamond Jubilee, used as museum.

LOWTHER

Ancient parish in West ward, Westmorland, including townships of Hackthorpe, Lowther, Melkinthorpe and Whale.

Acreage: 3,674 acres [1,487 ha], divided between constituent townships thus: Hackthorpe: 846 acres [342 ha]; Lowther: 1,667 acres [675 ha]; Melkinthorpe: 333 acres [135 ha]; Whale: 828 acres [335 ha]. Melkinthorpe Moor was enclosed 1810. ***Population:*** 373 in 1801; inflated to 560 in 1811 and 599 in 1821 during rebuilding of Lowther Castle. Stood at 494 in 1831 and 1851; then declined to range 400-450 across later 19th and early 20th century. Further decline to 307 by 1971, then rose again to stand at 402 in 2001. ***Landownership:*** *manor of Lowther* held in divided ownership in 13th and early 14th century. Sir Hugh de Lowther acquired one quarter of manor

1315; by 1422 Sir Robert Lowther held all of it and it descended through Lowthers, latterly earls of Lonsdale, to present. Lowther Castle rebuilt 1691-2 and again from 1806; uninhabited since 1936 and roof removed 1957. *Manor of Hackthorpe* held by Strickland family between 1361 and 1535, when purchased by Lowthers. *Manor of Melkinthorpe* held successively by Melkinthorpe, Musgrave, Fallowfield and Dalston families until sold to Sir John Lowther. *Manor of Whale* held by Whale family and later purchased by Lowthers. ***Economy:*** dominated by Lowther estate and castle. Farming; lime-burning and tile works at Hackthorpe mid-19th century; forestry (especially after 1950); tourism (bird of prey centre opened 1992; Lowther Castle gardens opened to public 2011). Attempts to establish carpet manufactory in former college at Newtown and linen factory in Lowther new village (laid out in 1760s) met with only limited success. ***Places of worship:*** medieval parish church of St Michael; rebuilt 1686, retaining medieval arcades; restored 1814-27 (tower altered 1824). Lowther family mausoleum in churchyard built 1857. ***Schools and other institutions:*** grammar school endowed 1638; school built 1640; rebuilt on new site near Hackthorpe in early 19th century; now Lowther Endowed Primary School. Girls' school endowed 1740; by 1817 income paid to schoolmistresses at Hackthorpe and Whale. College for gentlemen's sons established by Viscount Lonsdale 1697; closed 1740. Village hall (Lowther Parish Hall), Hackthorpe, built 1932.

LUPTON

Township in Kirkby Lonsdale parish, Lonsdale ward, Westmorland.

Acreage: 3,524 acres [1,426 ha], including 1,135 acres [459 ha] of commons, enclosed 1848. ***Population:*** in range 200-300 for most of 19th century with surge to 325 in 1891 associated with construction of Thirlmere aqueduct; declining across 20th century to 140 in 1981; stood at 165 in 2001. ***Landownership:*** *manor of Lupton* had been divided by c.1200; seems to have been consolidated in hands of Redmans of Levens from 14th to 16th centuries. Held by Sir Richard Hutton of Goldsborough 1639; sold to Sir Christopher Musgrave of Edenhall 1681; acquired by Lowthers in early 19th century. ***Economy:*** predominantly agricultural; small-scale quarrying and lime-burning mid-19th century. ***Places of worship:*** church of All Saints built 1868. ***Schools and other institutions:*** school shared with Hutton Roof (q.v.) from 1773. Lupton School built 1871; closed 1980.

MALLERSTANG

Chapelry and township in Kirkby Stephen parish, East ward, Westmorland.

Acreage: 8,380 acres [3,391 ha], consisting largely of unenclosed fell commons totalling 6,444 acres [2,608 ha]. ***Population:*** in range 200-250 for most of 19th and early 20th centuries (temporarily inflated to 585 in 1871 by influx of workers constructing Settle-Carlisle railway); declined across later 20th century to stand at 87 in 2001. ***Landownership:*** *Mallerstang chase* was hunting forest of barons of Westmorland and descended with barony (see Appleby). Pendragon Castle, guarding the valley, built by barons probably in 12th century; restored by Lady Anne Clifford 1660; dismantled 1685. ***Economy:*** pastoral farming (11 vaccaries in early 14th century); mining (lead, copper, tin, coal) in mid-19th century (had ceased by 1885); limestone quarrying and lime-burning; sandstone quarrying. ***Places of worship:*** chapel of ease at Outhgill (modern dedication St Mary); rebuilt by Lady Anne Clifford 1663; thatched until restoration 1768. Wesleyan Methodist chapel, also at Outhwaite, built 1878; closed late 20th century. ***Schools and other institutions:*** school at west end of chapel from 17th century (charitable endowment by Lady Anne Clifford); schoolroom built adjacent to chapel 1826; replaced by Board school in Outhwaite 1877; closed c.1969 and later used briefly as village hall; then sold. Village hall and reading room, the Travers Institute (named after Travers McIntire (d. 1920), vicar), built c.1920; closed 1970s; later demolished.

MANSERGH*

Township in Kirkby Lonsdale parish, Lonsdale ward, Westmorland.

Acreage: 2,668 acres [1,080 ha], including 950 acres [384 ha] of common land, enclosed 1848. ***Population:*** rising from 134 in 1801 to 247 in 1871, then declining steadily to 95 in 1971; rising again to 141 in 2001. ***Landownership:*** Mansergh was hamlet of Barbon in medieval centuries; deemed a manor in its own right by 1587, when sold by Marmaduke Redmayne (whose family had held it from early 15th century) to Henry Brabyn. By 1650s was in hands of Henry Ward, who sold it to Thomas Godsalve (who had bought Rigmaden 1661) before 1664. Descended in Godsalve family to Margaret (d. 1781), wife of Thomas Mawdesley of Mawdesley. Sold to John Satterthwaite 1781, whose trustees sold it to Christopher Wilson of Abbot Hall, Kendal, 1825 (who rebuilt Rigmaden Park

1825-8); descended in Wilson family to present. ***Economy:*** largely agricultural; small-scale quarrying 19th century. Races held annually in early/mid-19th century (memory preserved in name Racecourse Allotment). ***Places of worship:*** chapel of ease, dedicated to St Peter, built 1726 by Jacob Dawson on his property at Nether Hall; rebuilt 1879-80. ***Schools and other institutions:*** schoolmaster recorded 1646. School near chapel built and endowed by Christopher Wilson of Rigmaden 1839; closed 1946; building later used as parish hall.

MANSRIGGS

Township in Ulverston parish, Lonsdale hundred, Lancashire North of the Sands. Combined with Egton with Newland and Osmotherley CPs 2003.

Acreage: 569 acres [230 ha]. ***Population:*** 64 in 1801, rising to peak of 73 in 1871, then falling to 36 by 1951; remained fairly stable for rest of 20th century, standing at 42 in 2001. ***Landownership:*** part of Neville manor of Ulverston (q.v.). ***Economy:*** predominantly agricultural. ***Places of worship:*** none. ***Schools and other institutions:*** none.

MARDALE *see* Shap

MARTINDALE

Chapelry and township in Barton parish, West ward, Westmorland.

Acreage: 8,024 [3,247 ha]. Part of commons (1,403 acres [568 ha], including 103 acres [42 ha] of open field) were enclosed 1824 but extensive unenclosed fells of Martindale Common (4,182 acres [1,691 ha]) remain open. ***Population:*** rising from 165 in 1801 to peak of 208 in 1851; then falling to c.120-130 across first half of 20th century; further decline to stand at 49 in 2001. ***Landownership:*** Martindale was hunting forest, renowned for red deer, and descended with manor of Barton (q.v.). ***Economy:*** hill farming. Fulling mill at Howtown re-used as bobbin mill 1820s to 1850s. Deer-stalking in Martindale deer forest from 19th century: shooting lodge (The Bungalow) built 1910 by earl of Lonsdale for visit of Kaiser Wilhelm II. Tourism increasingly important from later 20th century. Outward Bound centre at Howtown opened 2008. ***Places of worship:*** chapel of St Martin in Howgrane valley; recorded from 13th century; rebuilt c.1633; repaired 1833. Replaced by St Peter's church, built 1880-2; St Martin's remains in occasional use. Wesleyan Methodist chapel, dated 1924; now converted to holiday cottage. 'Chapel in the Hause' (foundations visible at Boredale Hause) is of uncertain origin and function. ***Schools and other institutions:*** school endowed 1657; in mid-19th century schoolmaster lived and taught at Cotehow. National school built nearby 1834; closed 1940s. Reading room built near school 1913; converted to residential use. Former bobbin mill at Howtown leased by parish from 1930s; restored as community centre 1982.

MARYPORT

New town and port founded in mid-18th century, in Cross Canonby parish, Allerdale below Derwent ward, Cumberland. Maryport UD, created 1894, embraced CPs of Ellenborough & Ewanrigg and Netherhall, which were abolished 1929 to create Maryport CP. For pre-urban history, see Cross Canonby and Ellenborough & Ewanrigg.

Acreage. 1,515 acres [613 ha] in 1929.

Origins and growth of the town. In 1749 Humphrey Senhouse (1705-1770) of Netherhall started work building harbour and new town at Ellenfoot, which was already small anchorage (described by Thomas Denton 1688 as 'a kinde of port for small vessells'). Senhouse's motive was to gain share of coal trade, then dominated by Lowthers' port of Whitehaven. He called his new town 'Mary Port' in honour of his wife, Mary, daughter of Sir George Fleming, bishop of Carlisle. Grid of streets laid out, with over 100 rectangular plots being sold 1749-69. By 1770s Maryport contained c.340 families, perhaps 1,500 people. Town was founded on coal, iron and shipping. Coal mining expanded rapidly in Ellenborough during later 18th century. Iron blast furnace established beside River Ellen by group of Whitehaven merchants 1754; closed 1783. By 1829 town's industries also included three shipbuilding yards and manufactories of cotton and linen checks and sail cloth.

By 1801, combined population of Netherhall and Ellenborough & Ewanrigg townships stood at 3,403, from which it grew steadily to over 12,000 by 1891, as town boomed. Port expanded after control passed from Senhouse family to board of trustees 1833: new dock (basin of later Elizabeth Dock) had been constructed by 1837 and Maryport registered its own vessels from 1838. Arrival of Maryport and Carlisle railway in 1840 enabled port to increase coal traffic considerably as inland collieries expanded in its hinterland. Heavy industry continued. Two iron furnaces: Maryport Haematite Iron Co. opened 1868; closed 1882; Solway Iron Co. opened 1870; closed 1894; reopened until early 20th century. Rolling mills, iron and brass foundries and steam saw mills. Infrastructure of port grew: new 'floating dock', Elizabeth Dock, opened 1857, and Senhouse Dock in 1884, allowing larger vessels to trade. Shipbuilding expanded. By 1901 several earlier industries had ceased, including manufacture of glass, salt, cotton, paper, and lead pencils, but port, ironworks and shipyards continued. Other manufacturing industry included flour-milling, sail-making, tanning and brewing.

Maryport's fortunes declined across early and mid-20th century: coal and iron industries collapsed (metal and mining industries employed 1,390 men in 1921; fewer than 100 by 1932) and port was overshadowed by Workington. Population had dropped to under 11,000 by 1921. Most of collieries in vicinity had closed by 1950s – over 61% of town's insured population was unemployed in early 1930s. Coal shipments dwindled from over 200,000 tons p.a. in early 1920s (when Maryport was leading coal port in Cumberland) to around 30,000 tons by late 1930s. Port closed to traffic 1961; Senhouse Dock later converted into marina for pleasure craft. Traditional industries replaced by light industry after Second World War, on Solway and Glasson industrial estates. Town expanded in Netherton area, south of River Ellen, later in 20th century, leading to population growth to 12,393 in 1961. In 2001, population stood at 11,275.

Places of worship. Church for new town, St Mary's, a chapel of ease to Cross Canonby, founded by Humphrey Senhouse 1760; extended 1837; tower rebuilt 1847; nave rebuilt 1890-2. Christ Church, King Street, built 1872; closed 2013. Nonconformist places of worship proliferated: Quaker meeting house built 1772; rebuilt 1810; closed 1912. Presbyterian chapels in John Street ('English' chapel, built 1776; enlarged 1811; closed 1888) and Crosby Street (United Presbyterian, built 1831): they combined on Crosby Street site 1888. Wesleyan Methodist chapel, Well Lane, built 1806; replaced by new chapel at Back Brow, 1864; destroyed by fire c.1973; replaced by St Mark's Methodist Church, Kirkby Street, opened 1973. Primitive Methodist chapel, Kirkby Street, built 1839; enlarged 1870; closed c.1932. Baptist chapel, High Street, built 1834; replaced by larger building on Station Street, 'the Baptist cathedral of the North', opened 1891; demolished 1951; in turn replaced by Trinity Baptist Church, built 1968. Baptist chapel, Furnace Road, built 1861; now Gospel Hall. Congregational Church, Lawson Street, built 1883. Roman Catholic Church of St Mary and St Patrick, Crosby Street, built 1844; enlarged 1881-2. Cemetery, to north of town, opened 1856.

Schools. Town could boast 14 schools in 1818. School of Industry recorded 1829. National school founded 1840; moved to Eaglesfield Street 1847. British school, High Street, built 1845; replaced by new school, North Street, 1883. Church of England school built 1866, enlarged 1891 and 1896. Infant school, George Street, built 1876. 19th-century schools replaced by Maryport CE Junior School in modern building on Camp Road. Roman Catholic School built 1871; replaced by new Catholic Primary School, Ennerdale Road, Ellenborough, 1963. Netherhall School built as secondary modern school on edge of town 1955, becoming mixed comprehensive 1968 and specialist sports college 2003.

Other institutions. Customs house and harbour office built 1838; court house recorded 1847. Mechanics' Institute, library and reading room, Catherine Street, founded 1846. Public halls included Oddfellows' Hall, Eaglesfield Street, built 1841; The Athenaeum, built 1855; Cooperative and Central Halls by 1901. Senhouse Roman Museum opened 1990 in former Naval Reserve training battery (built 1885).

MATTERDALE

Chapelry and township in Greystoke parish, Leath ward, Cumberland. CP enlarged by absorbing Watermillock CP 1934.

Acreage: 7,288 acres [2,949 ha], including extensive areas of unenclosed fell grazings on Matterdale Common (2,605 acres [1,054 ha]), regulated 1882 under Commons Act 1976, when a further 2,500 acres [1,012 ha] of mossland and stinted pastures were enclosed. ***Population:*** 297 in 1801 rising steadily to 426 by 1871, then gradual decline to 229 by 1931 (last census year before boundary change). ***Landownership:*** part of barony of Greystoke (q.v.). ***Economy:*** predominantly hill farming; some quarrying and mining in 19th century; tourism from later 20th century. ***Places of worship:*** chapel of ease, licensed as parochial chapel 1573; probably rebuilt c.1685; thatched until 1848 when tower added. ***Schools and other institutions:*** school built 1716 and endowed by Rev. Robert Grisdale 1722; replaced by new larger building on adjacent site 1907; closed; both buildings used for community purposes.

MAULDS MEABURN *see* Crosby Ravensworth

MEALSGATE *see* Bolton (Cumberland)

MEATHOP AND ULPHA

Township in Beetham parish, Kendal ward, Westmorland.

Acreage: 1,991 acres [806 ha], growing to 3,147 acres [1,274 ha] after reclamation of marshes on Kent estuary in later 19th century. ***Population:*** in range 80-90 across first half of 19th century. Inflated by presence of isolation hospital, reaching peak of 328 in 1921; then declining to 143 by 2001. ***Landownership:*** part of manor of Witherslack (q.v.). ***Economy:*** agriculture; small-scale quarrying and lime-burning in 19th century; peat-digging on Foulshaw, Meathop and Ulpha Mosses; tourism (caravan parks) from later 20th century. ***Places of worship:*** none. ***Schools and other institutions:*** Westmorland Consumption Sanatorium (isolation hospital) at Meathop opened 1891; closed 1991; converted to private apartments.

MELMERBY

Ancient parish in Leath ward, Cumberland. Absorbed into Ousby CP 1934.

Acreage: 5,362 acres [2,170 ha], including extensive commons on Pennine edge. Melmerby Moor, Turn Moor, Mires common and Melmerby Low Fell (1,336 acres [526 ha]) enclosed 1858. ***Population:*** estimated at 270 in 1688; rising from 223 in 1801 to peak of 329 in 1841; then downward trend to 175 in 1931 (last census year for which separate figure available). ***Landownership:*** *manor of Melmerby* held in 13th century by Odard of Wigton, descending to his heirs, the Parvings, in 14th century. By 1380 had passed to Threlkeld family, who held it until death of Lancelot Threlkeld (d. 1673); it then passed through his granddaughter, Elizabeth Thirkeld, to Pattenson family, who held it until sold by John Edward Pattenson 1846. *Manor of Gale* held by Huttons of Hutton Hall, Penrith; then by Holme family in early 19th century. ***Economy:*** hill farming; small-scale quarrying in 19th century. Lead mining on Melmerby Fell by 18th century; had ceased by 1900. Bakery in village opened 1976. ***Places of worship:*** medieval parish church of St John; restored 1848 and 1895. Wesleyan Methodist chapel built 1848, closed 1975. ***Schools and other institutions:*** schoolhouse but no school taught 1818 'for want of the proper means of support'; school in operation by c.1830. New school opened 1862; closed 1974. Reading room founded 1882. Village hall rebuilt 2004 on site of earlier building.

MIDDLESCEUGH AND BRAITHWAITE

Township, mostly consisting of a detached portion of Carlisle St Mary's parish, Leath ward, Cumberland. Absorbed into Skelton CP 1934.

Acreage: 2,092 acres [847 ha], of which Middlesceugh and High Braithwaite (1,757 acres [711 ha]) formed detached portion of St Mary's parish, Carlisle, while Low Braithwaite lay in Hesket parish. Common land (338 acres [137 ha]) enclosed 1819. ***Population:*** in range 175-200 in first half of 19th century (peak 221 in 1821), then downward drift to 143 in 1931 (last census year for which separate figure available). ***Landownership:*** *manor of Middlesceugh* purchased from Allonby family by Sir Richard Fletcher (d. 1637) and descended thereafter with Hutton-in-the-Forest (q.v.). ***Economy:*** predominantly agricultural; quarry at Middlesceugh in mid-19th century. ***Places of worship:*** none. Inhabitants said to baptise, marry and bury at Sebergham in 17th century, but attended Ivegill chapel and buried their dead at Hesket-in-the-Forest in 19th. ***Schools and other institutions:*** none.

MIDDLETON

Chapelry and township in Kirkby Lonsdale parish, Lonsdale ward, Westmorland.

Acreage: 7,276 acres [2,945 ha], including 3,606 acres [1,459 ha] of unenclosed hill commons on Middleton Fell. ***Population:*** 288 in 1695; in range 250-300 for most of 19th century (temporary surge to 366 in 1861 due to railway construction); declining across 20th century to stand at 109 in 2001. ***Landownership:*** *manor of Middleton* exchanged mid-12th century by Gospatric son of Orm with William de Lancaster for Workington and Lamplugh. Held by Middleton family from 14th century for many generations; John Middleton enfranchised tenants in early 17th century. ***Economy:*** predominantly agricultural. ***Places of worship:*** chapel of ease recorded 1633; rebuilt 1634 and made parochial 1671. Rebuilt 1813 and again 1878-9; present church dedicated to Holy Ghost. ***Schools and other institutions:*** school taught in chapel 1633; endowment made 1670. Schoolhouse on south side of chapel built by 1829; replaced by new school 1899; closed 1952 and converted to dwelling. Hut for Women's Institute built near school 1921; derelict from 1980s; demolished c.2000.

MIDGEHOLME

CP created 1883 from extra-parochial moorland and detached portions of Banks and Naworth townships in Eskdale ward, Cumberland.

Acreage: 5,155 acres [2,086 ha], comprising following elements: Midgeholme (631 acres [255 ha]), a detached portion of Banks township; Midgeholme and Halton Lee West Fell (720 acres [291 ha]), an area of extra-parochial moorland; Bruthwaite Forest and Tindale Fell (3,803 acres [1,539 ha]), a detached portion of Naworth township. ***Population:*** in range 80-120 in mid-19th century, rising to peak of 557 in 1891; then declining across 20th century, particularly after 1930, to 47 in 1981; stood at 67 in 2001. ***Landownership:*** Tindale Fell given by Adam de Tindale to Hexham Priory, passing at Dissolution to barons of Gilsland, with which it descended thereafter. ***Economy:*** moorland grazing; extractive industries 19th and early 20th centuries, linked by tramways to Brampton & Hartleyburn Railway. Coalmines: colliery at Midgeholme recorded 1628; large-scale mining at from 1820s to later 19th century; Venture Drift (Howgill Colliery) reopened 1919 to 1934; several pits at Midgeholme reopened 1920s to 1953. Coke ovens at Midgeholme in later 19th and early 20th century. Quarries and lime works at Forest Head 19th century. Tindale zinc spelter works, in operation 1845-95; reopened for extraction of zinc residues 1928-31. ***Places of worship:*** Anglican services held in schoolroom, late 19th century. Primitive Methodist chapel (near Clesketts) opened 1870 (replacing tent chapel, established 1867); closed and sold 1932. ***Schools and other institutions:*** school at Midgeholme built 1885; transferred to Northumberland County Council 1925, becoming Hartleyburn First School; closed 1976. Reading room, recorded 1899 and 1946.

MILBURN

Chapelry and township in Kirkby Thore parish, East ward, Westmorland.

Acreage: 7,957 acres [3,220 ha], including extensive Pennine moorland of Milburn Forest (5,026 acres [2,034 ha]) which remains unenclosed common land. Milburn Fell Pasture (166 acres [67 ha]), enclosed 1857. ***Population:*** 237 in 1801, rising to peak of 348 in 1841; declining thereafter to 146 in 1981; stood at 171 in 2001. ***Landownership:*** *manor of Milburn* held by Patrick, earl of Dunbar early 14th century. Forfeited and granted to branch of Lancaster family, passing by marriage successively to Crackenthorpe (early 15th to mid-16th century), Sandford (mid-16th century to 1723) and

Honywood (from 1723) families, until sold by Philip Honywood to earl of Thanet 1780. Manorial rights sold to Corlands Minerals Ltd 1926. ***Economy:*** farming; quarrying and lime-burning in 19th century; coal and lead mining on Dun Fell and Milburn Forest. Worked by London Lead Co. in 19th century; main lead mine (with crushing mills) at Silverband on slopes of Dun Fell; closed after First World War. Silverband mine re-opened for barytes extraction 1939 to 1966 and 1980 to 1996. ***Places of worship:*** medieval chapel of ease of St Cuthbert recorded 1227; bellcote added 1894. Wesleyan Methodist chapel built 1834; closed 1988 and converted to dwelling. ***Schools and other institutions:*** schoolroom built on common; endowed 1790. New school built on village green 1851; now Milburn Primary School. Parish hall opened 1912, providing meeting room, supper room, reading room and library.

MILLOM

Industrial town in Millom parish, Allerdale above Derwent ward, Cumberland. Millom was UD from 1894 to 1934. For history of area before 1860, see Millom Rural.

Acreage. Millom UD, which covered much of Millom Below township, together with Haverigg (formerly in Chapel Sucken township), comprised 2,511 acres [1,016 ha].

Origins and growth of the town. Hodbarrow iron mine, which was to become 'the most productive haematite mine in the British Isles' developed in early 1860s. Annual output was c.120,000 tons in 1866; by 1880 it had risen to 343,000 tons; and by 1901 to c.400,000 tons. Iron works with two blast furnaces built 1866-7; number of blast furnaces had increased to six by 1874. Rapid influx of workers to mine and iron works was accommodated in a new town, formally inaugurated 1866. Combined population of Millom Above and Millom Below townships stood at around 900 until town of Millom was built but then rocketed: by 1871 population of Newtown and Holborn Hill had reached almost 3,000; by 1891 urban area contained 8,871 inhabitants and population of Millom UD peaked at 10,426 in 1901.

Employment in iron industry had begun to decline before First World War: by 1911 population of UD had fallen back to 8,612 and fell to a little over

7,000 by 1951; in 2001 it stood at 7,132. Iron-making continued until 1968 when Hodbarrow mine closed. Other 20th-century industrial activity included manufacture of nylon stockings from 1949, and tannery at Haverigg, making leather for shoe trade, established 1938; closed 1979. Tourism grew in late 20th century: flooded workings at Hodbarrow converted into nature reserve and marina.

Places of worship. Town church of St George built 1874-7, becoming separate ecclesiastical parish 1879. St Luke's church, Haverigg, built 1891. Primitive Methodist chapel at Holborn Hill built 1866. Wesleyan Methodist chapel, Queen Street, built 1872; closed 1993; replaced by adjacent former Sunday school building; still in use. Baptist chapel built 1884, replacing earlier chapel of 1867; still in use. Roman Catholic church built 1868; replaced by church of Our Lady & St James, built 1888. Salvation Army fortress, Nelson Street, built 1889; still is use. Bible Christian chapel, Newton Street; demolished. Spiritualist meeting house recorded 1938. Millom Community Church (Pentecostalist) established by 2011. At Haverigg: chapels for Baptists (opened 1869; short-lived as moved to Millom); Bible Christians (built 1873; demolished); Wesleyan Methodists (built 1878; demolished) and Primitive Methodists.

Schools. School Board formed 1876: built schools at Holborn Hill (c.1879 to 1970) and Lapstone Road (c.1876 to 1970; now Millom Infant School). Older schools closed on opening of Black Combe Junior School 1970. Catholic School in former church by later 19th century; replaced by new building, now St James Catholic Primary School, Lonsdale Road, 1971. Millom Institute (secondary school) built 1905; moved to new premises 1938 and became grammar school; enlarged 1950s to become core of Millom School, opened 1959. Board school at Haverigg by 1897; second school built early 20th century; now Haverigg Primary School.

Other institutions. Hodbarrow Mining Company hospital opened 1867; rebuilt 1888; closed 1934; demolished 1970s. Infectious diseases hospital, near Hodbarrow Pier, recorded 1897. Modern hospital (Millom Hospital), Lapstone Road. Market hall and local government offices built 1879; police station built 1894. Reading room and library built 1882. Library built 1887. Four public halls by 1901: Millom Public Hall (built 1873), Central Public Hall, Temperance Hall and Co-operative Hall. Recreation hall, St George's Road, built 1912 with theatre and cinema; now Millom Palladium Theatre and Studio. At Haverigg: reading room by 1897 and assembly rooms by 1901.

MILLOM RURAL

CP in Millom parish, Allerdale above Derwent ward, Cumberland, created 1894, covering rural parts of Millom Above and Millom Below townships, most of township of Chapel Sucken (including village of Kirksanton) and chapelry of Thwaites. Name changed to Millom Without 1934, when boundary changes occurred. This entry covers townships of Millom Above, Millom Below, Chapel Sucken and Thwaites.

MILLOM ABOVE and MILLOM BELOW. ***Acreage:*** Millom Below township (most of which became core of Millom UD – see Millom) covered 2,658 acres [1,076 ha]; Millom Above township covered 3,397 acres [1,375 ha]. Stinted pasture at Millom Mains (254 acres [103 ha]) enclosed 1824. Common marshland at Arnaby Marsh and Pool House Bank in Millom Above (totalling c.1,500 acres [c.600 ha]) enclosed 1824. ***Population:*** declining from 1,489 in 1901 to 1,132 in 1931 (last census before boundary changes). ***Landownership:*** Millom Castle was seat of lords of *seigniory of Millom*, which covered all land between rivers Esk and Duddon. Lords of Millom retained immediate lordship over much of Millom Above and Millom Below townships. Millom was held by Godard de Boyville in early 12th century, descending to Joan daughter of Adam de Boyville whose marriage to John de Hudleston (d. c.1252) brought estate to Hudleston family, with whom it descended to William Hudleston (d. 1745), whose daughter Elizabeth and her husband Sir Hedworth Williamson, sold it 1774 to Sir James Lowther. Manorial lordship remained with Lowthers into 20th century. ***Economy:*** charter for market and fair granted 1251, though no medieval urban growth followed; market was described as 'long discontinued' in 1777. Early iron mining at Waterblean (and Hodbarrow from c.1840). Charcoal burning (said to have destroyed woodland in deer park) and iron making (recorded in name 'Furnace Beck') by later 17th century. Small harbour at Borwick Rails, from which slate and corn were shipped in mid-19th century. Quarrying for roadstone at Ghyll Scaur Quarry (in Millom Park) from the 1930s; visitor centre (Millom Rock Park) established there late 20th century. ***Places of worship:*** medieval parish church of Holy Trinity (next to Millom Castle), probably pre-Conquest foundation (Viking Age sculptural fragments preserved in fabric); restored 1930. Mission room at Hill built c.1900. Wesleyan Methodist chapel at Hill by 1901; now village hall. ***Schools and other institutions:*** school in Millom Below endowed by Joseph Hudleston (d. 1700) but endowment had been lost by 1818. School in Millom Above endowed by William Atkinson of Thwaites 1811. Castle school (Millom and Hodbarrow

National School) by Holy Trinity church, built c.1858; closed c.1954. Elementary school at Hill, built 1883.

CHAPEL SUCKEN (including Kirksanton). ***Acreage:*** 2,371 acres [960 ha], excluding detached portion at Stainton, near Waberthwaite (713 acres [289 ha]), transferred to Muncaster CP 1886. Haverigg village was transferred to Millom (q.v.) 1894. Open fields and stinted pasture at Kirksanton Haws (556 acres [225 ha]) enclosed 1803-4. ***Population:*** separate figures only available 1821-1871. Population stood at around 250-300 until 1860s; then rose to 648 by 1871, reflecting start of Haverigg's growth into an industrial community as iron mining expanded. ***Landownership:*** manorial lordship descended with seigniory of Millom (see above). ***Economy:*** agricultural until 19th century. Lime-burning before mid-19th century. Iron mining north of Kirksanton 1879 to 1895; brewing and mineral water manufacture at Bank Springs Brewery, Kirksanton, by 1901. Airfield, RAF Millom, built on Kirksanton Haws 1939; closed to flying 1945; Haverigg wind farm built on site of airfield, 1992. ***Places of worship:*** place-name 'Kirksanton', recorded in Domesday Book, implies presence of early church: 'chapel of Kirksanton' recorded 1228 (no later records survive). Christ Church, Kirksanton, consecrated 1891; closed 1950. ***Schools and other institutions:*** HM Prison Haverigg opened 1967 on site of RAF Millom airfield training centre. Village hall at Kirksanton.

THWAITES. ***Acreage:*** 4,392 acres [1,777 ha], including 1,635 acres [662 ha] of unenclosed common on Thwaites Fell and 122 acres [49 ha] of common marsh at Lady Hall Marsh. ***Population:*** separate census figures only available 1821-1861: rising from 315 in 1821 to peak of 364 in 1851, falling back to 350 in 1861. ***Landownership:*** *manor of Thwaites* held by Thwaites family from 13th century; sold by John Thwaites of Ewanrigg to Sir John Lowther 1627; his son sold manor to Sir William Hudleston 1664 but regained possession 1667; it remained with Lowthers into 20th century. ***Economy:*** farming; lime-burning. Iron making at Duddon furnace, established 1736; closed 1867. Woollen mill complex (carding mill, weaving shed, dye works and fulling mill) at Hallthwaites from 1820s; closed 1935. Brushstock and handle mill, Duddon Bridge, mid-19th century. ***Places of worship:*** chapel of ease of St Anne built 1715 (consecrated 1725); rebuilt 1807; current church built 1852-4. ***Schools and other institutions:*** school built by inhabitants before 1722; replaced by Thwaites Board School; now Thwaites Primary School. Buckman Brow school for girls, endowed and built 1845 by Frances E. Millers of Duddon Grove; closed c.1924. Parish library established 1757 but lost by mid-19th century. Public hall at The Green by 1901; extended c.1940s; now Thwaites Village Hall.

MILNTHORPE

Market town and port in Heversham with Milnthorpe township, Heversham parish, Kendal ward, Westmorland. Milnthorpe became separate CP 1896.

Acreage: 1,125 acres [455 ha], including c.350 acres [c.140 ha] of commons enclosed under Heversham Enclosure Act 1803 ***Population:*** 1,099 in 1891 (first census year for which separate figure is available), rising steadily to stand at 2,106 in 2001. ***Landownership:*** see Heversham. ***Economy:*** small sea-port at Sandside, importing wines from France and other goods from Ireland, Isle of Man and west of England by 17th century, was principal port of Westmorland until 19th century. Four or five vessels berthed there by early 19th century. Weekly market established by charter 1334; lapsed 1920s. Noted in later 17th century for 'great' cattle fair in May. Iron forge (recorded 1671) had become paper mill by 1692. Several flax mills, wool carding mill, and two paper mills in 1829. Beela Mill, flax mill producing corn and flour sacks, rope and twine in mid-19th century; converted to Dobson's comb factory, 1868; still in operation. Arrival of railways killed port but fostered industries: London & North Western Railway (station at Ackenthwaite) opened 1846; closed 1963; Furness link (stations at Sandside and Heversham) opened 1876; closed 1963. Libby's dairy products factory opened c.1934; closed c.1995. Two annual fairs for sale of cattle in May and October, established c.1816; cattle fair at Milnthorpe station established 1849. Cattle fairs ceased 1929. Weekly market re-established 1966; by later 20th century Milnthorpe was focal village for surrounding area, with wide range of retailing and services. ***Places of worship:*** St Thomas' church built 1835-7; chancel added 1883. Independent chapel established 1820 in former cottages on Haverflatts Lane; taken over by Methodists late 1860s. New Wesleyan Methodist chapel, Beetham Road, opened 1904; closed 2014. Roman Catholic church of Christ the King, Haverflatts Lane, built 1970. ***Schools and other institutions:*** several dames' schools and girls' boarding schools in 19th and early 20th centuries. National school opened in former cottages on Milnthorpe Square 1819. New school on Mill Lane built 1834. St Thomas' Church School for Girls, Milnthorpe Square, opened 1867. Both closed when new county primary school (now Milnthorpe Primary School) built on Firs Road 1951. Secondary modern school opened on Milnthorpe playing fields 1951; moved to Haverflatts Lane c.1968; incorporated into Dallam Comprehensive School 1984. Gilbert Union workhouse at Ackenthwaite opened 1813; retained after formation of Kendal Poor Law Union 1836 and used for aged and infirm; became mental home and hospital 1918; closed 1960 and converted to housing. Kitching Memorial

Hospital opened 1881. Tattersall Almshouses, Church Street, built 1884. Market hall built 1843 with 'agricultural hall' for meetings upstairs; assembly hall added 1880; used as cinema 1920s-60s; premises bought by parish as Memorial Hall 1945. Kitching Memorial Institute, opened 1881, in memory of Dr John Kitching, London surgeon; contained library, magistrate's court, billiard and chess rooms; converted to Working Men's Institute 1921; closed and converted to flats 2007.

MOCKERKIN *see* Loweswater

MOORHOUSE *see* Burgh by Sands

MORESBY

Township in Moresby parish, Allerdale above Derwent ward, Cumberland.

Acreage: parish contained 2,193 acres [887 ha], of which 2,140 acres [866 ha] was in Moresby township and 52 acres [21 ha] in Parton (q.v.). Boundary changes resulted in small area ceded 1900 to Whitehaven MB (q.v.) and further 326 acres [132 ha] 1934. Common land on Moresby Moor (529 acres [214 ha]) enclosed 1763. ***Population:*** 371 in 1801 rising to 1,144 in 1891 (last census year before boundary changes). Moresby CP contained 1,280 inhabitants in 2001. ***Landownership:*** descended through de Moresby family from 12th century to death of Sir Christopher Moresby (d. 1499), when it passed successively to Pickering and Knyvett families, who sold it 1577 to William Fletcher. Descended in Fletcher family (William Fletcher built Moresby Hall c.1670) until sold to Sir John Brougham of Scales 1720, who sold to Sir James Lowther 1737. ***Economy:*** farming; coal mining by 17th century, continuing until 1961; open cast mining 1980s. Freestone quarrying in 19th century. Woollen textiles, recorded in name Walk Mill (i.e. fulling mill). ***Places of worship:*** medieval parish church of St Bridget on site of Roman fort of GABROSENTUM; rebuilt 1822; new chancel 1885. Primitive Methodist chapel at Moresby Parks built 1884; rebuilt on new site 1966; became community church 2011. Mission hall at Moresby Parks demolished 2009. Roman Catholic chapel 1932; closed by 2009. ***Schools and other institutions:*** Moresby Parks Board School built 1883; now Moresby Primary School. Reading room at Moresby Parks by early 20th century. Rosehill Theatre, founded by Sir Nicholas (Miki) Seker, opened 1959; extended 1988.

MORLAND

Township in Morland parish, West ward, Westmorland.

Acreage: 1,754 acres [710 ha]; common land on Morland Moor (413 acres [167 ha]) enclosed 1800. ***Population:*** 273 in 1801, rising to 426 in 1841, then gradually falling to 380 in 2001. ***Landownership:*** *manor of Morland* held of barony of Kendal. Part granted to Wetheral Priory 12th century, passing to Dean and Chapter of Carlisle at Dissolution. Remainder held by Windsor family 12th to 14th centuries; in late 16th century passed from Robert Bowes to Musgrave, Southaik and Backhouse families successively. ***Economy:*** farming. Quarrying and lime-burning in 19th century. Market and fair granted by charter 1362 but did not survive. Two linen mills in early 19th century, making coarse linen goods and tapes, threads and 'small wares'. Clothing company (Travelling Light) based in Morland 1980 to 2008. ***Places of worship:*** medieval parish church of St Lawrence (tower possibly pre-Conquest); chancel rebuilt 1600; north aisle rebuilt 18th century; major restoration 1896. Quaker meeting house built 1805; closed 1903. Wesleyan Methodist chapel built 1819; closed and converted to dwelling. ***Schools and other institutions:*** school in south transept of church founded 1653. Schoolhouse built c.1780; replaced by new building south-west of village 1875; in turn replaced by modern building nearby 1974; now Morland Area CE Primary School. Reading room and library established 1849. Morland Hall used as Red Cross hospital during Second World War. Village hall built 1949.

MOSEDALE

Township in Caldbeck parish, Allerdale below Derwent ward, Cumberland. Absorbed into Mungrisdale CP 1934

Acreage: 2,418 acres [979 ha], most of it unenclosed common land on slopes of Carrock Fell. ***Population:*** around 50 from 1841 to 1931, when separate figures were recorded. ***Landownership:*** subsidiary manor within manor of Caldbeck, held by lords of barony of Greystoke (q.v.). ***Economy:*** predominantly hill farming; mining at Grainsgill and Brandy Gill: lead and copper from 16th century; tungsten and wolfram between 1854 and 1981. ***Places of worship:*** Quaker meeting house built (or enlarged) 1702; discontinued 1865; reopened briefly until 1913; used as Anglican chapel of ease c.1936-70; reopened for Quaker worship 1973. ***Schools and other institutions:*** none.

MOSSER*

Township in Brigham parish, Allerdale above Derwent ward, Cumberland. Absorbed into Blindbothel CP 1934.

Acreage: 1,490 acres [603 ha], including 464 acres [188 ha] of common fell, enclosed 1867. ***Population:*** declining from 101 in 1801 to 56 in 1931 (last census year for which separate figure available). ***Landownership:*** granted to Adam de Mosser c.1203; manor had passed to Salkeld family by 1418. Tenants enfranchised by Thomas Salkeld of Corby, 1623. ***Economy:*** predominantly agricultural. Small-scale lead mining, late 18th/early 19th century. Tile-making at Aikbank, recorded 1840; had ceased by 1867. Handloom weaving early 19th century; fish smoking late 20th. ***Places of worship:*** chapel of ease, recorded 1546, originated as pre-Reformation chantry. Fell into decay in 17th century; rebuilt 1773 (and dedicated to St Philip); replaced by new church at Eaglesfield (q.v.) 1891 but continued in occasional use (dedication changed to St Michael by 1938). ***Schools and other institutions:*** school taught in chapel 1653. Small school in township recorded 1818.

MOTHERBY *see* Greystoke

MUNCASTER

Ancient parish in Allerdale above Derwent ward, Cumberland, containing townships of Muncaster (which included port village of Ravenglass) and Birkby, separated by River Esk.

Acreage: 6,496 acres [2,629 ha], divided between townships thus: Muncaster: 3,259 acres [1,319 ha]; Birkby: 3,237 acres [1,310 ha]. CP gained Stainton, detached portion of Chapel Sucken township (see Millom Rural), 1886. Birkby Fell, covering 1,383 acres [560 ha], remains unenclosed common land. ***Population:*** estimated at 620 in 1688; in range 570-630 across 19th century, with peak of 657 in 1831; declined from 522 in 1901 to 272 in 1981, rising to 335 in 2001. ***Landownership:*** *manor of Muncaster* held by Pennington family since 13th century, remaining with their descendants to present, their seat at Muncaster Castle originating as medieval tower house, enlarged 17th century and substantially redesigned 1860s-1880s. *Manor of Birkby* held by Stanley family of Dalegarth. ***Economy:*** farming; woollen textiles (fulling recorded in name 'Walk Mill' on northern boundary of parish); salmon fishery and mussel beds in Esk estuary; iron bloomery forge at Muncaster Head in 17th century; bobbin turning at Broad Oak, Birkby, in 19th century.

RAVENGLASS: harbour and market by Roman times; fort of GLANNOVENTA (name meaning 'market on shore') guarding harbour. Right to hold market and fair granted by charter 1208. By 17th century Ravenglass was small port with important cattle fair: by 1670s weekly market on Saturdays and cattle fair on 28 May had been added to ancient fair held in July. Attempts to revive market and fairs by Lord Muncaster in 1796 appear to have been short-lived. In 1870s became rail junction between Furness Railway and Ravenglass & Eskdale Railway (opened 1876); latter (and gardens at Muncaster Castle) became focus for tourism in 20th century.

Places of worship: medieval parish church of St Michael; restored 1873. Ruins at Chapel Garth, Birkby, excavated 1822, thought to be remains of religious building. Wesleyan Methodist chapel built 1833; reopened 1851. Independent chapel at Grove, near Ravenglass, recorded 1860 but not 1897. Baptist chapel, Ravenglass, built c.1845; closed before 1897. ***Schools and other institutions:*** endowed school established by Richard Brockbank 1696; school room built 1706; new school built 1876; closed 1980s; now hotel. Public hall at Ravenglass, built 1878, used (1901) as infant school on weekdays and for religious purposes on Sundays; part became school (closed and converted to dwelling); remainder now Muncaster Parish Hall. Reading room at Ravenglass established 1862.

MUNGRISDALE

Chapelry and township in Greystoke parish, Leath ward, Cumberland. Modern CP of Mungrisdale formed 1934 by amalgamation of Mungrisdale, Bowscale, Berrier and Murrah, and Hutton Roof CPs.

Acreage: 6,593 acres [2,668 ha], much of it fell grazings. 500 acres [202 ha] of Mungrisdale commons enclosed 1893; 4,400 acres [1,781 ha] of common remained unenclosed. ***Population:*** rising from 160 in 1801 to peak of 236 in 1921; then slow decline to 122 in 1931 (last census year before boundary change). ***Landownership:*** held by barons of Greystoke (q.v.). ***Economy:*** predominantly hill farming. Blue slate and flagstone quarrying in 19th century. Lead and plumbago mining at Bannerdale mid-19th century; closed 1870. Lime-burning on eastern side of township in 19th century. ***Places of worship:*** chapel of ease by 16th century; dedication to St Kentigern (or Mungo) thought to have given name to township. Rebuilt 1756; restored 1925. ***Schools and other institutions:*** public elementary school built 1835; closed 1960. Village hall originated as corrugated iron recreation room, a former mining office, bought from Carrock Mining Co. by Women's Institute 1920s; rebuilt in stone 2004.

MURTON

CP in Appleby St Michael parish, East ward, Westmorland, created 1894 and comprising townships of Murton and Hilton.

Acreage: Murton township: 5,730 acres [2,319 ha], including 3,331 acres [1,348 ha] of unenclosed common on Murton Fell; Hilton township: 4,984 acres [2,017 ha], including 3,558 acres [1,440 ha] of unenclosed common on Hilton Fell. ***Population:*** Murton contained around 200 inhabitants in earlier 19th century and Hilton around 300. Combined population stood at 441 in 1891, then declined to low of 290 in 1971, rising to 330 by 2001. ***Landownership:*** *manor of Murton* held by Cliffords in 13th century and by Musgraves of Hartley Castle from 1315 to 1614, when sold to Thomas Hilton of Hilton. *Manor of Hilton* was in divided ownership in 13th century; by 15th century share held by Bacon family ('Hilton Beacon') came to be treated as entire manor. Descended in Hilton family from 15th century until it and Murton were sold to Sir John Lowther 1696. ***Economy:*** pastoral farming; lead mining from medieval period; expansion in 19th century with large-scale mining in Scordale valley and smelt mill at Hilton. Operated by London Lead Co. from 1824 to 1876; re-opened 1896 for barytes; closed c.1919. Carding and spinning mill at Coupland Beck in 19th century. Bobbin mill at Murton. Military use: rifle range by 1897; commons taken over by Ministry of Defence for military manoeuvres 1980s. ***Places of worship:*** tradition of early chapel site in Murton township. Chapel of ease of St John the Baptist at Hilton built 1856. Wesleyan Methodist chapel at Murton built 1837; closed 1970. United Methodist chapel at Murton built 1841; closed 1928. ***Schools and other institutions:*** Thwaites School endowed 1738; rebuilt 1875; closed 1971. Murton Parish Institute established c.1923; rebuilt 1970s.

MUSGRAVE

CP in East ward, Westmorland, created 1894 by union of ancient parish of Great Musgrave and township of Little Musgrave in Crosby Garrett parish.

Acreage: Great Musgrave ancient parish contained 3,190 acres [1,291 ha];

Little Musgrave township, 1,208 acres [489 ha]. Commons in Great Musgrave (1,382 acres [559 ha]) enclosed 1859; Little Musgrave Intake (94 acres [38 ha]) and Little Musgrave common (378 acres [153 ha]) enclosed 1853. ***Population:*** Great Musgrave's population fairly stable, in range 160-190 across 19th century; Little Musgrave's population fell from peak of 80 in 1821 to 52 by 1891. Population of Musgrave CP stood at 240 in 1901; then declined across 20th century to low point of 128 in 1991 before rising to 152 in 2001. ***Landownership:*** *manors of Great Musgrave* and *Little Musgrave* held by Musgrave family from 12th century to 20th. ***Economy:*** predominantly agricultural. ***Places of worship:*** medieval parish church of St Theobold (Great Musgrave); rebuilt 1844-6. Wesleyan Methodist chapel at Great Musgrave built 1895; now demolished. ***Schools and other institutions:*** Great Musgrave endowed school (on common near Langrigg) established before 1778; rebuilt after new endowment by Rev. Septimus Collinson, provost of Queen's College, Oxford, 1828; closed late 20th century. Musgrave parish institute built 1920s.

NATEBY

Township in Kirkby Stephen parish, East ward, Westmorland

Acreage: 2,195 acres [888 ha], including 705 acres [285 ha] of unenclosed fell common. ***Population:*** in range 100-140 in first half of 19th century; inflated to 222 in 1871, probably as result of railway construction; stood at c.170 from 1881 to First World War, then declined to stand at 110 in 2001. ***Landownership:*** held by de Hastings family from 14th to 16th centuries, after which it passed to Whartons. Sold to Robert Lowther 1728 and descended to earls of Lonsdale. ***Economy:*** farming; quarrying and lime-burning in 19th century. Lead and coal mining on Nateby Common in 19th century. ***Places of worship:*** Congregational chapel built 1873; remains in use as Methodist chapel. ***Schools and other institutions:*** school recorded 1818. Board school for Nateby and Wharton built 1877; closed 1970; then used as village hall until 1990; converted to dwelling.

NATLAND*

Township in Kendal parish, Kendal ward, Westmorland. Part of CP, including Oxenholme, transferred to Kendal 1935.

Acreage: 1,155 acres [467 ha] before boundary change; 892 acres [361 ha] thereafter. ***Population:*** 204 in 1801; increased to 464 by 1891 as housing grew at Oxenholme. Stood at 690 in 1931 (last census year before boundary change). Population of reduced, largely rural, CP rose from 354 in 1951 to 747 in 2001 as Natland grew as dormitory village from 1960s. ***Landownership:*** held by Deincourt family from 1170s, from whom it descended with Sizergh to Strickland family (see Helsington); Sir Thomas Strickland and his brother Walter enfranchised the tenants 1669. *Reputed manor of Natland* held by descendants of Prickett family from later 17th century. ***Economy:*** agricultural; itinerant potters held annual pottery fair on Palm Sunday until c.1835. Helmside and Oxenholme village grew as result of opening of Lancaster and Carlisle Railway with station at Oxenholme 1846. Sandstone quarrying on The Helm, 19th-20th century; nursery and garden centre opened 1938; closed 2012. ***Places of worship:*** chapel recorded 1246; said to be ruinous and unused 1717; rebuilt c.1735; rebuilt again on new site nearby as St Mark's Church 1825; enlarged 1879; rebuilt again 1909-1910. Mission room at Helmside built 1906; closed 1951. ***Schools and other institutions:*** school taught in chapel, late 17th century and early 18th. Richard Frankland ran nonconformist academy at Natland 1674 to 1683. School built 1777 and endowed by will of Charles Shippard. New school built 1825; enlarged to accommodate growing population 1879 and again 1909; replaced and relocated 1967 (old school demolished 1969); now St Mark's CE Primary School. St Mark's Home for Waifs and Strays built 1884-5; closed 1974 and reopened as holiday centre for deprived children; closed 1994; Appleton Treatment Centre for traumatised children opened in grounds 1995. Reading room at Helmside built 1894; closed 1980s. Natland village hall opened 1956.

NENTHEAD *see* Alston Moor

NETHER DENTON *see* Denton, Nether

NETHERBY *see* Arthuret

NETHERHALL *see* Maryport

NETHER WASDALE

Township and chapelry in St Bees parish, Allerdale above Derwent ward, Cumberland. CP enlarged to include Wasdale Head 1934; name changed to Wasdale CP 2000.

Acreage: 8,574 acres [3,470 ha], including detached section at Easthwaite (131 acres [53 ha]), transferred to Irton CP 1934. Extensive unenclosed fells (4,994 acres [2,021 ha]) remain common land. ***Population:*** in range 150-200 across 19th and early 20th centuries (peak 211 in 1821); fell to 117 in 1931 (last census year before boundary change). ***Landownership:*** part of 'Middleward' of forest of Copeland, assigned to Joan Fitzwalter 1338 (see Egremont). ***Economy:*** hill farming; annual sheep fair in September recorded in 19th century. Medieval iron smelting suggested by names Smithy Beck and Cinderdale Beck. ***Places of worship:*** chapel of ease at Strands, recorded 1535; current building 17th-century, enlarged 1837. ***Schools and other institutions:*** schoolhouse built adjoining chapel yard c.1770. Parish school built by Stansfield Rawson (d. 1856) of Wasdale Hall; closed 1935; became village hall (Old School Community Hall).

NEWBIGGIN (near Ainstable) *see* Croglin

NEWBIGGIN (near Penrith) *see* Dacre

NEWBIGGIN (Westmorland)

Ancient parish in East ward, Westmorland.

Acreage: 1,196 acres [484 ha], including 728 acres [294 ha] of commons on Newbiggin High and Low Moor, enclosed 1850. ***Population:*** in range 110-140 across most of 19th and 20th centuries (peak 152 in 1821); declined across 20th century to stand at 96 in 2001. ***Landownership:*** *manor of Newbiggin* granted mid-12th century to Robert, steward of Appleby, later called 'de Newbiggin'; passed by marriage c.1332 to Crackanthorpe family, through which it descended until sold by D. R. F. Crackanthorpe to Church Commissioners 1956. Christopher Crackanthorpe rebuilt Newbiggin Hall 1533 (further rebuilt and extended in 19th century). ***Economy:*** largely agricultural; poultry breeding (19th century); market gardening (20th century). Sandstone quarrying in 19th century. ***Places of worship:*** parish probably carved out of Kirkby Thore. Medieval parish church of St Edmund; north chapel built 16th century; remainder rebuilt 1853-4. Wesleyan Methodist chapel built 1880; closed and converted to dwelling. ***Schools and other institutions:*** parish school built 1882 by lord of manor; closed. Reading room attached to school; now used as community hall.

NEWBIGGIN-ON-LUNE *see* Ravenstonedale

NEWBY (Cumberland) *see* Irthington

NEWBY (Westmorland)

Township in Morland parish, West ward, Westmorland.

Acreage: 2,986 acres [1,208 ha], including 1,239 acres [501 ha] of commons, enclosed 1820. ***Population:*** ranging from 234 to 338 in 19th century; declined to range 131 to 189 across 20th century; stood at 164 in 2001. ***Landownership:*** *manor of Newby or Newby Stones* passed from de Newby family to Vernons (by earlier 16th century) and by 1558 to Nevinsons; acquired by Lowthers 1722. ***Economy:*** predominantly agricultural: noted in later 17th century for 'excellent corn growing in plenty, wheat especially' (Thomas Denton). Limestone abundant and coal seam worked for lime-burning in 19th century. ***Places of worship:*** Wesleyan Methodist chapel built 1853; Primitive Methodist chapel built 1874; both closed and converted to dwellings. ***Schools and other institutions:*** none.

NEW HUTTON

CP in Kendal parish, Kendal ward, Westmorland, created 1897 by combining New Hutton township with Hay and Hutton-in-the-Hay areas of Scalthwaiterigg CP.

Acreage: new CP contained 4,759 acres [1,926 ha]; township of New Hutton had contained 2,591 acres [1,048 ha], including 1,280 acres [518 ha] of common land enclosed 1848. ***Population:*** New Hutton township stood in range 120-160 across 19th century (peak 172 in 1831). After 1897 boundary change, population of New Hutton CP was 234 in 1901; then fell to 201 in 1971 before rising to stand at 313 in 2001. ***Landownership:*** part of Richmond Fee of barony of Kendal (q.v.). ***Economy:*** predominantly agricultural. Killington reservoir, partly in township, built 1819 to supply water to Kendal-Lancaster canal. Fisher Tarn reservoir built 1899 to supply water to Kendal. Modern developments include industrial estate at Hill Top and caravan park. ***Places of worship:*** chapel of ease erected 1739; rebuilt as St Stephen's church 1828-9. ***Schools and other institutions:*** school built 1778; rebuilt c.1830 and again c.1873; closed 1946. Private preparatory school at Hill Top (Holme Park School) opened 1957; closed 2005. New Hutton Institute (village hall) opened c.1928.

NEWTON ARLOSH *see* Holme East Waver

NEWTON, HIGH *see* Allithwaite, Upper

NEWTON REIGNY

Township in Newton Reigny parish, Leath ward, Cumberland. Absorbed into Catterlen CP 1934.

Acreage: 1,013 acres [417 ha]; including 141 acres [58 ha] of commons, enclosed 1819. ***Population:*** rising from 128 in 1801 to 151 by 1831; then fluctuating in range 150-175 across rest of 19th and early 20th century; stood at 168 in 1931 (last census year for which separate figure available). ***Landownership:*** *manor of Newton Reigny* held by Reigny family in 12th and 13th centuries but in hands of Robert Burnell, bishop of Bath, by later 13th century; sold by Burnell to Hugh de Lowther 1290 and descended through Lowther family to earls of Lonsdale. ***Economy:*** predominantly agricultural. ***Places of worship:*** medieval parish church of St John (roof dated 1585); rebuilt 1876. ***Schools and other institutions:*** parish school was in Catterlen township (q.v.). Agricultural college at Newton Rigg established 1896 as Cumberland and Westmorland Dairy School and Farm; school of forestry added 1960s. Merged with University of Central Lancashire in 1990s; then University of Cumbria 2007; from 2011 part of Askham Bryan College. Village hall (Nichol Memorial Hall, in memory of F. G. Nichol, rector of Newton Reigny) built 1912.

NICHOLFOREST

Chapelry and township in Kirkandrews-upon-Esk parish, Eskdale ward, Cumberland.

Acreage: 8,497 acres [3,439 ha]. Nether Common enclosed 1761; Liddel Common (1,400 acres [567 ha]) enclosed 1810. ***Population:*** rising from 668 in 1801 to 907 by 1831; then steady decline (apart from temporary inflation to 1,216 in 1861 when Waverley railway line under construction) to 318 by 1971; rose again to 386 in 2001. ***Landownership:*** part of barony of Liddel (see Kirkandrews Moat Quarter). ***Economy:*** predominantly livestock farming until 20th century. Lime burning and quarrying in 19th century. Industrial camp built at Kershopefoot 1931 to aid long term unemployed return to work; became forestry base for Newfoundland Overseas Forestry Unit in Second World War; then local Forestry Commission office 1955-85. Large-scale coniferous afforestation from mid-20th century. ***Places of worship:*** chapel at Nicholforest recorded 1663; rebuilt 1678 and again 1726; rebuilt on new site at Kingfield 1817; rebuilt completely 1866-67. Small nonconformist chapel at Draught Sike late 19th century. ***Schools and other institutions:*** schoolmaster licensed to teach in chapel 1663. Endowed school recorded 1829. School at Warwicksland rebuilt 1870; closed 1989. School at Nookgate in mid-19th century. Wooden hut at Scuggate (known as 'Border Hall') served as village hall until replaced by wooden village hall (Nicholforest Hall) at Penton 1962; rebuilt 1964.

OLD HUTTON WITH HOLMSCALES

Township in Kendal ward, Westmorland, comprising Old Hutton in Kendal parish and Holmescales in Burton-in-Kendal parish.

Acreage: 3,976 acres [1,609 ha] including 1,274 acres [516 ha] of common land, enclosed 1848. ***Population:*** rising from 368 in 1801 to peak of 488 in 1841; then falling to 258 in 1961 before rising again to stand at 357 in 2001. ***Landownership:*** held by Curwen family of Workington in 13th and early 14th centuries and by Pickering family of Killington in 15th and 16th centuries. ***Economy:*** largely agricultural; small-scale quarrying at Holmscales. Tanning suggested by place-name Barkhouse Beck. Fulling mill recorded 1621. Mill at Bridgend used water power for threshing, winnowing, grinding and boring gun barrels in late 17th century (recorded by Thomas Machell). Bobbin mill and woollen worsted mill there in mid-19th century and Italian flax manufactory in 1885. Bobbin mill at Holmscales mid-19th century. Wind farm in operation from 2013. ***Places of worship:*** chapel of ease described as newly built 1470; rebuilt 1628. Rebuilt again 1699 and again as church of St John the Baptist 1873. Methodist chapel at Bridge End mill from c.1870 to 1902. ***Schools and other institutions:*** grammar school founded and endowed by Edward Milner 1613; rebuilt 1753 and again 1874; extended 1902, 1966 and 1976; now Old Hutton CE Primary School. Public hall built 1914.

ORMSIDE

Ancient parish in East ward, Westmorland, comprising townships of Great and Little Ormside.

Acreage: 2,713 acres [1,098 ha], divided between townships thus: Great Ormside: 2,118 acres [857 ha]; Little Ormside: 595 acres [241 ha]. Commons in Great Ormside enclosed 1773; commons in Little Ormside (259 acres [105 ha]) enclosed 1827. ***Population:*** around 200 throughout 19th century except during railway construction (population 686 in 1871); then fell to around 150 across 20th century, with low point of 91 in 1971; stood at 133 in 2001. ***Landownership:*** *manor of Ormside* held by Ormside family in 13th century; had passed to John de Derwentwater by 1310 and descended in Derwentwater and Radcliffe families (see Keswick) until 16th century. Held of Radcliffes by Barton family from 1436 or before until sold in 16th century by Thomas Barton to Sir Christopher Pickering (d.1620), passed by marriage to Hiltons and then Wyberghs until sold to George Stephenson of Warcop, whose heirs sold it to earl of Thanet 1770. ***Economy:*** predominantly agricultural; quarrying and lime-burning in 19th century; tourism (Wild Rose caravan park established late 20th century). ***Places of worship:*** medieval parish church of St James; extended early 16th century; Hilton chapel added 1723; restored 1885-6. Wesleyan Methodist chapel at Catherine Holme built 1879; closed c.2006. ***Schools and other institutions:*** school built by subscription 1852; closed 1960s; converted to village hall. Residential educational centres in former railway station (run by Ruffwood School, Knowsley, 1962 to 2008; continuing) and former mill (Ormside Mill Activity Centre, established 1999). Infectious diseases hospital opened 1899; closed 1949.

ORTON (Westmorland)

Ancient parish in East ward, Westmorland, embracing townships of Orton, Birkbeck Fells, Langdale, Raisbeck and Bretherdale and Tebay (for which see separate entry). Parish also shared extra-parochial common land of Birkbeck Fells Common with Crosby Ravensworth parish.

Acreage: ancient parish contained 24,513 acres [9,920 ha], distributed among constituent townships thus: Orton: 3,538 acres [1,432 ha]; Birkbeck Fells:

1,479 acres [599 ha]; Bretherdale: 2,264 acres [916 ha]; Langdale: 6,082 acres [2,461 ha]; Raisbeck: 4,572 acres [1,850 ha]; and Tebay: (q.v.) 6,855 acres [2,774 ha]. Birkbeck Fells Common (extra-parochial) contained 1,824 acres [739 ha]. Common land within manor (totalling 5,366 acres [2,172 ha]) enclosed 1779. Common cow pastures enclosed mid-19th century: Langdale Cowbound (320 acres [130 ha]) in 1844; Ellergill High Cowbound (46 acres [19 ha]) in 1867. ***Population:*** rising from 1,230 in 1801 to around 1,500 in mid-19th century, peaking as result of influx of railway workers at 1,917 in 1881. After Tebay became separate CP 1897, population of remainder of parish fell from 832 in 1901 to 552 in 1971; stood at 594 in 2001. ***Landownership:*** *manor of Orton* (including Raisbeck) held by Gamel de Pennington in 12th century; had been divided into moieties by 1277 when in hands of Dacre and Musgrave families. Dacre moiety descended in family until 1614 when it was sold to tenants. Musgrave moiety, subdivided between Blenkinsopps and Warcops, was also sold to tenants; thereafter manor held in trust by four nominal lords; courts continue to be held. *Manor of Bretherdale* held by Byland Abbey (Yorks.) and *manor of Langdale* by Watton Priory (Yorks.); at Dissolution both acquired by Wharton family and later passed to earls of Lonsdale. ***Economy:*** principally livestock farming; small-scale quarrying and lime-burning in 18th and 19th centuries. Stocking knitting, heyday in later 18th and early 19th century. Market charter granted 1336 with fair in October; revived 1658 when further fairs granted. Never major market centre: described as 'small market town' in 1670s; market said to be 'of little moment' in mid-19th century but was revived 1863. Livestock fairs (on 3 May, early September and mid-October) continued to be held in late 19th century. Service stations on M6 motorway, established by local farmers John and Barbara Dunning: Tebay West opened 1972; Tebay East opened 1993. Chocolate factory (Kennedy's Fine Chocolates) established 1991; moved to former school building 1996. ***Places of worship:*** medieval parish church of All Saints; chancel rebuilt 1878-9. Wesleyan Methodist chapel built 1833; still in use. Primitive Methodist mission chapel at Greenholme opened 1901; closed 1994. ***Schools and other institutions:*** school built by subscription c.1730 (endowed as grammar school 1750); moved to new building 1809; relocated 1963; now Orton CE Primary School. Greenholme school, serving Bretherdale and Birkbeck, endowed 1733; rebuilt 1861 and also used as chapel; closed c.1963. Langdale school opened 1873; closed 1960s. Market Hall built by subscription 1865; now used as village hall. Temperance hall built 1858; disused. Reading and Recreation Society, formed 1883, built reading room.

ORTON, GREAT (Cumberland)

Ancient parish in Cumberland ward, Cumberland, embracing townships of Great Orton and Baldwinholme.

Acreage: 4,287 acres (1,735 ha), divided between townships thus: Great Orton: 1,686 acres [682 ha]; Baldwinholme: 2,601 acres (1053 ha). ***Population:*** estimated at 340 in 1688; rising from 378 in 1801 to peak of 519 in 1851. In range 400-450 for much of later 19th and 20th century; stood at 383 in 2001. ***Landownership:*** *manor of Orton* held by Orton family in 13th and 14th centuries, passing by marriage of Joan, daughter and heiress of Giles Orton (d. 1369) to Sir Clement Skelton of Armathwaite; then divided between their four daughters. John Brisco (d. 1583) of Crofton bought two-thirds of manor; his grandson, John Brisco (d. 1632) bought final third and manor descended in Brisco family, who built Orton Park 1830s. ***Economy:*** agricultural; 20th-century specialisation in market gardening and poultry farming. ***Places of worship:*** medieval parish church of St Giles; restored 1886. Methodist chapel recorded 1899. ***Schools and other institutions:*** school built at west end of church c.1703; replaced by new parish school, endowed by Pattinson's charity, built on common land on boundary between townships by c.1814. New school, south of village, built 1859 by Sir Wastell Brisco; now Great Orton Primary School. Parish room erected 1899, used as reading room. Village hall (Orton and District Public Hall) built 1954.

OSMOTHERLEY

Township in Ulverston parish, Lonsdale hundred, Lancashire North of the Sands. Combined with Egton with Newland and Mansriggs CPs 2003.

Acreage: 1,929 acres [781 ha], including 750 acres [304 ha] of common land on Osmotherley Common, enclosed 1862. ***Population:*** rising from 218 in 1801 to peak of 474 in 1881; then falling across 20th century to 224 in 1981; stood at 234 in 2001. ***Landownership:*** *manor of Osmotherley* held by Osmotherley ('Asmunderlaw') family from Redmaynes of Levens from 13th century to c.1480s. In hands of earls of Derby in 16th century. ***Economy:*** predominantly agricultural; small-scale quarrying (slate; flagstone) in 19th century. Sickle mill at Broughton Beck mid-19th century. ***Places of worship:*** chapel of ease of St John the Evangelist built 1874 (services previously held in school, using portable altar rails). Wesleyan Methodist chapel recorded 1831; no trace survives. ***Schools and other institutions:*** school built near Broughton

Beck 1770; became public elementary school; closed 1927. Village hall at Broughton Beck (serving Osmotherley and Mansriggs), converted from water mill, donated to parish 1937.

OUGHTERSIDE AND ALLERBY

Township in Aspatria parish, Allerdale below Derwent ward, Cumberland.

Acreage: 2,194 acres [888 ha] before boundary changes 1934, when 139 acres [56 ha] gained from Aspatria & Brayton CP, and 57acres [23 ha] from Hayton & Mealo CP, while smaller acreages were lost to them. Common field, Oughterside Townfield (295 acres [119 ha]) enclosed 1776 and Aspatria West Moor (part of which lay in township) enclosed 1825. ***Population:*** rising from 269 in 1801 to 705 in 1861, reflecting growth of mining village of Prospect; then falling to stand at 586 in 1931 (last census year before boundary changes). ***Landownership:*** *manor of Oughterside* held by Orfeur family before passing to Lawsons of Brayton and Isel in 18th century. *Manor of Allerby* acquired by Thomas Porter of Bolton by marriage with heiress of Thomas Lowther of Allerby, probably in 15th century; passed by widow of their descendant Thomas Porter (living 1578) to Eaglesfield family, from whom it was bought by Lamplughs of Dovenby in late 17th century. ***Economy:*** farming. Coal mining, recorded by 1681, expanded after building of Maryport & Carlisle railway in 1840s; in decline by 1900. Oughterside Colliery closed 1943 but opencast mining stripped out remaining deposits in 1970s-80s. Red sandstone quarries in late 19th century. ***Places of worship:*** Primitive Methodist chapel at Prospect built 1860; closed 1980s. Congregationalists met in schoolroom 1851. ***Schools and other institutions:*** school at Oughterside founded by Sir Wilfred Lawson before 1850; rebuilt as Board school 1875; enlarged 1896; now Oughterside Foundation School. Oughterside and Prospect village hall, at Prospect, built c.1964.

OULTON

Township in Wigton parish, Cumberland ward, Cumberland. Absorbed into Woodside CP 1934.

Acreage: 2,878 acres [1,165 ha], including 1,527 acres [618 ha] of common land, enclosed 1835, and part of Wedholme Flow, which survives as unenclosed peat moss. ***Population:*** 294 in 1801, rising to peak of 421 in 1851, then falling to stand at 271 in 1931 (last census year for which separate figure available). ***Landownership:*** *manor of Oulton* held by Dalstons of Dalston for many generations, until death of Sir George Dalston 1765, after which it was sold. In hands of Taylor family of Dockray Hall in late 18th/early 19th century. ***Economy:*** predominantly agricultural; commercial peat extraction on Wedholme Flow in 20th century. ***Places of worship:*** Baptist chapel erected 1722; rebuilt 1832; later used as Primitive Methodist chapel; disused by 1984. Anglican services sometimes held in school in late 19th century. ***Schools and other institutions:*** school, recorded in name School House, established by 1847; new CE school built 1875; closed 1992. Parish Institute built 1924.

OUSBY

Ancient parish in Leath ward, Cumberland. CP enlarged by absorbing Melmerby CP 1934.

Acreage: 7,472 acres [3,042 ha], including extensive commons on Pennine edge. Ousby In Fell, Low Moor, Star Howes, Fell Green, The Mires, Castle Slack and School Green (totalling 1,670 acres [675 ha]) enclosed 1863; Ousby Fell (3,581 acres [1,449 ha]) remains unenclosed common land. ***Population:*** estimated at 365 in 1688. Rising from 253 in 1801 to 329 in 1871, followed by decline to 225 by 1931 (last census year before boundary change). ***Landownership:*** said to be named from its Danish owner, Ulf son of Haldane. Complex descent through medieval period: manor had apparently been divided into eighths by 1289. Ownership remained divided but manorial rights had been obtained by Crackanthorpes of Newbiggin by 17th century. ***Economy:*** farming; lead and coal mining by late 17th century; had largely ceased by mid-19th century. Crushing mill at Smittergill Head c.1851 to 1875. ***Places of worship:*** medieval parish church of St Luke; rebuilt late 16th century and again 1858, further restoration 1896. Wesleyan Methodist chapel of 1838, replaced by larger building 1872; had closed by 1985. ***Schools and other institutions:*** school taught in church by 1703. National school built 1856; closed c.1966; converted for use as community centre.

PAPCASTLE

Township in Bridekirk parish, Allerdale below Derwent ward, Cumberland.

Acreage: 1,240 acres [502 ha]; reduced to 1,108 acres [448 ha] after loss of the Goat area, beside River Derwent, to Cockermouth UD 1934. Papcastle and Dovenby commons (totalling 262 acres [106 ha]) enclosed 1842. ***Population:*** rising from 283 in 1801 to 736 in 1861, as result of industrial growth at Goat; then declining to 549 in 1931 (last census year before boundary change). Reduced CP, containing village of Papcastle, grew from 266 in 1951 to 449 in 1971; stood at 406 in 2001. ***Landownership:*** site of Roman fort of DERVENTIO said to have been caput of Allerdale barony before Cockermouth (q.v.) in 12th century. *Manor of Papcastle* descended with barony until mid-16th century, when it was bought by William, Baron Dacre of Gilsland. In Crown hands after attainder of Leonard Dacre in 1571; sold to Lamplughs of Dovenby Hall by the early 17th century and descended with Dovenby (q.v.) thereafter. ***Economy:*** Goat took its name from mill-race (Old English *gota* = watercourse), suggesting early industrial activity. Flax weaving started early 19th century at Goat mills, developing to major enterprise by Harris family at Derwent Mill, built 1834; closed 1934; used as shoe factory from 1940 to 1990. Other industrial activity at Goat in 20th century included James Walker's (gasket manufacturers) and other small industrial units. ***Places of worship:*** chapel of St Osyth (location unknown) recorded 1475-1521; Anglican mission church built in village 1875; closed 2002. ***Schools and other institutions:*** day school planned for new mission 1877, but never opened. Reading room, now village hall, opened 1895.

PARTON

Coastal settlement in Moresby parish, Allerdale above Derwent ward, Cumberland, which became separate administrative township c.1726.

Acreage: 53 acres [21 ha]. ***Population:*** rising from 360 in 1801 to 858 in 1871, then steeply to peak of 1,479 in 1881, after which it fell back slowly to 1,395 in 1951, continuing to fall to 924 in 2001. ***Landownership:*** see Moresby. ***Economy:*** fishing and harbour by 1680s (Thomas Denton considered Parton 'as fit for a harbour as Whitehaven and safer rideing at anchor') but coal trade ceased after pier was washed away 1795. Brewing and tanning from 1750s to 19th century. Lowca iron works (see Harrington) also provided employment from early 19th century. ***Places of worship:*** United Free Methodist chapel built 1863; rebuilt 1896; closed 2006. Independent chapel built 1861; closed c.1954. ***Schools and other institutions:*** school for boys built and endowed by Joseph Williamson 1818. School of Industry for girls and infants, established by Mary Robinson, built 1837 and enlarged 1893; now St Bridget's CE Primary School. Former endowed school converted for use as village hall c.1984.

PARTON AND MICKLETHWAITE *see* Thursby

PATTERDALE

Chapelry and township in Barton parish, West ward, Westmorland, including village of Glenridding.

Acreage: 16,735 acres [6,775 ha], including extensive fell commons. Open field (Overdale common field; 56 acres [23 ha]) enclosed 1861. Hartsop High and Low Pastures (1,988 acres [805 ha]) enclosed 1865. ***Population:*** 261 in 1801, rising to 961 in 1921, then falling to 460 in 2001. ***Landownership:*** *manor of Patterdale* held by Threlkeld family in early 17th century when Thomas Threlkeld sold it to Joan, widow of George Mounsey (d. 1624); descended through Mounsey family (known as 'kings of Patterdale') until sold to John Marshall, mill-owner of Leeds, 1825, who rebuilt Patterdale Hall 1845-50. Remained in Marshall family until sold 1937. *Hartsop* was held by Lancaster family from 13th to 17th century, passing by marriage of Frances, daughter of Christopher Lancaster to Sir Christopher Lowther, subsequently descending to earls of Lonsdale. ***Economy:*** hill farming. Lead mining by later 17th century: major expansion at Greenside lead mine, Glenridding, from c.1830, employing 300 people in 1849; smelter built 1834; mine closed 1961. Smaller lead mines in Hartsop, worked sporadically until 1940s. Small ancient copper mine at Hartsop. Slate quarrying at Hartsop and Place Fell in 19th century. Tourism from 19th century: boarding houses and 4 hotels by 1906; steamers on Ullswater; outdoor pursuits with centres at Glenridding from later 20th century. ***Places of worship:*** chapel of ease of St Patrick recorded 1348; rebuilt 1852-3. Chapel shared by Methodists and Quakers built 1842. Methodists split off and built new chapel at Glenridding 1890; closed early 1990s. Quakers retained old chapel which closed 1965. ***Schools and other institutions:*** 'ancient' grammar school recorded 1818; rebuilt 1836 and again 1873; enlarged 1895 and modernized 1970, 2003 and 2008; now Patterdale CE Primary School. Infant school at Glenridding by 1860; reading room adjacent by 1897; became Glenridding Public Hall 1917.

PATTON

Township in Kendal parish, Kendal ward, Westmorland. Absorbed into Whinfell CP 1986.

Acreage: 637 acres [258 ha]. ***Population:*** rising from 77 in 1801 to peak of 98 in 1871; falling back to 60 in 1981 (last census year for which separate figure available). ***Landownership:*** part of Patton granted to Conishead Priory in 13th century. Half of hamlet of Patton held by Thornburgh family by mid-15th century; they leased Conishead's land there in early 16th century and acquired it after Dissolution. Descended through Thornburgh family until sold to the tenants 1723. ***Economy:*** predominantly agricultural; small-scale quarrying. ***Places of worship:*** none. ***Schools and other institutions:*** none.

PENNINGTON

Ancient parish in Lonsdale hundred, Lancashire North of the Sands, including mining village of Swarthmoor.

Acreage: 2,845 acres [1,151 ha], including 1,163 acres [471 ha] of common land on Pennington Moor and Swarthmoor, enclosed 1827. ***Population:*** rising from 273 in 1801 to 489 by 1851, then rapid increase to 1,698 by 1881 as a result of expansion of iron mining. Declined from 1,501 in 1901, to 1,333 in 1961 but recovered to 1,794 by 2001. ***Landownership:*** *manor of Pennington* held by Pennington family from 12th century, descending through that family to present (see Muncaster). ***Economy:*** farming; small-scale slate quarrying; lime-burning; gravel digging in 19th century. Iron ore mining at Lindal Moor and Carkettle by 1799; rapid expansion in mid-19th century; all mines had closed by 1914. Reservoirs supplying water to Barrow-in-Furness constructed 1879 (now used for fishing). ***Places of worship:*** medieval parish church of St Michael (deemed chapel under Urswick in 12th century); rebuilt 1826-7; heavily restored 1924-6. St Leonard's mission chapel, Swarthmoor, opened 1883; lower floor used as working men's institute and reading room; mission church no longer in use. Bethel Primitive Methodist chapel at Swarthmoor, built 1864; now Swarthmoor Methodist Church. Bible Christian chapel by 1888; closed (building later used as garage). ***Schools and other institutions:*** parish school recorded 1675. Replaced by public elementary school at Row Head, opened 1876; enlarged 1885; now Pennington CE Primary School. Parish poor house acquired under bequest from James Fell 1743; sold 1866 and inmates transferred to Ulverston Workhouse. Hospital for infectious diseases established at High Carley 1884; County Sanatorium added 1916; closed 1984. Pennington Memorial Hall built c.1947.

PENNY BRIDGE *see* Egton with Newland

PENRITH

Market town and ancient parish in Leath Ward, Cumberland, including village of Carleton.

Acreage. 7,587 acres [3,070 ha], including extensive common land on Beacon Fell, enclosed under Forest of Inglewood enclosure award 1819.

Landownership. *Honour of Penrith* created 1242 and granted to king of Scots; seized by Edward I 1295 and conferred on Antony Beck, bishop of Durham, 1298, who soon forfeited it to Crown. Granted by Richard II to John duke of Brittany 1378 and to Ralph Neville, 1st earl of Westmorland, 1397, who is said to have built Penrith Castle; his grandson Richard earl of Warwick (d. 1472) forfeited honour to Crown. In 1696 William III granted honour of Penrith to William Bentinck, duke of Portland, whose descendant, 3rd duke, sold it to his brother-in-law William Cavendish, 5th duke of Devonshire, 1787. Penrith contained three inferior manors: *manor of Bishop's Row*, held by bishops of Carlisle from 12th century; *manor of Penrith and Hutton Hall*, held by Hutton family until sold to John Gaskarth 1734, whose son sold it to Lord Lonsdale 1790; and *manor of Carleton*, held by Carleton family until 1707, then passing through several families in 18th century until bought by John Cowper 1828; his descendant G. T. M. C. Cowper sold Carleton Hall 1947.

Origins and growth of the town. Penrith area, at meeting of north-south and east-west routeways, was probably focal point from an early date. It developed as market centre for Eden valley and was on a par with Cockermouth in importance as urban settlement by later middle ages. From later 18th century Penrith became centre for textile manufacture and place of social gatherings on race days and when assize courts were held. Its position close to 'picturesque' scenery of Lake District made it an increasingly fashionable resort during 19th century. Not a major tourist centre in itself, railway (from 1846) and road connections (M6 Junction 40 crossroads with A66, opened 1970) contributed to Penrith's importance in 20th and 21st centuries.

Market function probably developed by 12th century around early church of St Andrew, which became core of medieval town. Gained borough status by early 13th century (burgages recorded in 1220s), though this was later lost. Market charter (with fair) granted 1222. In addition, special great fairs for cattle and horses were held. Industries included dyeworks and weaving shops (recorded 1310) and tanneries (recorded from 1379). By 17th century it had become major market centre: Sir Daniel Fleming described it 1671 as 'a large

Towne, well built and accounted the second in [Cumberland] ... Here is a very great market every Tuesday for all sorts of provisions'. Both he and Thomas Denton, writing 1688, reported that up to 400 'beeves' were slaughtered on market days in autumn and Denton noted importance of livestock and hiring fair on Whit Tuesday. By 1688 there were four guilds: merchants, tanners, shoemakers and skinners. Penrith also specialised in pewter manufacture during later 17th and earlier 18th century. Outside urban area stone quarrying on Penrith fell (Beacon Hill). Population estimated at 1,350 in 1688.

By 1769 population estimated to have grown to c.2,000. From later 18th century Penrith began to be known as manufacturing town, specialising in 'checks for aprons and bed hangings, linen cloth for shirting and sheeting and a few ginghams'. By 1829, however, textile trade was much reduced. Tanning, brewing (two breweries recorded 1787) and clockmaking also important. Altham's iron works, Albert Street, established early/mid-19th century. Market continued to flourish on Tuesdays and Saturdays, with Devonshire Arcade built 1807. Population almost doubled from 3,801 in 1801 to 7,387 in 1851. Gas light company founded 1830; piped water supply from River Eamont came with opening of water works 1854.

Arrival of railway further consolidated town's prosperity: Lancaster and Carlisle railway opened 1846, followed by Eden Valley railway 1862 and Cockermouth, Keswick & Penrith railway 1865. In 1850s 'a rage for improvement was manifest': Penrith Building Society formed 1850 (dissolved 1878) acquired land for new streets on slopes of Beacon Fell, leading to expansion of built-up area to north-east of medieval urban core. In early 20th century Penrith still boasted brewery, foundry, agricultural implement makers, and some basket making. Railways transformed transportation of farm produce: Agricultural Hall, built 1870, enabled livestock sales to be conducted under cover. Penrith Farmers & Kidd auction company formed 1922; moved to new premises in Skirsgill 1987. Modern agricultural service businesses include BOCM Paul, animal feed mill, and Alba Proteins, processing animal by-products for industrial use. Population continued to grow across later 19th and 20th century, to stand at 14,756 in 2001.

Places of worship. Medieval parish church of St Andrew, a pre-Conquest foundation to judge from important Viking Age sculpture in churchyard ('Giant's Grave' and 'Giant's Thumb'). Medieval tower survives; body of church rebuilt 1721-2. Augustinian Friary in existence from c.1299 until Dissolution. Second Anglican church, Christ Church, built 1848-50, to serve increased population and also to provide new burial ground. Town cemetery opened 1872. Presbyterian chapel, Rowcliffe Lane, built 1785, replacing earlier place of worship; in turn replaced by United Reformed Church, Lowther Street, 1884. Quaker meeting house, originally farmhouse, purchased 1699; enlarged 1730, 1803 and 1992. Roman Catholic chapel in Bishop's Yard, blessed 1833; replaced by St Catherine's Church, Drover's Lane, 1850; enlarged 1860. Wesleyan Methodist chapel, Sandgate Head, built 1815; replaced by present Methodist Church, Wordsworth Street, built 1873. Primitive Methodist chapel, Arthur Street, built 1857; converted to Temperance Hall 1873 when Primitives moved to former Wesleyan chapel at Sandgate Head; which closed 1967; used by Pentecostalist church late 20th century; converted to housing 2014. Ebenezer Independent chapel, Duke Street, built 1824; replaced by Congregational Church on same site 1865; closed c.1990; converted to housing. Plymouth Brethren Gospel Hall, Queen Street, built 1873. Salvation Army barracks built 1882 in Castletown; later moved to Hunter Lane. Church in the Barn community established 1980s; moved 1997 to Wetheriggs. Kings Church Eden established 2002-3. Bangladeshi Eden Mosque established c.2009.

Schools. John de Eskeheved licensed by bishop of Carlisle 'to teach the art of grammar' 1340. Chantry in parish church founded by Bishop Strickland 1395 with priest to teach music and grammar; chantry dissolved 1547. Grammar school in St Andrew's churchyard and Bishops Yards, endowed by Elizabeth I 1564; moved to new premises on Ullswater Road 1915; now Queen Elizabeth Grammar School. Robinson's school for education of poor girls founded 1661; spinning and knitting school later added; closed 1971 (now museum and tourist information centre); replaced by new school on Scaws housing estate, now Beaconside CE Primary School. Girls' National school of industry, Drovers Lane, established 1813; rebuilt 1853; enlarged 1882 with classroom for infants. Boys' National school, Benson Row, established 1816; rebuilt 1871; closed 1979; demolished. CE infant school established 1828; moved 1833 to house in Meeting House Lane; now used by pre-school nursery and toddler group. Wesleyan schools established in Meeting House Lane 1844. British School, Castlegate Head, built 1847. Ragged (mission) school erected 1853 in Drovers Lane, for children from Townhead district; discontinued 1874. St Catherine's Catholic primary school, Drovers Lane, built 1882; enlarged 1899; rebuilt 1970; continues. Brunswick Road Board schools opened 1894; extended 1901; now Brunswick Infant School. New schools built in post-War period: secondary modern schools, Tynefield (girls) and Ullswater (boys), established 1950s; merged 1980 to become Ullswater High School, now Ullswater Community College; and Wetheriggs Junior School;

now North Lakes Primary School. Independent preparatory school (Hunter Hall) moved from Great Salkeld to Frenchfield, Penrith, 1986.

Other institutions. Moot Hall, where manorial court leet held, dates from 1678, probably replacing earlier structure; occupied by draper's shop since 1850. Town Hall, Stricklandgate, converted from houses 1905-6. Penrith Beacon, in use before 1363, an important landmark in times of border warfare; stone tower erected 1719. Lock-up (now known as 'The Old Jail'), Scotland Road, recorded 1547; replaced by new house of detention at Town Head, built 1825, along with constable's residence; in turn replaced by police station, Hunter Lane, built 1859. Carleton Hall became headquarters of Cumbria Constabulary, formed 1974. Penrith magistrates' court building opened 1973; closed 2011. Parish workhouse, Queen Street, replaced by Penrith Poor Law Union Workhouse, a mile west of town, built 1838; closed 1960s. Isolation hospital at Fairhill built 1895; closed and demolished later 20th century. Jubilee Cottage Hospital, Beacon Edge, built 1898-9; closed 1987 (converted to housing); replaced by new Penrith Community Hospital, Bridge Lane.

Book society and general news room formed c.1829 at Crown Inn; a number of subscription libraries also existed. Mechanics' Institute founded 1830-1; reading room added 1846. YMCA public library founded 1853. Working men's reading room built 1853; converted into free library and museum 1883, which was moved to new Town Hall building 1906 and to St Andrew's churchyard 1992. Newspapers published in Penrith: *Cumberland and Westmorland Advertiser* and *Penrith Beacon Advertiser*, both established 1854; and *Cumberland and Westmorland Herald*, established 1860 under different title. Assembly room at George Hotel, described 1847 as 'one of the finest ... in the north of England'. Parish rooms (now Masonic Hall) built 1882. Drill hall and concert hall, Portland Place, built 1893. Alhambra Buildings housed skating rink and hall in 1911; later became cinema. Albion Hall, Castlegate, formerly dance hall, converted to theatre 1925-6. Race course built to north of Beacon Fell 1814; converted to golf course 1890. Castle Park opened 1923.

PENRUDDOCK *see* Hutton Soil

PLUMBLAND

Ancient parish in Allerdale below Derwent ward, Cumberland, including villages of Arkleby and Parsonby.

Acreage: 2,568 acres [1,039 ha]. Plumbland Common (139 acres [56 ha]) and Parsonby Common (40 acres [16 ha]) enclosed at unknown date. Parsonby Hill and other small areas (36 acres [15 ha]) enclosed 1817. Arkleby Green (13 acres [5 ha]) enclosed 1803. Warthole Common (157 acres [63 ha]) remains open pasture but common rights extinguished. ***Population:*** estimated at 410 in 1688. Rising from 330 in 1801 to peak of 800 in 1851 (attributed to opening of collieries and lime works and establishment of free school); then gradual decline across later 19th and 20th century to stand at 367 in 2001. ***Landownership:*** parish contained four manors. *Manor of Plumbland,* in hands of Orfeur family by early 14th century; sold to Lawson family 1709. *Manor of Parsonby,* held by rector of Plumbland. *Manor of Arkleby,* held by Martindale family until forfeited to Crown 1568 and granted to Penruddock; sold to Gustavus Thompson 1725, left by his widow to Gale 1772; sold to Craik 1788; to Harrison then to Saterthwaite 1790; bought by Lawson 1803. *Manor of Warthole (or Wardhall),* held by Calder Abbey; Dykes family there by 15th century; manor descended in Dykes (later Ballantine-Dykes) family to 20th century. ***Economy:*** agriculture; fulling mill and colliery at Wardhall, recorded 1688. Coal pits in Plumbland and Arkleby in early 19th century; small drift mine worked by unemployed miners c.1930. Limestone quarrying; large quarries at Thrushgill (1837-1891) and Wardhall Guards (1847; extended 1887-8; quarrying ended 1942), with waggonway to Maryport & Carlisle railway, provided lime for West Cumberland iron works. Lime-burning: two lime works in mid-19th century; large limekilns at Wardhall lime works late 19th century (had closed by 1900). High Close quarry, worked by T. Armstrong Ltd 1938-1957, produced aggregate and road stone. Brick- and tile-making at Plumbland mid-19th century. ***Places of worship:*** medieval parish church of St Cuthbert at Parsonby; rebuilt 1869-71. Independent (Congregational) chapel built 1847; closed c.1961; bought by mission hall congregation 1962 (now Plumbland Evangelical Chapel). Plumbland mission hall (a wooden hut) set up by breakaway group of Congregationalists 1946; dismantled and sold 1963. ***Schools and other institutions:*** school recorded 1660; grammar school at Parsonby opened 1800 (endowed by John Sibson (d. 1760) but not established until after his widow's death); now Plumbland CE Primary School. Reading room with small library established by curate of Plumbland c.1861; closed c.1930. Church rooms built c.1919; demolished and rebuilt as village hall 1984.

PLUMPTON STREET *see* Hesket-in-the-Forest

PLUMPTON WALL

Township in Lazonby parish, Leath Ward, Cumberland. Absorbed into Hesket CP 1934.

Acreage: 3,052 acres [1,220 ha]; commons enclosed under Inglewood Forest enclosure award 1819. ***Population:*** rising across early 19th century from 206 in 1801 to lie in range 300-320 across later 19th and early 20th century (peak of 345 in 1881); stood at 320 in 1931 (last census year for which separate figure available). ***Landownership:*** Plumpton park was royal demesne, part of Inglewood Forest. Disparked by Henry VIII; leased to Musgraves in later 16th century; then to earl of Annandale 1622; sold to Lowther family 1653. ***Economy:*** agricultural. ***Places of worship:*** chapel of ease of St John at Salkeld Gate built 1767; replaced by present building 1907-8. Methodist chapel at Brockley Moor, bought from Presbyterians 1888; closed. ***Schools and other institutions***: endowed school at Salkeld Gate, built 1759; rebuilt 1898; remodelled and extended 1976; now Plumpton Primary School. Institute near Salkeld Gate in early 20th century.

PONSONBY

Ancient parish and township in Allerdale above Derwent ward, Cumberland.

Acreage: 2,427 acres [982 ha] including 202 acres [82 ha] of surviving common land on Ponsonby Fell. Commons at In Fell, Low Moor and Pikehow Moor (totalling 96 acres [39 ha]) enclosed 1853; further 54 acres [22 ha] at Gaterigg Bank and Calder Gills enclosed 1855. ***Population:*** estimated at 375 in 1688; in range 150-180 for most of 19th and early 20th centuries, with peak of 190 in 1851; rose to 218 in 1951, then fell back to 92 by 2001. ***Landownership:*** *manor of Ponsonby* held from 12th century by descendants of Punzun (from whom it was named), who took surname de Ponsonby. By 1388 had passed to Stanley family, who held it until mid-20th century. Nicholas Stanley sold Ponsonby Hall to Home Office 1951. ***Economy:*** largely agricultural until 20th century; some freestone quarrying. Calder Hall became site of world's first commercial nuclear reactor: construction started 1953; power station opened 1956; closed 2003. ***Places of worship:*** medieval parish church (dedication unknown); tower and spire added 1840; restored 1874. ***Schools and other institutions:*** no parish school; parish was served by schools at Calder Bridge (see Beckermet St Bridget) and Gosforth. Approved school for boys opened at Pelham House 1941; closed in 1980s.

POOLEY BRIDGE *see* Barton

PORT CARLISLE *see* Bowness-on-Solway

PORTINSCALE *see* Above Derwent

PRESTON PATRICK

Township in Burton-in-Kendal ancient parish, Kendal ward, Westmorland.

Acreage: 3,657 acres [1,480 ha]; act to enclose commons (estimated at 1,800 acres [c.725 ha]) 1814, but no award known. ***Population:*** rising from 335 in 1801 to 514 by 1851; surge to 782 in 1891; then around 450 in early 20th century, falling to 293 in 1981; rose again to stand at 438 in 2001. ***Landownership:*** *manor of Preston Patrick* held by Thomas son of Gospatrick, ancestor of Curwens of Workington, in late 12th century. Suffix 'Patrick' records ownership by his son Patrick de Culwen or Curwen. Part granted to Premonstratensian house founded at Preston Patrick c.1191, which moved to Shap. Former Shap Abbey lands granted by James I to Philip Lord Wharton; sold in 1730 to Robert Lowther, descending to earls of Lonsdale. Remainder of manor descended through Preston family and was sold to Francis Charteris of Hornby Castle 1717. ***Economy:*** farming; bobbin manufacture at Park End and extensive bleach works at Gatebeck in mid-19th century. Gunpowder manufacture at Gatebeck from 1852 to 1936, with associated hoop-making and cooperage, linked by tramway (opened 1876; closed 1936) to Milnthorpe station. ***Places of worship:*** medieval chapel of ease of St Gregory recorded 1331; made parochial 1701; replaced by St Patrick's Church on same site, built 1852. Quaker meeting house built 1691; rebuilt 1869; in use. ***Schools and other institutions:*** curate taught school in early 18th century. School built by public subscription 1780; rebuilt 1869; closed 1949. Memorial Hall built 1923 as reading room and for social and public gatherings.

PRESTON QUARTER

Township in St Bees parish, Allerdale above Derwent ward, Cumberland. This entry covers rural parts of township, excluding areas transferred to Whitehaven MB in 1890s. Preston Quarter CP was abolished 1934.

Acreage: 3,156 acres [1,277 ha] before boundary change in 1894, when northern part of CP was transferred to Whitehaven MB and remainder renamed Preston Rural Quarter CP. After transfer of further land to Whitehaven and to St Bees CP 1896, name reverted to Preston Quarter CP until abolished 1934, when northern section was transferred to Whitehaven, and southern to Rottington CP. ***Population:*** as result of Whitehaven's expansion, population rose rapidly from 1,886 in 1801 to stand at 7,551 in 1891, before transfer of industrial urban areas to Whitehaven. Population of remaining rural area lay in range 175-200 between 1901 and 1931 (last census year for which separate figure available). ***Landownership:*** granted to St Bees Priory on its foundation c.1120, forming core of priory's endowment. After Dissolution it passed, as part of *manor of Kirkby Begock*, through Challoner and Wybergh families, and was mortgaged to Lowthers. Sir John Lowther of Whitehaven gained possession 1665 and manor descended to earls of Lonsdale. ***Economy:*** largely agricultural; coal mining and tile making in Prestonhows area mid-19th century. ***Places of worship:*** ancient religious site associated with pre-Conquest cult of St Bega ('St Bees') lay in Preston Quarter until transfer to St Bees CP 1896. Benedictine priory founded on site c.1120; on Dissolution, nave of priory church used as parish church. ***Schools and other institutions:*** St Bees Grammar School, established by Archbishop Edmund Grindal 1583, evolved into independent public school. Major expansion of buildings from 1842; school chapel dates from 1907; school closed 2015. St Bees Theological College (first Anglican training college outside Oxford and Cambridge) founded 1816, using restored chancel of priory church as lecture hall; closed 1895. Public library belonging to school established before 1700.

PRESTON RICHARD

Township in Heversham parish, Kendal ward, Westmorland, including village of Endmoor.

Acreage: 2,134 acres [864 ha]. ***Population:*** rising from 296 in 1801 to 718 in 1891; then declining in early 20th century to 605 by 1911; rising again (as result of housing development at Endmoor and Gatebeck) to stand at 1,307 in 2001. ***Landownership:*** *manor of Preston Richard* held by Preston family (several generations of whom bore name Richard) from 13th century, passing by marriage of Margaret, daughter of Richard Preston (d. c.1390) and Alan Pennington of Muncaster, to Penningtons. By 17th century manor (or a moiety of it) was held by Prestons of Holkers who had inherited from Bensons of Hugill. George Preston sold it to Sir John Lowther c.1650. ***Economy:***

farming; lime-burning; flax-spinning at Crooklands Mill in early 19th century; had become bobbin mill by mid-19th century. ***Places of worship:*** none. ***Schools and other institutions:*** school built at Endmoor 1862; replaced by new school on new site 1991; now St Patrick's CE Primary School. Working men's institute at Endmoor in 1890s, containing 'well stocked library and reading room'. Endmoor village hall built c.1984.

PROSPECT *see* Oughterside & Allerby

RAISBECK *see* Orton (Westmorland)

RAVENGLASS *see* Muncaster

RAVENSTONEDALE

Ancient parish in East ward, Westmorland, which included village of Newbiggin-on-Lune.

Acreage: 16,407 acres [6,640 ha], including 6,806 acres [2,755 ha] of surviving common land on Ravenstonedale Common and Ravenstonedale Moor. Parish divided into four 'angles' for local administration: Town Angle (2,355 acres [953 ha] of rateable land), covering Ravenstonedale village and north-east of parish; Bowderdale Angle (c.942 acres [381 ha]), covering Bowderdale and Weasdale valleys; Fell-End Angle (1,545 acres [625 ha]), covering south-east corner of parish; and Newbiggin Angle (1,425 acres [577 ha]), covering Newbiggin-on-Lune. ***Population:*** downward drift from 1,138 in 1801 to 838 in 1901 and 570 in 2001. Temporarily inflated to 1,262 in 1861 as result of railway construction. ***Landownership:*** *manor of Ravenstonedale* granted by Torphin to Gilbertine Priory of Watton (Yorks.) in 12th century. At Dissolution bought (after grant for life to archbishop of York) by Sir Thomas Wharton, descending in Wharton family until sold after attainder of duke of Wharton 1728 to Robert Lowther of Maulds Meaburn, from whom it descended to earls of Lonsdale. ***Economy:*** predominantly agricultural; quarrying and lime-burning in 19th century. Lead mining at Fell End Clouds in 19th century. Market at Ravenstonedale (obsolete by mid-19th century); Whitsuntide fair and three cattle fairs in 19th century. South Durham & Lancashire Union railway station at Newbiggin (opened 1861; closed 1952) fostered growth of market function there: Tuesday butter market at Newbiggin by 1885. ***Places of worship:*** medieval parish church of

St Oswald; tower rebuilt 1738, remainder 1744. Footings of Gilbertine cell to north of church. Supposed medieval chapel near St Helen's Well, Newbiggin. St Aidan's church, Newbiggin, built 1892; closed and converted to dwelling early 1980s. Protestant nonconformity strong in parish: High Chapel, built 1727 for Presbyterian congregation; became Independent (Congregational) 1811; then United Reformed; closed 2006 (converted to community and heritage centre). Quaker meeting house, Fell End, built 1705; closed 1790s (when meeting moved to Narthwaite); demolished 1899. Quaker meeting house at Narthwaite built 1823; closed 1907. Wesleyan Methodist chapel (Low Chapel) built 1839; still in use. Primitive Methodist chapel at Newbiggin built 1837; closed. Wesleyan Methodist chapel at Fell End built 1861; still in use. New Methodist Church at Newbiggin built 1939; still in use. ***Schools and other institutions:*** grammar school founded by Thomas Fothergill 1688; rebuilt as public elementary school 1872; extended 1895; now Ravenstonedale Endowed Primary School. British school at Newbiggin built by subscription 1872; closed. Fell End school built by subscription 1872; closed c.1956. Reading rooms at Ravenstonedale (built 1891; still open) and Newbiggin. Newbiggin Public Hall built 1925.

REAGILL *see* Crosby Ravensworth

RENWICK*

Ancient parish in Leath ward, Cumberland. Absorbed into Kirkoswald CP 1934.

Acreage: 4,299 acres [1,740 ha]. Moorland at Middle Moor and Long Moor, totalling 850 acres [350 ha], enclosed 1818; further 2,525 acres [1,004 ha] of common land on Renwick Fell enclosed 1864. ***Population:*** around 150 in mid-17th century (though estimated at 265 in 1688); rose from 189 in 1782 to peak of 375 in 1831; then fell to 174 in 1931 (last census year for which separate figure available). ***Landownership:*** *manor of Renwick* granted to Adam de Staveley, lord of Sedbergh and Dent, early 12th century; sold 1296-7 by Thomas de Staveley to Michael de Harcla; forfeited on attainder of Sir Andrew Harcla 1323; granted to Robert de Eaglesfield 1328, who used it to endow Queen's College, Oxford, 1341, with whom lordship of manor remained. ***Economy:*** predominantly agricultural: sheep grazing on fells said to be important in later 17th century. Coal mining recorded 1620; continued until mid-19th century. Quarrying and lime-burning from late 18th century to

1880s. ***Places of worship:*** medieval parish church of All Saints; rebuilt 1737 and 1845-6. Wesleyan Methodist chapel built 1818; rebuilt 1905; still in use. ***Schools and other institutions:*** school near church recorded 1675; disused by 1711. Schoolhouse built 1758 and endowed by Thomas Tallentire (d. 1767); replaced by new National school on waste near church 1838; in turn replaced by Board school 1876, closed 1987. Second school opened by Thomas Robinson, curate, at Linghouse (near Scalehouses) c.1790; closed c.1830. Reading room established in former (1838) school building 1879; closed 1950s; reopened 1977.

RIBTON

Township in Bridekirk parish, Allerdale below Derwent ward, Cumberland. Absorbed into Camerton CP 1934.

Acreage: 616 acres [249 ha], including former common land on Ribton Moor (date of enclosure not known). ***Population:*** declining from c.50 in early 19th century to under 25 by 1851; stood at just 17 in 1931, last year for which separate census figures available. ***Landownership:*** *manor of Ribton* held by Ribton family by 1308; sold by William Ribton to Christopher Curwen and others 1597; bought by Thomas Lamplugh before 1636; descended through Lamplugh family (who built Ribton Hall – now demolished – c.1680) until sold to Lowthers 1722. ***Economy:*** agriculture; salmon fishery in River Derwent, recorded 1688. Royal Navy Armaments Depot covered much of township in mid-20th century (see Broughton). ***Places of worship:*** none. ***Schools and other institutions:*** none.

RICKERBY *see* Stanwix

ROCKCLIFFE

Ancient parish in Cumberland ward, Cumberland, embracing townships of Rockcliffe Castletown and Rockcliffe Churchtown, latter including Rockcliffe village and settlements of Harker and Todhills.

Acreage: 4,939 acres [1,999 ha], divided between townships thus: Castletown: 3,453 acres [1,381 ha]; Churchtown: 1,486 acres [601ha]. Rockcliffe Common enclosed 1815. ***Population:*** estimated at 330 in 1688. Parish population rose from 497 in 1801 to 949 in 1861. Drop to 784 in 1871 attributed to pulling down of dilapidated cottages and amalgamation of farms. Population continued to fall, reaching low point of 545 in 1911, then rising again to lie in range 700-750 for most of 20th century (733 in 2001). ***Landownership:*** *manor of Rockcliffe* held under barony of Burgh by Rockcliffe ('Routheclive') family until sold to Ralph de Bray 1205, whose descendants conveyed it to John le Franceis 1248-9; after complex descent, sold to Dacre family in late 14th century, reuniting it with barony of Burgh (q.v.), with which it descended. Castle and demesnes sold 1682 to Rev. Charles Usher, whose granddaughter devised them to William Strong, mid-18th century; sold to Robert Mounsey, a Carlisle lawyer, who built Castletown House 1809-11. ***Economy:*** predominantly agricultural until 20th century. Inns on main routes through parish: fords across River Eden at Peat Wath (disused by mid-19th century) and Rockcliffe Wath (ferry from Rockcliffe village by 19th century) and major roads (A74/M74 and A7). Parish crossed by railways north from Carlisle: Caledonian railway (opened 1847), with station at Rockcliffe; North British railway (opened 1862), with station at Harker. Industrial estates on southern edge of parish from late-20th century. ***Places of worship:*** medieval parish church of St Mary; rebuilt 1848; tower and spire destroyed by lightning 1899 and rebuilt 1901. Mission room at Todhills by 1920s; now demolished. ***Schools and other institutions:*** unendowed school recorded 1814 and another by c.1830. School at Castletown maintained by G. G. Mounsey by 1860. Parochial school built 1871; enlarged 1890; replaced by modern building on new site, now Rockcliffe CE Primary School. Reading room in Rockcliffe village built 1882; converted to Mounsey Arms public house c.1916, replacing two village pubs closed by Central Control Board. Memorial Hall erected 1921; replaced in 1980s by new village hall (The Rockcliffe Centre) on edge of village. Recreation hut at Harker in 1920s.

ROSLEY *see* Westward

ROTTINGTON

Township in St Bees parish, Allerdale above Derwent ward, Cumberland. CP enlarged by addition of parts of Preston Quarter and Sandwith CPs 1934. Absorbed into St Bees CP 1973.

Acreage: 643 acres [260 ha], before boundary changes. ***Population:*** fluctuating in range 45-70 across 19th and early 20th centuries (peak of 78 in 1901); stood at 60 in 1931 (last census year before boundary changes). ***Landownership:*** *manor of Rottington* passed by marriage from Rottington family to Sandes family 1421; sold to Curwens of Workington 1579. Henry Curwen (d. 1725) devised it to his cousin Sir Michael Warton (d. 1725), on whose death it passed to Charles Pelham (d. 1763), who sold it to Sir James Lowther 1762. ***Economy:*** predominantly agricultural; small-scale sandstone quarrying in late 19th century. ***Places of worship:*** none. ***Schools and other institutions:*** none.

RUSLAND *see* Colton

RYDAL AND LOUGHRIGG

Township in Grasmere parish, Kendal ward, Westmorland.

Acreage: 4,889 acres [1,979 ha], including 936 acres [379 ha] of surviving common land on Loughrigg Fell. ***Population:*** rising steadily from 230 in 1801 to 503 in 1921; then levelling off to stand at 534 in 1971 (last census year for which separate figure available). ***Landownership:*** *manor of Rydal* granted to Roger de Lancaster before 1274; passed to Fleming family by marriage of Isabel daughter of Sir John Lancaster to Sir Thomas Fleming 1408-9, and descended in Fleming family of Rydal Hall until 20th century. ***Economy:*** hill farming; fisheries (recorded in name Fishgarths Wood; 'fish arks' in rivers Rothay and Brathay in 17th century); small-scale quarrying in 19th century. Bobbin manufacture at Skelwith Bridge mid-19th century; had ceased by 1900. Tourism from earlier 19th century (early attraction was William Wordsworth's home at Rydal Mount 1813 to 1850). ***Places of worship:*** church of St Mary at Rydal built 1823-4 by Lady le Fleming; extended 1884. ***Schools and other institutions:*** two dame schools recorded 1818. School supported by Lady le Fleming recorded 1860. Rydal Hall leased by Carlisle Diocese as retreat and conference centre from 1962. Reading room or institute at Rydal, built before 1897; now disused.

ST BEES

Township in St Bees parish, Allerdale above Derwent ward, Cumberland. CP was enlarged by transfer of part of Preston Quarter CP, including parish church and St Bees School, 1896. Further enlarged by transfer of part of Hensingham CP 1934 and by absorbing Rottington CP 1973.

Acreage: 1,814 acres [734 ha], excluding 34 acres [14 ha] of detached portions in Preston Quarter, absorbed into St Bees CP 1882; thereafter 1,848 acres [748 ha] until 1896 boundary changes. St Bees Common (320 acres [129 ha]) enclosed 1816. ***Population:*** rising from 406 in 1801 to 1,041 in 1891 (last census year before boundary changes). Population of enlarged CP fluctuated seasonally: numbers in 1921 (1,609) said to be inflated by influx of summer visitors and in 1931 (1,028) to have been reduced because of school vacations. Population stood at 1,717 in 2001. ***Landownership:*** *manor of St Bees or Kirkby Beacock* granted to St Bees Priory on its foundation c.1120. After Dissolution passed through Challoner and Wybergh families, and was mortgaged to Lowthers. Sir John Lowther of Whitehaven gained possession 1665 and manor descended to earls of Lonsdale, who remained lords in 20th century. ***Economy:*** farming; sandstone quarrying in mid-19th century. Tourism from mid-19th century: hotel erected 1847 'for visitors and students'; station on Whitehaven & Furness Junction Railway opened 1849. ***Places of worship:*** for medieval priory and parish church, see Preston Quarter. United Methodist Free Church built 1866; still in use. ***Schools and other institutions:*** for St Bees School, see Preston Quarter. Ladies' boarding school recorded 1847. National school for girls built in St Bees village before 1860; replaced by mixed Board school on new site 1875; now St Bees Village Primary School. Village hall, founded as 'Hodgett's Club' c.1882, containing reading room, billiard room and dance hall; used as cinema in 1940s. Recreation ground by late 19th century; outdoor swimming pool, built by unemployed men, opened 1934.

ST CUTHBERT WITHOUT

CP in St Cuthbert's parish, Carlisle, Cumberland ward, Cumberland, comprising townships of High and Low Blackwell, Botcherby, Brisco, Carleton, Harraby, and Upperby. Carlisle city boundary extended 1887 and 1894, resulting in some loss of territory; Botcherby, Harraby and Upperby lost when city boundaries extended again 1912. CP absorbed Wreay CP (q.v.) 1934. This entry treats St Cuthbert Without CP as it existed between 1894 and 1912 and is arranged by constituent townships.

Acreage: 8,522 acres [3,449 ha] before boundary changes. Commons and waste grounds totalling 1,673 acres [677 ha] in Carleton, Brisco (and Wreay, in St Mary's parish, Carlisle) within manor of Botchergate enclosed 1778. ***Population:*** 1,039 in 1801, rising steadily to 1,872 in 1861 (last census year for which population of constituent townships is recorded separately).

BLACKWELL (or Blackhall). ***Acreage:*** Blackwell High township: 2,563 acres [1,037 ha]; Blackwell Low township: 974 acres [394 ha]. Blackhall commons (described 1688 as narrow, barren and of little use other than for turbary) enclosed 1697. ***Landownership:*** *manor of Blackwell* granted by Henry I to Odard de Logis, whose descendants took surname de Wigton. Margaret de Wigton (1293-1349) granted it 1334 to Robert Parving. Sold to William Stapleton who sold it 1429-30 to Dacres, through which family it descended until sold 1716 to Sir Christopher Musgrave of Edenhall, remaining with that family to 20th century. ***Economy:*** agricultural until 20th century; Carlisle race course opened 1904 (replacing former course at The Swifts). ***Places of worship:*** Wesleyan Methodist chapel at Stoneraise built 1891; closed c.2010. ***Schools and other institutions:*** bequests to school at Blackwell 1785 and 1798; Durdar school at Stoneraise built by 1862; rebuilt 1871; now Stoneraise Primary School.

BOTCHERBY. ***Acreage:*** 514 acres [208 ha]. ***Landownership:*** *manor of Botcherby* granted to Guy the hunter early 12th century, descending through his heirs, who took surname de Botcherby, to mid-13th century; later descent unclear. ***Economy:*** agricultural until 20th century; housing development since c.1930. Laundry near Botcherby Bridge, late 19th century. ***Places of worship:*** St Andrew's mission church built 1890. ***Schools and other institutions:*** Board school built 1899; closed mid-1920s; re-opened 1939; converted to community centre. St Joseph's Home for elderly (with chapel attached) built 1892; closed 1977; destroyed by fire 2002.

BRISCO. ***Acreage:*** 1,956 acres [792 ha]. ***Landownership:*** part of *manor of Botchergate*, held by Carlisle Priory, passing to Dean and Chapter of Carlisle at Dissolution. ***Economy:*** largely agricultural; quarrying (notably at Newbiggin Quarry) later 19th century. Calico print works established before 1797; converted to iron works for making engines 1846; bought by Cowans Sheldon who moved to St Nicholas works (see Carlisle) later 19th century. ***Places of worship:*** none. ***Schools and other institutions:*** none.

CARLETON. ***Acreage:*** 1,448 acres [586 ha], including part of Wragmire Moss, enclosed 1854. ***Landownership:*** see Brisco. ***Economy:*** largely agricultural; brick and tile works at Scalesceugh mid-19th century. ***Places of worship:*** chapel of Garlands Hospital opened 1875; closed 1997. ***Schools and other institutions:*** bequest to schoolmasters in Brisco, Carleton and Harraby 1785; in 1860 only Carleton had school; new school built 1872; closed 1984. Garlands mental hospital, built as Cumberland and Westmorland Lunatic Asylum, opened 1862; closed 1999 and site redeveloped for housing; replaced by Carleton Clinic.

HARRABY. ***Acreage:*** 616 acres [249 ha]. ***Landownership:*** see Brisco. ***Economy:*** skinnery in 19th century. Lakeland laundry established 1935. Major expansion of housing 1948 to 1955 and later. Industrial estates from later 20th century. ***Places of worship:*** Harraby Methodist church opened 1948; closed 2009. Mission hall c.1954, replaced by St Elisabeth's church, built 1966-7. Roman Catholic church of Christ the King opened 1954. ***Schools and other institutions:*** schools for new housing estates: Petteril Bank School (opened in pre-fabricated huts 1943; rebuilt 1972); Inglewood Infant and Junior Schools (opened 1953-4); Pennine Way Primary School (opened 1958); Harraby Secondary School (opened 1956; became North Cumbria Technology College; renamed Richard Rose Academy 2008). Harraby community centre opened 1963.

UPPERBY. ***Acreage:*** 455 acres [184 ha]. ***Landownership:*** see Brisco. ***Economy:*** alabaster mill manufacturing plaster of Paris mid-19th century. Clay-digging and brick-making late 19th century. Steam laundry from 1890s to mid-20th century with wadding/dyeing works adjacent c.1900. Major housing development 1920s to 1950s. ***Places of worship:*** St John's church built 1843; chancel added 1889-90; municipal cemetery adjacent from 1881. Primitive Methodist chapel built 1901; still in use. ***Schools and other institutions:*** school built c. 1828; rebuilt and enlarged 1854; became parish room when new school built 1912; now Upperby Primary School. Institute, Lamb Street, built 1922. Upperby Park, on former brick fields, opened 1923. Petteril Bank (villa built 1829) extended to accommodate Cumbria Archive Centre, opened 2011.

ST JOHN'S IN THE VALE *see* Castlerigg, St John's and Wythburn

SALKELD, GREAT

Ancient parish in Leath ward, Cumberland.

Acreage: 3,774 acres [1,527 ha], including common moorland to west of village enclosed under Inglewood Forest enclosure award 1819. ***Population:*** estimated at 345 in 1688; rapid increase from 289 in 1811 to 403 in 1821 attributed to enclosure of commons; rose to stand in range 475-500 from 1851 to 1901; then declined to 348 in 1971 before rising again to 445 in 2001. ***Landownership:*** part of honour of Penrith (q.v.). ***Economy:*** agricultural; fishery in River Eden in 17th century; small-scale quarrying 17th-19th centuries. ***Places of worship:*** medieval parish church of St Cuthbert (fortified tower of c.1380). Presbyterian chapel in fields to west of village built c.1710; rebuilt 1750; replaced by new building in village 1876; closed 1953 and converted to dwelling. Wesleyan Methodist chapel at Salkeld Dykes built 1832; still in use. ***Schools and other institutions:*** endowed school founded c.1515; school taught in chancel of church until schoolhouse built 1696. Endowment lost; school taught by curate in early 19th century but had ceased by mid-19th century. New parish school built 1856; closed 2004. Hunter Hall, an independent primary school, opened 1985; moved to Carleton, Penrith, 1992. Reading room and library built 1895; village hall added 1911; whole building now village hall.

SALKELD, LITTLE

Township in Addingham parish, Leath ward, Cumberland. Absorbed into Hunsonby CP 1934.

Acreage: 1,149 acres [465 ha]. Common enclosed 1849. ***Population:*** in range 100-130 across most of 19th century, except for surge to 173 in 1871; declined to 91 in 1931 (last census year for which separate figure available). ***Landownership:*** *manor of Little Salkeld* granted to Carlisle Priory in 12th century, passing to Dean and Chapter of Carlisle at Dissolution. Little Salkeld Hall, seat of Salkeld family until George Salkeld sold it 'for a song' to Col. Thomas Cholmley during Civil War; passed to Charles Smalwood by 1680s in whose family it descended until sold c.1790 to Col. Samuel Lacy, who excavated grotto known as Lacy's Caves on bank of River Eden in late 18th century. ***Economy:*** largely agricultural. Gypsum mine and Long Meg plaster works established 1880; closed 1975. ***Places of worship:*** chapel at Little Salkeld recorded 1282. Barn converted into Anglican chapel in mid-19th century; closed 1970. ***Schools and other institutions:*** none.

SALTER AND ESKETT

Extra-parochial township in Allerdale above Derwent ward, Cumberland. Absorbed into Lamplugh CP 1934.

Acreage: 638 acres [258 ha]. ***Population:*** said to be only 4 farms and 5 families in 1829. Population in range 35-45 until 1860s, then rose rapidly to 196 in 1881 with development of iron-ore mines and quarries; remained high until 1920s (166 in 1921) but dropped as industries declined, to stand at 64 in 1931 (last census year for which separate figure available). ***Landownership:*** granted to St Bees Priory in 12th century; purchased after Dissolution by Dr Leigh, subsequently passing to Salkeld, Patrickson, Robertson, Fryer and Dickinson families. ***Economy:*** extensive limestone quarrying, lime-burning and iron ore mining from mid-19th century to later 20th: Eskett Pit, other mines and quarries recorded 1865; Salter Hall Quarry closed 1926 (now flooded); Eskett Quarry (limestone) remains open. ***Places of worship:*** none. ***Schools and other institutions:*** none.

SANDFORD *see* Warcop

SANDWITH

Township in St Bees parish, Allerdale above Derwent ward, Cumberland. CP abolished 1934 and divided between Whitehaven MB and Rottington CP.

Acreage: 1,400 acres [567 ha]. ***Population:*** rose from 180 in 1801 to stand in range 300-350 for most of 19th and early 20th centuries (peak of 422 in 1881); 332 in 1931 (last census year for which separate figure available). ***Landownership:*** held by St Bees Priory until Dissolution; granted c.1560 to Sir Thomas Chaloner (d. 1565); reverted again to Crown; granted to St Bees School 1604. ***Economy:*** agriculture; sandstone quarrying. Gypsum mine and alabaster quarry in cliffs at Barrowmouth by 1860s; expansion of anhydrite mining for Marchon Chemical Works (see Whitehaven) from 1955 to c.1987. ***Places of worship:*** Anglican mission from St Bees: services held in school room from before 1847 until 1972. ***Schools and other institutions:*** school built 1804; closed 1972. Lighthouse at St Bees Head since 1718 (coal-burning); rebuilt after fire 1822 and again 1865; automated 1987: formed a self-contained community when manned.

SANTON *see* Irton

SATTERTHWAITE

Township in Hawkshead parish, Lonsdale Hundred, Lancashire North of the Sands.

Acreage: 7,322 acres [2,963 ha], divided into four sections: Satterthwaite (2,051 acres [830 ha]), Dale Park (1,113 acres [450 ha]), Graythwaite (1,648 acres [667 ha]) and Grizedale (2,510 acres [1,016 ha]). ***Population:*** rising from 274 in 1801 to 452 in late 19th century (increase of 96 between 1821 and 1831 attributed to establishment of two bobbin mills) then falling to 214 by 1931; slight increase by later 20th century to stand at 257 in 2001. ***Landownership:*** part of Furness Fells granted to Furness Abbey in 12th century. After Dissolution, *manor of Satterthwaite* was included in sale of lands 1614. *Grizedale* estate held by Rawlinson family in 16th and 17th centuries; then Irtons, who sold it to Rev. Thomas Bowman, whose son held it 1850. Montague Ainslie, built New Hall, Grizedale, 1836; estate sold to Harold Brocklebank 1903 and purchased by Forestry Commission 1937. *Graythwaite*: Sandys family owned Graythwaite Hall with some interruptions from c.1555 to early 20th century. Low Hall held by Sawrey family in 16th century, passing by marriage to Rawlinsons in whose hands it remained in 1914, though most of land had been purchased by Colonel Sandys. ***Economy:*** woodland industries important from medieval times, producing charcoal for smelting iron. Bloomery sites thought to be medieval; bloom-smithy in operation 1564. Force Forge (bloomery-forge) opened pre-1680. Cunsey Furnace, charcoal-fired blast furnace, built 1711; sold 1750 and ceased working about then. Bobbin mills at Force Forge and Force Mill established between 1821 and 1831; one converted into electricity works by 1914. Afforestation: Montague Ainslie of Grizedale Hall planted c.1,200 acres [c.500 ha] with larch; Forestry Commission plantations on 8,000 acres [3,238 ha] purchased 1937, initially solely for timber production but from 1960s encouraging public access. Grizedale Vistor Centre and cafe, way-marked walks and sculpture since 1977; also mountain biking and Go Ape aerial assault course (opened 2003). ***Places of worship:*** chapel of ease at Satterthwaite since late 16th century; rebuilt c.1675 and again 1754; replaced by All Saints church, built 1837. Chapel at Graythwaite disused long before 1722. ***Schools and other institutions:*** school at Satterthwaite recorded 1848; replaced by new school at Bowkerstead c.1849 to serve both Satterthwaite and Rusland; closed 2006. Parish room at Satterthwaite built 1883; reading room added 1900; now village hall. Theatre in the Forest at Grizedale opened 1970; closed 1999.

SAWREY *see* Claife

SCALEBY

Ancient parish in Eskdale ward, Cumberland, comprising townships of West Scaleby and East Scaleby.

Acreage: 3,647 acres [1,476 ha], divided between townships thus: West Scaleby: 1,965 acres [795 ha]; East Scaleby: 1,683 acres [681 ha]. Common land at Scaleby Common Wood meadow (62 acres [25 ha]) enclosed 1843, and 200 acres [81 ha] of peat bog on Scaleby Moss enclosed 1853. ***Population:*** estimated at 340 in 1688. Rising from 486 in 1801 to peak of 618 in 1821; gradual decline thereafter to 374 in 1901, falling to low of 275 in 1991; stood at 349 in 2001. ***Landownership:*** *manor of Scaleby* granted to Richard de Tilliol ('Richard the rider') early 12th century and descended in Tilliol family until 15th century, passing by marriage to Musgrave family. Sir Edward Musgrave restored and extended Scaleby Castle 1597-1606. Sold in 1650s to Richard Gilpin, Puritan minister of Greystoke, whose grandson mortgaged it to Edward Stephenson, who foreclosed 1741; descended through Stephenson (from 1834 Standish) family until sold 1946. ***Economy:*** predominantly agricultural; much meadow land: noted for horse-breeding in later 17th century. Commercial peat digging on Scaleby Moss by mid-19th century; tile-making in 19th century. ***Places of worship:*** medieval parish church of All Saints; restored 1828 and 1861-2. Wesleyan Methodist chapel at Scaleby Hill opened 1828; closed; now private house. ***Schools and other institutions:*** school endowed 1773; rebuilt as National school adjoining church 1850s; closed 1933 and became parish hall. Reading room and public hall built 1895, containing library (established 1898); now village hall.

SCALES (Lancs.) *see* Aldingham

SCALTHWAITERIGG

CP in Kendal parish, Kendal ward, Westmorland, created 1897 from part of township of Scalthwaiterigg, Hay & Hutton-in-the-Hay. For remainder of former township, see New Hutton.

Acreage: CP contained 1,242 acres [503 ha] before loss of territory (c.200

acres [c.80 ha]) to Kendal 1919 and 1935. Former township of Scalthwaiterigg, Hay & Hutton-in-the-Hay contained 3,434 acres [1,390 ha]. Common land at Mint's Feet (105 acres [42 ha]) enclosed 1815. ***Population:*** township population rising throughout 19th century from 250 in 1801 to 653 in 1891 (last census year before boundary changes). Scalthwaiterigg CP contained 585 inhabitants in 1911, most in built-up area on edge of Kendal; after loss of those areas to Kendal, remaining population fell from 182 in 1921 to low of 81 in 1971; stood at 104 in 2001. ***Landownership:*** part of Marquis fee of barony of Kendal (q.v.). ***Economy:*** farming; industrial colony at Mealbank since 18th century: fulling mill and woollen manufacture from 1748; Braithwaite & Co. woollen mill from 1830s; closed 1966. Snuff manufacture (Samuel Gawith & Co.) from 1792 to 1921. Oil mill (extracting oil from mill waste) late 19th/early 20th century. Industrial estate on site of former mills. ***Places of worship:*** hospital of St Leonard (site now Spital Farm) founded early 13th century; dissolved at Reformation. ***Schools and other institutions:*** school established at Mealbank 1846 by F. & C. Braithwaite for children of workers at their woollen mill; rebuilt 1868 with reading room and library; later known as Scalthwaiterigg Mealbank Council School; closed mid-20th century and converted to housing.

SCOTBY *see* Wetheral

SEASCALE *see* Gosforth

SEATON

Township in Camerton parish, Allerdale below Derwent ward, Cumberland. Siddick and industrial area on coast transferred to Workington MB 1899.

Acreage: 2,626 acres [1,063 ha] before boundary change. Seaton common and townfield enclosed 1826. ***Population:*** rising from 562 in 1801 to 835 in 1851; then growing rapidly to stand at 3,229 in 1891, last census year before boundary change, which resulted in 1,884 inhabitants being transferred to Workington. Population of remaining area of Seaton grew again after Second World War, rising from 2,649 in 1951 to 4,861 in 2001. ***Landownership:*** Gospatric son of Orm, ancestor of Curwens, said to have had his seat at Seaton (supposedly on site of Roman fort at Burrow Walls) in mid-12th century, before moving it across River Derwent to Workington. *Manor of Seaton* descended with Workington thereafter. ***Economy:*** market charter granted

1280; salmon fishery in River Derwent from medieval period (memory preserved in name Salmon Hall). Rabbit warren (and race course) on coastal common land in 17th century. Coal mining: Seaton colliery (pits at St Helens and Moorhouse Guards in north of township and at Seaton Banks in south) recorded by 1673, expanding during 1720s and 1730s; waggonway (first railway in Cumberland) from colliery to Workington harbour, completed 1732. Mining expanded in 19th century and continued, exploiting undersea seams, well into 20th: mining at Moorhouse Guards ended 1945; Siddick pit remained open until 1966. Iron making: Seaton ironworks, charcoal-powered blast furnace, built at Barepot 1762; closed 1857. Major blast furnaces built by Workington Haematite Ironworks 1856. Associated industries included coke-making (24 coke ovens built at St Helens Colliery 1894) and fireclay brick-making, with brick and tile works of Seaton Fire Brick Co. at Moor House Guards from 1840. 20th-century industrial development: textile mill on former iron works site, built 1937; paper and board manufacture (Thames Board Mill established 1966); chemicals (Ectona Fibres, established 1968); construction industry (Thomas Armstrong, from 1986); retail park (Dunmail Park) on site of textile mill opened 1988; wind farm along coast, built 1996. ***Places of worship:*** possible early chapel site preserved in name St Helens in north of township. Schoolroom licensed for services 1852; Anglican church of St Paul built 1881-3. Holy Trinity church, Northside (otherwise West Seaton), built 1893. Wesleyan Methodist chapel in Low Seaton in later 19th century. ***Schools and other institutions:*** parochial school, Low Seaton, by 1850s. National school, High Seaton; replaced by Seaton CE Primary school on new site; old building became Seaton Infant School (now Seaton Academy). Primary school, Northside (West Seaton) since 1878. Siddick School built 1902; closed 1967. Reading and recreation room at Seaton, built late 19th century. Good Templars' Hall built 1884.

SEBERGHAM

Ancient parish in Allerdale below Derwent ward, Cumberland, containing two townships or 'bounds': High Bound (covering Sebergham Churchtown, Warnell and Parkhead) and Low Bound (covering Upper and Nether Welton).

Acreage: 5,557 acres [2,249 ha], of which 3,724 acres [1,507 ha] was in High Bound and 1,833 acres [742 ha] in Low Bound. Common land on Sebergham Common and Warnell Fell (2,896 acres [1,172 ha]) enclosed 1765. ***Population:*** estimated at 495 in 1688; rising from 434 in 1801 to peak of

745 in 1861, then slowly falling to 361 in 2001. ***Landownership:*** part of forest of Inglewood, with which superior *manor of Sebergham* descended. Another estate called *manor of Sebergham* granted to Carlisle Priory in 13th century, passing to Dean and Chapter of Carlisle after Dissolution. Smaller subsidiary manors within parish included *Warnell* (acquired by Denton family from Dacres 1506; sold to Lowthers 1774); *Parson's Park* (belonging to rector of Caldbeck) and *Hartrigg* (held by Dalston family before being acquired by John P. Fletcher in mid-19th century). ***Economy:*** largely agricultural. Fulling mill at Warnell, recorded 1688. Coal mining and limestone quarrying at Warnell by 17th century; lime-burning and tile-making in 19th. ***Places of worship:*** medieval parish church of St Mary; tower added 1825; restored 1880 and 1905. St James' Church (Anglican) at Welton built 1874. Wesleyan Methodist chapel at Welton (by 1861; closed and converted to cottage). ***Schools and other institutions:*** school taught in west end of church early 18th century; schoolhouse near church by early 19th century (housing circulating library and reading room, established 1858); closed and sold to church 1952; now used as meeting room and accommodation for groups. School at High Welton (later Stony Cross Board School; then Welton Primary School) built 1745; rebuilt 1865 and again in 1960s; closed 2011. Village halls at Welton (built mid-20th century) and Churchtown (built 1983).

SEDBERGH

Market town and township in Sedbergh parish, Ewecross wapentake, West Riding of Yorkshire.

Acreage: 19,603 acres [7,933 ha], divided between four subdivisions thus: Howgill & Bland: 5,443 acres [2,203 ha]; Frostrow & Soolbank: 3,292 acres [1,332 ha]; Marthwaite: 2,570 acres [1,040 ha]; and Cautley & Dowbiggin: 8,297 acres [3,358 ha]. Over 15,000 acres [6,100 ha] survives as registered common land, largest commons being Brant Fell (6,758 acres [2,735 ha]) and Bluecaster Fell (7,556 acres [3,060 ha], partly in Garsdale). ***Population:*** rising from 1,639 in 1801 to around 2,200-2,300 by mid-19th century, remaining fairly stable across 20th century. Stood at 2,705 (including Sedbergh School) in 2001. ***Landownership:*** *manor of Sedbergh* held by de Staveley family until mid-13th century, passing by marriage of Alice, daughter of Adam de Staveley, to Henry Fitz-Randulph; their descendants, who took name Fitz-Hugh, held it until 15th century. Divided into moieties through 16th century: one moiety granted by Henry VII to Edward Stanley, Lord Mounteagle; other moiety passing through Scrope and Dacre families. Both parts bought 1601 by Sir Thomas Strickland, in whose family manor remained until sold to John Upton of Ingmire Hall in late 19th century. Ingmire estate was finally broken up 1928.

Origins and growth of the town. Tuesday market and annual fair on 8 September granted 1254; market may have failed as re-established 1526. Remained small in 19th century, though livestock fairs still held in 1880s. Quarrying for limestone, Silurian 'Blue Rag' stone and sandstone; lime-burning in 19th century. Textile industry: hand knitting and some hand loom weaving in 17th and 18th centuries. Water-powered textile mills from late 18th century. Birks Mill (cotton, woollen) established 1790; had become bobbin mill with dyeing and cleaning works by 1880s; site used for egg packing later 20th century; now food depot. Hebblethwaite Mill built 1793, initially as carding mill; later spinning; sold 1812. Millthrop Mill established 1796, initially as cotton mill, later making horse blankets; burnt down 1967; now private housing. Farfield Mill built 1837 as woollen mill by Joseph Dover; specialised in horse blankets, tweeds; water wheel replaced by turbine 1896 and by steam 1911; rebuilt after fire 1909; closed c.1995; reopened as arts and heritage centre. At Howgill, fulling mill had become wool spinning mill by 1816; disused by 1890s. Tourism grew in 20th century, with specialism in book-selling as a 'Book Town'. Sedbergh School remained major employer.

Places of worship. Medieval parish church of St Andrew; restored 1886. Chapel of ease at Howgill built and endowed by John Robinson 1685; replaced by Holy Trinity Church, built 1838. St Mark's Church, Cautley, built 1845-7. St Gregory Vale of Lune built as private chapel ('Weathercock chapel') by Uptons of Ingmire Hall 1861; remodelled c.1907 and consecrated 1918; closed 1984. Mission room at Millthrop built 1888. Sedbergh School chapel 1890, rebuilt 1895-7. Quaker meeting house at Brigflatts built 1675; still in use. Wesleyan Methodist chapels in Sedbergh (built 1805; rebuilt on new site 1864 and 1914; still in use); Cautley (built 1845; still in use) and Frostrow (built 1886; still in use). Primitive Methodist chapel, Millthrop, built 1889; closed 1993. Independent chapel, Sedbergh, built 1821; rebuilt on new site 1828 and enlarged 1871; now United Reformed Church. Roman Catholic chapel in Millthrop from 1956 to 1975, sharing parish church since then.

Schools. Sedbergh School founded 1525 as grammar school combined with chantry; became Free Grammar School of King Edward VI 1551; re-organised under new scheme of governance 1874, becoming independent boarding school. Rebuilt 1716 (building now school library); most school buildings date from period 1878-1907. Parish school adjoining church replaced by National school, built 1842. British school to north of town near Castlehaw built 1843; rebuilt 1901. These combined and replaced by Sedbergh Primary School 1967. Secondary modern school built 1949; became comprehensive 1980; now Settlebeck High School. Several dame schools in 19th century, including one run by Julia Green converted to preparatory school 1897-1938. Baliol School established as girls' school 1900; closed 1932; used as military training school during Second World War; converted to local authority special school 1970s; closed c.2011. Schools outside town included that attached to Howgill chapel (built late 17th century; rebuilt on new site 1850; closed c.1964) and Cautley and Dowbiggin School (opened 1877; closed 1955).

Other institutions. Parish workhouse opened 1732; Sedbergh Poor Law Union (covering whole ancient parish, including Garsdale and Dent) formed 1840; new Union Workhouse built on southern edge of town 1853-4; closed 1930; converted to housing 1976. Almshouses (Widows' Hospital) founded by Thomas Palmer, built 1849; renovated 1960; still open. Market hall and reading room given by Rev. J. H. Evans, headmaster of Sedbergh School, 1858; now library. Public hall built 1865; later became a cinema; now bookshop. Peoples Hall built 1956 as village/community hall. Village hall at Howgill in former school building. Cemetery opened 1890 (given to town by Florence Upton Cottrell-Dormer of Ingmire Hall); Queens Gardens opened 1902.

SEDGWICK

Township in Heversham parish, Kendal ward, Westmorland.

Acreage: 495 acres [200 ha]. ***Population:*** rising from 161 in 1801 to 247 by 1851, then dropping to 194 in 1861 as result of relocation of gunpowder works; rose again to 291 in 1901, then fell back to low of 152 in 1951 before housing development led to increase to 380 by 2001. ***Landownership:*** held with Sizergh (of which manor it was often termed a 'hamlet') from 13th century (see Helsington). Wakefield family, bankers and gunpowder manufacturers, acquired large landholding through purchases in 19th century, building Sedgwick House c.1810 (rebuilt 1868-9). ***Economy:*** largely agricultural until establishment of gunpowder works ('Old Sedgwick' works) by John Wakefield 1764; closed 1854 when relocated to 'New Sedgwick' works on other side of River Kent (see Helsington). New incorporating mill for gunpowder established at Basinghyll 1790; closed 1936. Small-scale quarrying of limestone 'marble' 18th and 19th centuries. ***Places of worship:*** none.

Schools and other institutions: school for girls near Sedgwick House from c.1820; replaced by new school in village c.1870; closed c.1906 when pupils transferred to Crosscrake school. Village hall/club room built by Wakefield family c. 1870; rebuilt 1889; still in use. Reading room established by 1868.

SELLAFIELD *see* Beckermet St Bridget; Ponsonby

SELSIDE AND WHITWELL

Township in Kendal parish, Kendal ward, Westmorland.

Acreage: 3,388 acres [1,371 ha]. Whiteside Common (642 acres [260 ha]) enclosed 1829. ***Population:*** rising from 192 in 1801 to peak of 335 in 1841, then declining to 202 by 1891; in range 170-200 across 20th century; stood at 198 in 2001. ***Landownership:*** *manor of Selside* held by Thornburgh family by 14th century; *Whitwell* held by Bellingham family. Two manors united in 16th century by marriage of Thomasin Bellingham (d. 1582) to Sir William Thornburgh (d. 1552). Descended in Thornburgh family, but sold off in parts early 17th century. Selside Hall passed on death of Francis Thornburgh (d. 1769) to his sister Mary, wife of Ralph Riddell, in which family it remained until 2007. ***Economy:*** predominantly agricultural. Coaching inn (Plough Inn) on main Kendal-Shap road (A6); closed 2011. ***Places of worship:*** chapel at Selside Hall may have served community; rebuilt at distance from Hall (as Thornburghs were recusants) 1707; rebuilt again as St Thomas' church 1838; restored 1894. ***Schools and other institutions:*** schoolmaster licensed 1691 and curate taught school 1695 but no endowment. School endowed by John Kitchen 1730; rebuilt by subscription 1793, and again 1837; rebuilt after fire 1897; now Selside Endowed CE Primary School. Village hall (Selside Memorial Hall) built c.1950.

SETMURTHY

Township in Brigham parish, in Allerdale above Derwent ward, Cumberland.

Acreage: 2,780 acres [1,125 ha], excluding small detached portion at Wythop Mill, annexed to Embleton CP 1887. Common land in Setmurthy and Embleton, totalling 3,150 acres [1,275 ha], enclosed 1824. ***Population:*** fluctuating in range 140-190 across 19th and earlier 20th century; then declining in later 20th century to stand at 108 in 2001. ***Landownership:*** part of honour of Cockermouth (q.v.). Subsidiary *manor of Hewthwaite* held by Huthwaite family from 13th century, passing by marriage to Swinburne family in later 14th century and descending with Brigham (q.v.). John Swinburne rebuilt Hewthwaite Hall 1581. ***Economy:*** largely agricultural; Hay Fell noted as sheep heaf in 17th century, when there was also race course on common. Tanning implied by name Barkhouse. Small-scale quarrying 19th century. Forestry plantations, tourism and leisure businesses in 20th century. ***Places of worship:*** medieval chapel of ease, recorded early 13th century; present church of St Barnabas built 1794; rebuilt 1870. Quaker meeting house at Ruddings built 1799; closed 1828. ***Schools and other institutions:*** reader in chapel was also schoolmaster in early 18th century; schoolroom built 1795. New school built 1896; closed 1934; school building became parish hall.

SHAP

Ancient parish in West ward, Westmorland, including chapelries of Swindale and Mardale. Divided 1904 into Shap Urban and Shap Rural CPs; this entry treats whole of ancient parish.

Acreage: 27,177 acres [10,998 ha] before Shap Rural CP gained 159 acres [64 ha] of Crosby Ravensworth CP in 1935. Acreage includes extensive unenclosed fell commons of Ralfland Forest (3,838 acres [1,553 ha]) and Mardale Common (3,345 acres [1,354 ha]). Shap Common (2,463 acres [997 ha]) enclosed 1813; Shap rough intake (35 acres [14 ha]) in 1836; Wadshaw and Hazel Bank, Fawcett Forest (307 acres [124 ha]) in 1867. ***Population:*** rising across 19th century from 828 in 1801 to peak of 1,416 in 1881. Population of Shap Urban CP in range 1,000-1,200 across 20th century, standing at 1,221 in 2001. Population of Shap Rural CP fluctuated in range 250-400 in earlier 20th century, then declined to stand at 119 in 2001. ***Landownership:*** *manor of Shap* held by Curwens of Workington before being granted to Shap Abbey. At Dissolution passed to Lord Wharton and was later sold to Robert Lowther, from whom it descended to earls of Lonsdale. *Mardale* and part of *Swindale* lay in Thornthwaite Forest (see Bampton). *Manor of Rosgill* held by Rosgill family in 13th and 14th centuries, passing by marriage c.1362 to Salkeld family, who held it until 1631, when it passed by marriage to Christians, from whom it was purchased by Lowthers in later 18th century. *Manor of Hardendale and Wasdale* held by Byland Abbey (Yorks.), passed to Crackanthorpes of Newbiggin at Dissolution. ***Economy:*** village of Shap, strung out along main Kendal-Penrith road (A6) developed into small market centre: charter for market and 3 annual fairs obtained 1687; market hall built soon thereafter. Market and fairs had declined almost to extinction by mid-19th century. Boosted by arrival of Lancaster-Carlisle railway 1846, with station at Shap (station closed 1968). Market revived for farm produce, butter and eggs: new market house built 1861. Rural parts dominated by livestock farming and quarrying. Slate quarries by mid-19th century, at Thornship and Rosgill and at Mosedale (roofing slate quarried until 1920s). Granite quarrying, both pink granite in Wasdale (for building) from 1864 and blue at Wickersgill (for aggregate) from 1882, when Shap Silicated Granite Pavement Co. formed. Extensive limestone quarrying and lime-burning by mid-19th century. Large-scale commercial quarrying and lime works from 20th century: limestone quarry at Shap Beck opened 1930s; Shap Fell (or Hardendale) quarry opened 1962 with lime works nearby. ***Places of worship:*** Shap Abbey founded c.1201. Medieval parish church of St Michael; tower built 1828; rest rebuilt 1897-8. Keld chantry chapel founded c.1350. Chapel of ease at Mardale built 1596; consecrated for burials 1728; demolished 1937 on construction of Haweswater reservoir. Chapel of ease at Swindale recorded 1703; rebuilt 1749; restored and re-opened 1874; closed 1937. Quaker meeting house in Shap built 1704; closed by 1778. Wesleyan Methodist chapel, Shap, built 1845; rebuilt 1935; closed 2009. ***Schools and other institutions:*** school taught in parish church in 1703; separate schoolhouse built by 1815. Replaced by parish school for boys built 1838; Board school for girls and infants opened 1882; these combined 1955 on site of boys' school; now Shap Endowed CE Primary School. Former girls' school site became Shap Secondary School, open from 1955 to 1970. Endowed school attached to Swindale chapel founded 1703; still in use in 1898. Board schools at Rosgill (opened 1878; closed 1955) and Wickersgill (opened 1882 for children of granite quarry workers; closed 1960s; converted to dwelling c.2009). Workhouse for West Ward Poor Law Union built 1877; had closed by 1924; later used as children's home, then sold to Shap Granite Co. for housing. Reading room opened 1872; enlarged 1924 as Memorial Hall. Former secondary school became magistrate's court 1970; then youth centre and now Shap Library. Library and reading room at Mardale in late 19th century.

SILLOTH

Port and coastal resort in Holm Cultram parish, Allerdale below Derwent ward, Cumberland.

Silloth became separate CP 1934, its territory (1,324 acres [536 ha]), including old village of Skinburness, carved from Holme Low CP (q.v.), and Skinburness Marsh.

Origins and growth of the town. Silloth was creation of Carlisle & Silloth Bay Railway & Dock Co., established 1854 with intention of providing Carlisle with port for ocean-going ships. Railway linking Carlisle to Silloth built 1856 and wet dock constructed 1857-9. From early on, promoters recognised that Silloth's potential as seaside resort could fill gap left by shaky revenues from port. New town was laid out on grid plan by 1860 and by 1871 Silloth was described as a 'watering place [to which] many visitors resort during the summer months'.

Floating dock and harbour, built 1857-9, not only enabled trade to grow but also linked Silloth to wider Irish Sea region: steamers daily to Liverpool (until 1917), and twice-weekly to Dublin (still operating 1929); as well as regular services to Isle of Man. New dock opened 1885 and port flourished between then and First World War. Early industries included salt-making, chemical works (from c.1868, processing imported phosphate and guano) and mineral water manufacture. Solway Chemical and Manure works, a second fertilizer manufacturer using imported phosphate, established 1878; closed 1962. Carr's Flour Mill, built 1887, processing grain imported from North America, ensured port's survival. Sand dunes along shore used for weapons testing at Battery from 1886 and marsh towards Skinburness used as training ground for Royal Artillery.

By 1866 two hotels had been built on sea front: Solway Hotel and Queens Hotel. Public baths, drawing sea water by steam engine, opened 1856 and resort also sported pavilion on dunes and cricket ground. By 1899 there were bowling green, golf links, tennis courts and refreshment room on the Green (Silloth Cocoa House Co. Ltd). By 1901 there were 10 hotels and 117 lodging houses/apartments in Silloth and further tourist accommodation, including Skinburness Hotel (hydropathic hotel built in 1880s), at Skinburness. Population had risen to c.2,000. In its heyday in early 20th century, Silloth drew on middle-class visitors from border counties and excursionists from industrial workforce in West Cumberland and Carlisle. By 1951 town's population stood at 3,081.

Economic stagnation during 1950s and 1960s, as port declined and railway station closed (in 1964), resulted in fall in population to 2,662 by 1971. Port underwent revival in late 20th century, with increased flour imports, and revivals of cattle trade and fertilizer manufacture. In early 21st century it handled c.80 ships yearly, importing grain, raw materials for fertilizer manufacture, paper pulp and molasses for animal foodstuff. Tourism revived with growth of caravan parks along coast in late 20th century. Population rose again in later 20th century to stand at 2,932 by 2001.

Places of worship. Christ Church (Anglican) built 1869-70; tower and spire added 1878. Congregational church, Wampool Street, built 1862; closed 1952 and sold to Roman Catholics; now Our Lady of the Assumption RC Church. Wesleyan Methodist chapel erected 1875; now Silloth Methodist Church. Primitive Methodist chapel, Esk Street, erected 1877; closed c.1932 and used as church hall; sold 1970s; demolished 1983. St Andrew's Presbyterian (now United Reformed) church, Solway Street, built 1886. Mission room, Solway Street, built 1887.

Schools. St Paul's National School, Liddel Street, built 1857 (building converted into Solway Coast AONB Discovery Centre, opened 2002). Infant school built 1896; closed 1982 and converted into Silloth Community Hall. Silloth Primary School opened on new site 1982. Secondary school opened 1939; became comprehensive 1977-8; now Solway Community Technology College. Several private schools in earlier 20th century.

Other institutions. Silloth Baths opened 1856; closed after First World War and converted to tea room in 1920s and to amusement arcade in 1950s. Sanatorium for Cumberland and Westmorland (now Silloth Nursing Home) built 1862; children's ward added 1882. Oddfellows' Hall built before 1886; now converted to supermarket. Theatre, Silloth Pavilion, built on part of the Green in 1920s but demolished 1950s.

SKELSMERGH*

Township in Kendal parish, Kendal ward, Westmorland. Part of CP was transferred to Kendal 1935.

Acreage: 2,093 acres [847 ha] before boundary change; 1,939 acres [784 ha] after loss of 154 acres [62 ha] to Kendal 1935. Common land on Skelsmergh Fell (196 acres [79 ha]) enclosed 1829. ***Population:*** rising from 247 in 1801 to 367 in 1881, then falling back to 309 in 1931 (last census year before loss of territory to Kendal). Reduced CP had population of under 300 in later 20th century; stood at 271 in 2001. ***Landownership:*** *manor of Skelsmergh* held by Leyburne family from 13th to 16th centuries; then sold in parcels. Parts acquired by Bellingham family became *reputed manor of Skelsmergh*, descending with Levens (q.v.). Leyburnes retained Skelsmergh Hall and some land until forfeited 1715. ***Economy:*** farming. Fulling mills at Beck Mills (Spital Mills) on River Mint and on River Sprint in 16th century. New fulling mill at Beck Mills, built 1723, had become flax mill by 1804 and worsted spinning mill 1829 to 1865; other uses thereafter; converted to housing 1948. Logwood and sickle mill at Scarfoot by 1743; woollen mill there in 19th century, also grinding raw materials for dyeing; closed 1920. Bobbin mill built at Oakbank late 1840s; later wood turning and paper waste; closed 1981. Stocks Mill a tannery in 1743; bobbin mill in earlier 19th century. Tannery at Garth Row in 17th century. Industrial estates on area transferred to Kendal 1935 with wide range of manufacturing, distribution and servicing activities in late 20th century. ***Places of worship:*** chapel of ease dedicated to St John the Baptist recorded 1578 but fell into decay in mid-17th century; modern Anglican church of St John the Baptist built 1871. Roman Catholic chapel of Ss Robert and Alice at Dodding Green originated in mass centre there, recorded from 1680s; room converted to chapel in 18th century; closed 1990; buildings taken over by Cenacolo Community 2004. Christian Brethren mission room at Garth Row by 1858; closed 1955. Kingdom Hall for Kendal Jehovah's Witnesses built 2006. ***Schools and other institutions:*** National school at Garth Row built 1874; closed 1949. Parish room built 1897, becoming village hall (Skelsmergh Community Hall).

SKELTON

Ancient parish in Leath ward, Cumberland, containing townships of Skelton, Lamonby (with Ellonby) and Unthank (with Laithes). CP enlarged by absorbing Hutton-in-the-Forest and Middlesceugh & Braithwaite CPs 1934.

Acreage: ancient parish covered 7,417 acres [3,002 ha], divided between constituent townships thus: Skelton: 2,155 acres [872 ha]; Lamonby: 2,630 acres [1,064 ha]; Unthank: 2,632 acres [1,065 ha]. Parish included c.4,000 acres [1,619 ha] of common pastures, enclosed 1767. Open fields enclosed 1787. ***Population:*** estimated at 615 in 1600, in 1801 it stood at 729, divided between townships as follows: Skelton: 270; Lamonby: 244; Unthank: 215. Remained fairly stable in range 600-700 from 1871 to 1931 (last census year before boundary change). ***Landownership:*** *manor of Skelton* held by Boyvill family from 12th century; divided after death of Helewise (d. 1272), wife of Eustace Baliol, between six heiresses. One part descended to Southaik family, who were of Hardrigg Hall, Lamonby. ***Economy:*** predominantly agricultural; limestone quarrying and lime-burning in 19th century. Several inns serving traffic on Penrith-Wigton road. ***Places of worship:*** medieval parish church of St Mary or St Michael; nave rebuilt 1794; chancel rebuilt 1879. Supposed private chapel at Lamonby Hall in 16th century. Primitive Methodist chapels at Skelton (built 1865; now Skelton Methodist Church) and Lamonby (recorded 1901). ***Schools and other institutions:*** school at Skelton built 1750 by Isaac Milner; further endowment by Joseph Nelson 1818; rebuilt and enlarged 1849; replaced by modern building on new site; now Skelton Primary School. School at Lamonby by 1899; closed and converted to dwelling. Library established 1846 (with reading room, recorded 1901). Toppin Memorial Hall opened 1923, built in memory of J. C. Toppin of Musgrave Hall (d. 1915 aged 80); now housing library.

SKELWITH BRIDGE *see* Hawkshead and Monk Coniston with Skelwith

SKIDDAW FOREST

Extra-parochial fell pasture, which formed separate CP until absorbed into Underskiddaw CP 1934.

Acreage: 2,969 acres [1,202 ha], almost entirely open fell. ***Population:*** probably uninhabited until Skiddaw House was built c.1829 as keeper's lodge and shepherd's house (inhabited until 1969): 7 to 10 inhabitants in later 19th century. ***Landownership:*** demesne land of Fortibus share of honour of Cockermouth in 13th century. After reunification of two halves of honour 1323, it descended with honour of Cockermouth (q.v.). ***Economy:*** demesne cattle pasture in 13th century; fell grazing thereafter. Grouse shooting in 19th century. Hostel for walkers (Skiddaw House) since 1980s. ***Institutions:*** none.

SKIRWITH

Township in Kirkland parish, Leath ward, Cumberland. Absorbed into Culgaith CP 1934.

Acreage: 5,249 acres [2,124 ha]; common land on Skirwith Moor (691 acres [280 ha]) enclosed 1858. ***Population:*** 189 in 1801, rising (with some fluctuation) to 314 in 1861; slow decline thereafter to 227 in 1931 (last census year for which separate figure available). ***Landownership:*** *manor of Skirwith*, part of barony of Adam son of Swein in 12th century; held by Middleton family in 16th century, passing to Huttons who sold it 1606 to Agnes, widow of William Fleming; then descended in Fleming family of Rydal (q.v.) to 20th century. Other estates included *manor of Bank*, held by Salkeld family until

1399 when passed by marriage to Crackanthorpes who held it until 20th century, and *Skirwith Abbey*, built by John Orfeur Yates 1768, passing to his son Francis Aglionby (Yates) who sold it 1822 to William Parker, in whose family it descended to late 20th century. ***Economy:*** farming. Coal mining (ceased by 1860); quarrying and lime-burning; lead mining on Cross Fell. ***Places of worship:*** church of St John the Evangelist built 1856. Wesleyan Methodist chapel built 1819; rebuilt 1868; sold 1974 and converted to dwelling. ***Schools and other institutions:*** school built 1828; rebuilt on new site 1878; closed 1974; used as hostel by Buttershaw School, Bradford, 1976-86. Cottage reading room recorded 1897, with library, recorded 1901. Village hall opened 1929.

SLEAGILL

Township in Morland parish, West ward, Westmorland.

Acreage: 1,390 acres [563 ha], including c.600 acres [243 ha] of common land enclosed 1813. ***Population:*** in range 120-150 for most of 19th century (peak of 184 in 1831); then drifted down across 20th century (low point of 81 in 1971) to stand at 100 in 2001. ***Landownership:*** part of manor of Newby (q.v.) in 14th century; held by de Sleagill family in 13th century. Later descent uncertain. ***Economy:*** predominantly agricultural. Quarrying and lime-burning in 19th century. ***Places of worship:*** Methodist church built 1954; in use. ***Schools and other institutions:*** school recorded 1818. Endowed school for boys and girls built 1858; closed c.1974 and converted to dwelling. School for girls and infants recorded 1860.

SMARDALE

Township in Kirkby Stephen parish, East ward, Westmorland. Absorbed into Waitby CP 1895.

Acreage: 1,664 acres [673 ha], including 643 acres [260 ha] of common land on Smardale Fell, enclosed 1849. ***Population:*** in range 35-55 across 19th century, standing at 36 in 1891 (last census year for which separate figure available). Numbers swelled to 180 in 1871 by railway navvy 'shanty town' during railway construction. ***Landownership:*** purchased from Helbeck family by Smardales in 1292; passed by marriage to Warcops c.1385 and, after 1580, to Dalston family of Dalston, who built (or rebuilt) Smardale Hall. W. H. Wakefield was lord of manor 1885. ***Economy:*** farming; sandstone quarries and extensive limestone quarrying and lime-burning in mid-19th century. ***Places of worship:*** Chapel Well (spring) said to be site of early chapel. ***Schools and other institutions:*** none.

SOCKBRIDGE

Township in Barton parish, West ward, Westmorland, including village of Tirril.

Acreage: 1,159 acres [469 ha], including several small detached portions and one large one at High Winder (51 acres [21 ha]) but excluding intermixed detached portions of High Barton township. Sockbridge CP contained 1,195 acres [484 ha] after absorption of detached portions into respective CPs. Until 1934 Barton Fell (1,718 acres [695 ha]) was common to townships of Sockbridge, High Barton and Low Winder. Common land at Sockbridge and Yanwath & Eamont Bridge (228 acres [92 ha]) enclosed 1818. ***Population:*** in range 200-250 across 19th and first half of 20th century (peak of 263 in 1831 attributed partly to establishment of school); dropped to 196 in 1961, then rose again to stand at 397 in 2001. ***Landownership:*** *manor of Sockbridge* held by Lancaster family from 13th century, until passing to Lowther family by marriage of Frances, daughter of Christopher Lancaster, to Sir Christopher Lowther 1638; descended in Lowther family thereafter. ***Economy:*** farming; limestone quarrying and lime-burning; chair-making in earlier 19th century; brewing from mid-19th century: brewery at Tirril closed 1899 when taken over by Penrith brewery. ***Places of worship:*** Quaker meeting house at Tirril built 1731 (porch dated 1733); rebuilt 1801; closed 1862. Used as village reading room from 1902; sold 1932 and converted to dwelling. Wesleyan Methodist chapel built 1879; closed by 1980; converted to dwelling. ***Schools and other institutions***: 'mathematical academy' established in Tirril by John Slee (d. 1828) and continued by his son Thomas (d. 1849); gained renown as boarding establishment, attended by 'university gentlemen'. Public elementary school built 1854 in corner of township close to Yanwath; became Sockbridge & Tirril Yanwath Council School; now Yanwath Primary School. Reading room and library at Tirril built 1914; now village hall.

SOLPORT

Township in Stapleton parish, Eskdale ward, Cumberland. CP enlarged by absorbing Trough CP 1934.

Acreage: 3,237 acres [1,310 ha] before enlargement. Common land at Bableton Bogs and Nethershawhead (470 customary acres [c.750 statute acres; 300 ha]) enclosed 1760; Shawhill and Blackbush Rigg commons enclosed 1761. ***Population:*** 259 in 1801, rising to peak of 360 in 1821; declined thereafter, falling to 152 by 1931 (last census year before boundary change). ***Landownership:*** *manor of Solport* appears to have been held with Stapleton by Tilliol family until passing by marriage in early 15th century to Colvilles and from them to Musgraves of Hayton. Sir Edward Musgrave sold it in 17th century to Grahams of Netherby, in which family it descended. ***Economy:*** largely agricultural; limestone quarrying and lime-burning; tile-making; coal mine at Hagg Beck in mid-19th century. ***Places of worship:*** place-names Chapel Brow/Well said to record site of early chapel. Quaker meeting house at Thorney Land built 1698; rebuilt 1773; closed 1870 but occasional use until 1926 (used by Wesleyan Methodists c.1890). United Presbyterian church at Holywell built 1837; restored 1882; now disused. ***Schools and other institutions:*** 'old school room' at Shank Hill, recorded 1860 when about to be rebuilt; rebuilt 1879; enlarged 1960s, 1970s and 2007; now Shankhill CE Primary School.

SOULBY (Cumberland) *see* Dacre

SOULBY (Westmorland)

Township in Kirkby Stephen parish, East ward, Westmorland.

Acreage: 2,644 acres [1,070 ha]; approximately 1,090 acres [440 ha] of commons (including 90 acres [36 ha] of open field) enclosed 1817. ***Population:*** rising from 237 in 1801 to 309 in 1851; temporarily inflated to over 450 in 1861 and 1871 by labourers building railways; then drifting downwards from 275 in 1881 to 164 by 1991; stood at 286 in 2001. ***Landownership:*** *manor of Soulby* held by Soulby family before passing to Musgraves in early 14th century and descending in that family to 20th century. ***Economy:*** predominantly agricultural; two annual cattle fairs established c.1797 and 1825 (recorded 1860 but had ceased by 1885). Quarrying and lime-burning in 19th century. ***Places of worship:*** chapel of ease (now St Luke's

Church) built 1662-3 (burial ground consecrated 1849); restored 1873. Wesleyan Methodist chapel built 1833; replaced by new chapel 1893; still in use. ***Schools and other institutions:*** school endowed 1768; replaced by new building on different site 1876; closed 1971. Reading room built 1903. Former school building became village hall 1976.

STAFFIELD

Township in Kirkoswald parish, Leath ward, Cumberland. Absorbed into Kirkoswald CP 1934.

Acreage: 5,636 acres [2,281 ha]; including 3,206 acres [1,320 ha] of commons enclosed 1816. ***Population:*** rising from 276 in 1801 to 309 1811 and 1821; then declining to 193 in 1931 (last census year for which separate figure available). ***Landownership:*** *manor of Staffield* held by Thomas de Moulton of Gilsland by c.1300, passing to Dacres and descending with manor of Kirkoswald (q.v.). *Reputed manor of Staffield*: properties purchased from Alanby family by Sir Richard Fletcher of Hutton c.1607 claimed 1702 to be one-third of manor; courts held by Fletchers (later Fletcher-Vanes) with whom it descended to 19th century. ***Economy:*** predominantly agricultural. Small-scale freestone quarrying; lime-burning at Clints Quarries; coal mines at Burned Edge and elsewhere on Scarrowmanwick Fell, mid-19th century. ***Places of worship:*** none. ***Schools and other institutions:*** none.

STAINBURN

Township in Workington parish, Allerdale above Derwent ward, Cumberland. CP abolished 1934, part being added to Workington MB; part to Winscales CP.

Acreage: 1,228 acres [497 ha]. Stainburn townfields enclosed by private agreement 1794. Stainburn Common (900 acres [364 ha]) enclosed 1814. ***Population:*** in range 140-150 across early 19th century; then rising to 256 by 1881 and to 270 in 1911 (last census year for which separate figure available). ***Landownership:*** *manor of Stainburn* granted by Waldeof, lord of Allerdale and honour of Cockermouth, to St Bees Priory on its foundation c.1125. After Dissolution came into hands of Curwens of Workington; later held by Lowther family. ***Economy:*** salmon fishery on River Derwent, with fish coops at Stainburn Pool by 1278; continuing to 20th century. Coal mining by later 17th century. Tannery at Sunnyside, probably by 1829; closed mid-20th century (still in operation 1929). Market gardening recorded 1920s. ***Places of worship:*** medieval chapel (oratory of St Bees Priory) recorded 1406 (location perhaps recorded in name Chapel Close). ***Schools and other institutions:*** Board school (mixed) built 1883; closed mid-20th century.

STAINMORE

Chapelry and township in Brough parish, East ward, Westmorland.

Acreage: 16,329 acres [6,608 ha], including extensive moorland commons. Capel Rigg Intake (37 acres [15 ha]) enclosed 1857; Stainmore Common East (4,075 acres [1,649 ha]) enclosed 1890. Regulated pastures of East Stainmore South Moor (2,577 acres [1,043 ha]) and North Moor (3,726 acres [1,508 ha]) remain registered common land. ***Population:*** rising from 530 in 1801 to peak of 707 in 1831; then downward drift (501 by 1901) to low point of 218 in 1971; rising again late 20th century to stand at 253 in 2001. ***Landownership:*** Stainmore was hunting forest of barony of Westmorland and descended with barony (see Appleby). ***Economy:*** pastoral farming (11 vaccaries recorded 1392). Quarrying, both flagstones (recorded in name Slate Quarry Moss) and limestone with lime-burning in 19th century. Lead mining (with smelt mills at Augill Bridge and Smeltmill Beck) and coal mining in 18th and 19th centuries. Cabbish lead mine reopened for barytes mining 1905-30. Grouse shooting on moorland 19th-21st centuries. ***Places of worship:*** chapel of ease, now St Stephen's church, South Stainmore, endowed as school by Christopher Buckle 1594; consecrated as chapel 1608; repaired 1699; rebuilt 1842-3. St Mary's church, North Stainmore, built 1861; made redundant 1972 and converted to dwelling. Primitive Methodist chapel at Mouthlock, South Stainmore, built 1831; rebuilt nearby 1909; converted into residential centre 1980s. Primitive Methodist chapel, North Stainmore, built 1860; closed 2010. ***Schools and other institutions:*** school at South Stainmore founded 1594; schoolhouse built adjacent to chapel 1699; closed c.1970; subsequently used as outdoor pursuits centre. School near North Stainmore church opened 1880; closed c.1958; now North Stainmore Village Hall. South Stainmore Parish Institute opened 1932.

STAINTON (near Carlisle) *see* Stanwix

STAINTON (near Penrith) *see* Dacre

STAINTON (Lancs.) *see* Urswick

STAINTON (Westmorland)

Chapelry and township in Heversham parish, Kendal ward, Westmorland.

Acreage: 1,735 acres [702 ha]. ***Population:*** in range 350-400 across 19th century, then declining to 249 in 1971; rising again to stand at 301 in 2001. ***Landownership:*** passed by marriage of Helen daughter of Anselm de Furness to Ralph Deincourt of Sizergh in early 13th century; descended with Sizergh to Stricklands who held it until 1674, when Sir Thomas Strickland enfranchised the tenants. ***Economy:*** predominantly agricultural. Mills on St Sunday's Beck: Halfpenny Mill (flax by mid-19th century; later uses: coconut matting, paper and sweet manufacture; closed early 20th century); Low Bridge (bobbin making, 1825 to c.1865); Stainton (woollen cloth by mid-19th century; closed 1925); and Stainton Cross (flax until 1886). ***Places of worship:*** medieval chapel of ease at Crosscrake, recorded 13th century; said 1717 to have been ruinous and unused for many years; rebuilt 1773; replaced by church of St Thomas 1874-5 (tower removed 1963-4). Independent chapel built 1698; closed 2007. ***Schools and other institutions:*** chapel used as school (recorded 1692 and 1733-1755). School rebuilt 1828; enlarged 1905; remodelled 1973; now Crosscrake CE Primary School.

STANWIX

Ancient parish in Eskdale ward, Cumberland, embracing townships of Cargo, Etterby, Houghton, Linstock, Rickerby, Stainton, Stanwix and Tarraby. Brunstock township, detached portion of Crosby-on-Eden parish, transferred to Stanwix CP 1887. Each of the nine townships is treated separately below.

Acreage: ancient parish extended to 6,251 acres [2,500 ha]. Part of parish transferred to Carlisle MB 1887 and further transfer took place 1951. Conversely Stanwix CP gained territory by absorbing part of Crosby-on-Eden 1934. Renamed Stanwix Rural CP 1966. ***Population:*** Stanwix and Etterby townships became suburbs of Carlisle from later 19th century. Population of ancient parish grew from 1,039 in 1801 to 2,877 in 1881, last census year before boundary changes (for population of individual townships, see below). Population of Stanwix Rural CP stood at 3,043 in 2001.

CARGO (1,166 acres [472 ha]). ***Population:*** rising from 237 in 1801 to 292 in 1851. ***Landownership:*** *manor of Cargo* held by Robert de Ross, lord of Wark, 1274; descended in Ross family until 1338 when it passed by marriage to Parrs of Kendal. Exchanged for other lands with Elizabeth I c.1573. Granted to Sir John Whitmoor early 17th century, descending in that family until after 1688; later purchased by Dacres, who sold it to Joseph Lamb of Newcastle-upon-Tyne 1793. ***Economy:*** farming; fishing. Fording point (Doudle Ford) across River Eden. Industrial estates by railway later 20th century. ***Places of worship:*** Wesleyan Methodist chapel built 1857; closed 1968. ***Schools and other institutions:*** school built 1856; used for worship on summer Sundays 1860; closed c.1977. Recreation hut in village in 1920s.

ETTERBY (302 acres [122 ha]). ***Population:*** rising from 49 in 1801 to 452 by 1871 (last census year for which separate figure available). ***Landownership***: part of manor of Westlinton (q.v.), held under barony of Burgh. ***Economy:*** farming; nurseries mid-19th century; brick-making late 19th century. ***Places of worship:*** Presbyterian mission hall built 1881; extended 1890; closed 1940s and sold for use as youth centre 1948; demolished c.1970. Convent of the Sacred Heart built 1891-2. St Mark's mission church (Anglican), on Belah housing estate, built 1951-2. ***Schools and other institutions:*** school intended for girls built at convent 1891-2; opened as Chadwick Memorial Industrial School for Roman Catholic boys 1903; closed 1924 and became Nazareth House orphanage and home for aged poor; sold 1951 and became Austin Friars School, now co-educational independent secondary day school.

HOUGHTON (1,516 acres [614 ha]). ***Population:*** rising from 226 in 1801 to 381 in 1851; stood at 339 in 1871 (last census year for which separate figures available). ***Landownership:*** descended with Tarraby (see below). ***Economy:*** farming; early 19th-century villas for Carlisle's élite (Houghton Hall, Houghton House and Knells). ***Places of worship:*** Anglican church of St John the Evangelist, built 1840-2. Wesleyan Methodist chapel built 1893; closed 1953. ***Schools and other institutions:*** school close to church rebuilt 1841; repaired and improved 1897. Reading room recorded 1865; replaced by village hall built 1903, used as reading and recreation room and for public meetings.

LINSTOCK (1,127 acres [456 ha]). Linstock and Crosby Commons enclosed from 1691; Crosby Holme, The Tofts, Little Hill and Willows common fields enclosed 1772. ***Population:*** comparatively stable across early and mid-19th century: 167 in 1801; 207 in 1871 (last census year for which separate figure available). ***Landownership:*** *manor or barony of Linstock or Crosby* granted by Henry I to Walter his chaplain, who gave it to Carlisle Priory. Assigned to bishop of Carlisle at partition of priory's lands 1249, descending with bishopric thereafter. Linstock Castle was seat of bishops in 13th century. ***Economy:*** farming; tile manufacture mid-19th century. ***Places of worship:*** chapel of St Peter, near Linstock, recorded 1362. Mission room recorded 1910 may have been 'small place of worship' built by G. H. Head, recorded 1860. ***Schools and other institutions:*** Women's Institute Hall opened 1922.

RICKERBY (623 acres [252 ha]). ***Population:*** in range 80-110 in earlier 19th century, standing at 113 in 1871 (last census year for which separate figure available). ***Landownership:*** sub-manor under barony of Linstock (see above), held with Scaleby (q.v.). Came to George Head Head 1834 and to his heir Miles McInnes in 1870s; estate largely split up in 1900s. ***Economy:*** farming; dominated by Rickerby House estate and park from 1830s. ***Places of worship:*** none. ***Schools and other institutions:*** school built by G. H. Head 1836; separate boys' and girls' schools recorded 1865. Rickerby House became children's home by 1920s; converted to boys' preparatory school c.1935; closed c.1952; building bought by Cumberland Education Committee and re-opened as Eden School 1953; closed 1993. Rickerby Park bought by Carlisle Citizens League 1920 as war memorial; Memorial Bridge across River Eden opened 1922.

STAINTON (604 acres [244 ha]). ***Population:*** in range 60-70 from 1801 to 1841. ***Landownership:*** held under manor of Westlinton by Musgrave family of Crookdake until sold to Sir John Lowther 1686. ***Economy:*** agricultural. ***Places of worship:*** none. ***Schools and other institutions:*** none.

STANWIX (419 acres [170 ha]). ***Population:*** rising from 337 in 1801 to 1,013 in 1871 (last census year for which separate figure available). ***Landownership:*** part of manor of socage of Carlisle Castle. ***Economy:*** Knowefield Nurseries (covering 150 acres [60 ha]) belonging to Little & Ballantyne, established 1860s; said to be one of largest in country; lost to housing development from 1930s. ***Places of worship:*** medieval parish church of St Michael, on site of Roman fort of *UXELODUNUM;* rebuilt 1841-5; apse added 1893. ***Schools and other institutions:*** National school built 1843, near Knowefield; replaced by Board school off Church Street, built 1886; enlarged 1898; now Stanwix Primary School. Cumberland Reformatory for Boys built by G. H. Head 1854; closed 1882; site taken over by Carlisle College of Art 1951; now University of Cumbria Brampton Road Campus. Reading room by 1860. Stanwix Community Centre, refurbished 2012.

TARRABY (493 acres [200 ha]). ***Population:*** rising from 110 in 1801 to 150 in 1851; had fallen back to 86 by 1871 (last census year for which separate figure available). ***Landownership:*** *manor of Tarraby* held of barony of Levington (Kirklinton) by Aglionby family until exchanged with Sir John Lowther for Nunnery 1690s; exchanged again with Dalston family who enfranchised tenants 1764. Drawdykes Castle remained seat of Aglionby family until 1789. ***Economy:*** farming. Military training camp (Hadrian's Camp) opened 1939; closed 1969; became travellers' camp 1980s. ***Places of worship:*** none. ***Schools and other institutions:*** none.

BRUNSTOCK (637 acres [258 ha]), detached portion of Crosby-on-Eden parish, transferred to Stanwix CP 1887. ***Population:*** rising from 65 in 1801 to 108 in 1831, then falling back to 67 in 1871 (last census year for which separate figure available). ***Landownership:*** part of bishop of Carlisle's manor of Crosby on Eden (q.v.). Brunstock House built 1828-30 for Carlisle lawyer, George Saul. ***Economy:*** agricultural. ***Places of worship:*** none. ***Schools and other institutions:*** none.

STAPLETON

Township in Stapleton parish, Eskdale ward, Cumberland. CP enlarged by absorbing Bellbank CP 1934.

Acreage: 4,744 acres [1,920 ha]. Common land at Longrigg Moss, Bolton Fell and Kirkcambeck Common (totalling 1,800 acres [728 ha]) enclosed 1800. ***Population:*** rising from 239 in 1801 to 550 in 1841; gradual decline thereafter to 289 in 1931 (last census year before boundary change). ***Landownership:*** moiety of *manor of Stapleton* held by Stapleton family by early 14th century; sold by John Stapleton to Thomas Lord Dacre 1525, descending thereafter with barony of Gilsland (see Brampton). ***Economy:*** predominantly agricultural; tile works and lime-burning in 19th century. ***Places of worship:*** medieval parish church of St Mary; rebuilt 1829-31. Wesleyan Methodist chapel at Roweltown; restored 1891. ***Schools and other institutions:*** school endowed 1778; schoolhouse on common pulled down 1814 and new parish school, Longrigg School, built near church; rebuilt 1871; closed and converted to dwelling. Reading room and library established in parish school late 1850s. Public hall opened 1909.

STAVELEY-IN-CARTMEL

Chapelry and township in Cartmel parish, Lonsdale hundred, Lancashire North of the Sands.

Acreage: 4,295 acres [1,738 ha], including commons, enclosed after 1796. ***Population:*** rising from 315 in 1801 to peak of 438 in 1871, then falling back to 329 in 1931; rising again to 428 in 2001. ***Landownership:*** part of endowment of Cartmel Priory; not a separate manor. Appears to have been divided into small estates in post-medieval centuries. ***Economy:*** largely agricultural; limestone quarrying and lime-burning in mid-19th century; charcoal burning 16th to 19th centuries; tourism increasingly important from later 20th century. ***Places of worship:*** chapel of ease recorded 1618; restored 1678; had become ruinous by 1779; major restoration 1793 (when church was dedicated to St Mary); restored again 1897. ***Schools and other institutions:*** school taught in chapel (recorded 1618 and early 18th century). Boys' school endowed 1779; rebuilt 1847. Girls' school built and endowed by Mary Dixon 1801; closed. National school beside church built by subscription 1875; closed mid-20th century.

STAVELEY, NETHER

Township in Kendal parish, Kendal ward, Westmorland. CP merged with Over Staveley and Hugill CPs 2004.

Acreage: 2,563 acres [1,037 ha]. Common enclosed under Act of 1816. ***Population:*** rising from 131 in 1801 to 299 by 1851 and 399 by 1901. Opening of railway station 1847 led to expansion of Staveley village into township, with major growth mid-20th century: population grew from 387 in 1931 to 677 by 2001. ***Landownership:*** see Over Staveley. ***Economy:*** predominantly agricultural; coppice woodland. Expansion of Staveley village in 20th century included commercial development. ***Places of worship:*** Roman Catholic church of Sacred Heart opened 1957. ***Schools and other institutions:*** Temperance Hall opened 1887 (now converted to dwelling).

STAVELEY, OVER

Township in Kendal parish, Kendal ward, Westmorland. CP merged with Nether Staveley and Hugill CPs 2004.

Acreage: 2,580 acres [1,044 ha]. Commons (1,228 acres [497 ha]) enclosed 1854. ***Population:*** rising across 19th century from 324 in 1801 to 736 by 1881; temporarily inflated to 885 in 1891 by influx of workmen constructing Thirlmere aqueduct; declined across 20th century from 702 in 1901 to 516 in 2001. ***Landownership:*** part of Marquis fee of barony of Kendal. Divided at early date into 'Staveley Godmund' (William Godmund granted Staveley late 12th century) and 'Staveley Gamel'. Godmund family continued to hold land there in 16th century. Moiety of *manor of Staveley* held by John Hotham by 1413; descended in Hotham family until sold to Bellinghams 1569, descending thereafter with Levens (q.v.). ***Economy:*** market charter granted 1335; market did not survive; livestock fairs held in 19th century. Water-powered mills on River Kent: fulling mill recorded 1274; sickle mill recorded 1689; cotton mill in 1784. By mid-19th century three mill sites: Scroggs Mill (fulling mill by 1689; sickle mill in 18th century; converted to bobbin manufacture early 19th century; bone mill 1885; turning pick and hammer shafts 1897; disused by 1912); Old Mill, later Kentmere Mills (cotton mill built 1784; closed 1810 and became woollen mill; now packaging manufacture) and Low Mill or Chadwick's Mill (bobbin manufacture; wood turning for tool handles from mid-20th century; closed 1990s). Further mills nearby in Hugill (q.v.). Business park developed on site of Chadwick's Mill 1993. ***Places of worship:*** chapel of ease of St Margaret endowed 1338; gained parochial status 1587; only tower survives. New church of St James built adjacent to school 1865. Primitive Methodist chapel, School Lane, built 1834; replaced by new chapel on Main Street 1866; closed 1939. Wesleyan Methodist chapel built 1836; rebuilt on Kentmere Road 1876; closed 1994. ***Schools and other institutions:*** school in chapel by 1530s. Endowment for school 1696 but school still taught in chapel 1717, as no schoolhouse had been built. School built by subscription before 1779; rebuilt 1841 and again 1859; now Staveley CE Primary School. Working Men's Institute and reading room built 1874; new village hall and institute built 1936.

STORTH *see* Beetham

STRICKLAND, GREAT

Township in Morland parish, West ward, Westmorland.

Acreage: 2,339 acres [947 ha]. Act to enclose commons passed 1830. ***Population:*** in range 240-280 for most of 19th and early 20th centuries (peak of 345 in 1851); then declining to 170 in 1971, before rising again to 246 in 2001. ***Landownership:*** *manor of Strickland* (which included Little Strickland) held by Strickland family from 13th century; had passed to Fallowfield family by 1430, descending by marriage of Lucie Fallowfield to John Dalston (d. 1692) of Acorn Bank; Christopher Dalston sold it to Sir John Lowther late 17th century, from whom it descended to earls of Lonsdale. ***Economy:***

predominantly agricultural; limestone quarrying and lime-burning in 19th century. ***Places of worship:*** church of St Barnabas, built as chapel of ease 1872. Quaker meeting house built 1681; closed 1806; became private house. Independent meeting house at Strickland recorded 1684. Wesleyan Methodist chapel built 1887; closed 1980s. ***Schools and other institutions:*** endowed school built 1790; replaced 1848 at expense of Mrs Sarah Plumer; closed c.1967. Village hall, timber and corrugated iron building which had been munitions store at Gretna, erected early 1920s.

STRICKLAND, LITTLE

Township in Morland parish, West ward, Westmorland.

Acreage: 790 acres [320 ha]; common enclosed 1811. ***Population:*** rising from c.100 in early 19th century to peak of 135 in 1851; then drifting back to 102 by 1931, declining to stand at 66 in 2001. ***Landownership:*** see Great Strickland. ***Economy:*** predominantly agricultural; limestone quarrying and lime-burning in 19th century. ***Places of worship:*** St Mary's chapel of ease built 1814 (repaired and extended 1895-6) to serve Thrimby chapelry, replacing older chapel at Thrimby (q.v.). ***Schools and other institutions:*** endowed school, replacing that at Thrimby, built c.1814; closed 1976. Village hall, a corrugated iron building perhaps dating from 1920s; still in use.

STRICKLAND KETEL *see* Burneside

STRICKLAND ROGER *see* Burneside

SUBBERTHWAITE

Township in Ulverston parish, Lonsdale hundred, Lancashire North of the Sands. Merged with Blawith to form Blawith and Subberthwaite CP 1986.

Acreage: 1,236 acres [500 ha] in two parts: Lower Subberthwaite (397 acres [160 ha]), covering Stennerley and Gawthwaite, separated from rest by Gawthwaite Moor and Combs (330 acres [134 ha]), moorland intercommoned with Lowick. ***Population:*** rising from 90 in 1801 to peak of 163 in 1831; remaining at c.150 until 1880s, then falling to 70 in 1981 (last census year for which separate figure available). ***Landownership:*** part was held of manor of Broughton in Furness (q.v.) but Subberthwaite was termed a manor in 1769-70. ***Economy:*** farming; blue slate quarrying: extensive quarry at Gawthwaite Moor from early 19th century to mid-20th. ***Places of worship:*** mission room at Gawthwaite (also used as reading room) built 1868; closed later 20th century and converted to dwelling. ***Schools and other institutions:*** school said to have been founded c.1778 but appears not to have survived.

SUNDERLAND

Township in Isel parish, Allerdale below Derwent ward, Cumberland. Absorbed into Blindcrake CP 1934.

Acreage: 806 acres [326 ha]. Sunderland Common enclosed, with other commons in manor of Isel, 1813. ***Population:*** fluctuated in range 55-80 across 19th and early 20th century (peak of 99 in 1851); stood at 60 in 1931 (last census year for which separate figure available). ***Landownership:*** part of manor of Isel (see Blindcrake, Isel and Redmain). ***Economy:*** predominantly agricultural; limestone quarrying, lime-burning and drainage tile-making in 19th century. ***Places of worship:*** early Quaker meetings held in open on Setrah Hill from 1653 to 1673, when moved into local houses. Wesleyan Methodist chapel built 1862; closed 1930s (now dwelling house). Mission hall established by 1920s. ***Schools and other institutions:*** none.

SWARTHMOOR *see* Pennington

TALKIN *see* Hayton

TALLENTIRE

Township in Bridekirk parish, Allerdale below Derwent ward, Cumberland. Absorbed into Bridekirk CP 1934.

Acreage: 1,992 acres [806 ha]. Tallentire Common enclosed by private agreement 1811. ***Population:*** in range 200-250 for most of 19th and early 20th century (peak of 250 in 1851), but had fallen back to 184 by 1931 (last census year for which separate figure available). ***Landownership:*** *manor of Tallentire* said to have been granted to Odard son of Lyulph in 12th century; in hands of John Radcliffe, lord of Castlerigg and Derwentwater, in 1578. Bought by George Fletcher (d. 1637), son of Lancelot Fletcher of Cockermouth, and descended in Fletcher family, passing by marriage of Anne

Fletcher (d. 1721) to Matthias Partis, a Newcastle tobacco merchant. Sold 1776, after death of Henry Fletcher Partis, to William Browne, descending in that family (who enlarged Tallentire Hall 1863) until death of William Browne (d. 1894). ***Economy:*** largely agricultural; limestone quarries and lime-burning in 19th century. Wind farm built 2012-13. ***Places of worship:*** Congregational chapel, built 1876; closed and converted to dwelling house. ***Schools and other institutions:*** school built by William Browne 1863; closed during Second World War, when it was used as hostel for Land Army.

TARRABY *see* Stanwix

TEBAY

Township in Orton parish, East ward, Westmorland, which became separate CP 1897. Tebay CP absorbed part of Grayrigg CP 1935.

Acreage: 6,855 acres [2,774 ha] before boundary change; including 1,217 acres [493 ha] of registered common land on Tebay Fell and 946 acres [383 ha] at Roundthwaite common. Roundthwaite Cowbound pastures (274 acres [111 ha]) enclosed 1858. ***Population:*** stood at c.400 in mid-19th century before arrival of railway, growing rapidly to 684 by 1871 and to peak of 1,093 in 1901; declined across 20th century to low of 594 in 1981; stood at 728 in 2001. ***Landownership:*** *manor of Tebay* appears to have been in divided ownership in middle ages; whole acquired by Wharton family in 16th century, passing to Lowthers, probably in 1730s. ***Economy:*** hill farming. Tebay village grew on arrival of railway with Tebay station (opened 1852; closed 1968) at convergence of Lancaster & Carlisle Railway (opened 1846) and South Durham & Lancashire Union Railway (opened 1861). Market established by 1880s with market hall near station. Small industrial estate at Tebay goods yard and engine sheds after railway closure. For service stations on M6, see Orton. ***Places of worship:*** St James' church built 1878-80. Primitive Methodist chapel built 1865; still in use. Gaisgill Wesleyan Methodist chapel built 1841; restored 1895; closed 2003. ***Schools and other institutions:*** school endowed 1672; schoolhouse at Old Tebay replaced by new school at Mount Pleasant 1863; enlarged 1892; closed 1960s (now guest house). Infant school built near Methodist chapel early 20th century; now Tebay Primary School. Victoria Institute (wooden building, formerly Sedbergh School chapel, moved to Tebay by train c.1899) opened 1900; now closed. Gaisgill reading room built 1909; now closed.

TEMPLE SOWERBY

Chapelry and township in Kirkby Thore parish, East ward, Westmorland.

Acreage: 1,241 acres [502 ha.]. Common land at Temple Sowerby Moor, Down Moors, Whinns and Parson's Close enclosed 1774. ***Population:*** rising from 299 in 1801 to stand in range 370-440 across mid- and later 19th century (peak of 476 in 1871); downward drift across 20th century to low point of 279 in 1971, rising again to 330 in 2001. ***Landownership:*** *manor of Sowerby* given to Knights Templar *c.*1228, passing on their suppression 1312 to Robert de Clifford, who granted it to Knights Hospitaller 1323. Sold to Thomas Dalston 1544, who settled it on his son Christopher, whose seat was at Acorn Bank; then descended through Dalston family (who rebuilt Acorn Bank in 17th century and remodelled it in mid-18th) until late 18th century, when it passed by marriage through several families to Boazman family by mid-19th century. ***Economy:*** farming; livestock fairs established early 19th century; tanning in mid-19th century. Brick and tile works near Acorn Bank by late 19th century; gypsum quarrying there by 1913 and nearby at Houtsay Park across 20th century. ***Places of worship:*** medieval chapel of ease (now St James' church) recorded 1227-8; rebuilt 1754; enlarged 1770; tower built 1807-8; restored 1870s. Independent chapel converted from existing building 1813; survived until 1850s. Wesleyan Methodist chapel built 1872; now closed and converted to dwelling. ***Schools and other institutions:*** school taught in chapel in 17th century (recorded 1672); schoolhouse built 18th century; extended 1872 and 20th century; now Temple Sowerby CE Primary School. Infant School recorded 1885; appears to have been converted to reading room (with small library attached) by 1900. Free library in village recorded 1860. Victory Memorial Hall built as First World War memorial c.1920.

THORNTHWAITE *see* Above Derwent

THRELKELD

Chapelry and township in Greystoke parish, Leath ward, Cumberland.

Acreage: 5,956 acres [2410 ha]. Small open field ('townfield' of 11 acres [4 ha]) enclosed 1842. Threlkeld Common, covering 1,102 acres [446 ha] of fells around Saddleback/Blencathra, remains unenclosed common land. ***Population:*** rising from 260 in 1801 to 419 by 1881; then more rapid increase to peak of 614 by 1931; followed by decline to 454 by 2001. ***Landownership:***

manor of Threlkeld held by Threlkeld family from 13th century until death of Sir Lancelot Threlkeld (d. 1506-10), when it passed by marriage to Pickering family; manorial rights purchased by Lowther family before 1632. ***Economy:*** hill farming; fair for sheep and cattle on first Thursday in September in early nineteenth century. Coaching inns on Penrith-Keswick turnpike road (turnpiked 1762). Station on Cockermouth, Keswick & Penrith Railway, opened 1864. Mining and quarrying: lead and zinc mining at Gategill from 17th century to 1928. Threlkeld Granite Quarry (in Castlerigg, St John's & Wythburn township) opened 1864; closed 1982; museum on site opened 1996. Granite, concrete and pre-cast flagstone works opened near railway station in 1890s. Tourism from late 18th century. ***Places of worship:*** medieval chapel of St Mary, recorded c.1225. Present church built 1776, replacing thatched church. Wesleyan Methodist chapel at Scales built 1842; closed and converted to dwelling c.1970. Mission room in Blease Road built 1885; used as Methodist chapel until 1959; converted to dwelling. Methodist chapel at granite quarry opened 1903; closed mid-1960s. ***Schools and other institutions:*** schoolmaster recorded 1659. School taught in church until schoolhouse built 1776; rebuilt on new site 1849; now Threlkeld CE Primary School. Granite Quarry School opened 1897, built and maintained by quarry company to serve employees' families; closed 1952. Public room (now village hall) built 1901. Library recorded 1860. Tuberculosis sanatorium at Far End Row Farm opened 1904 (used as hostel from 1976 and field study centre from 1993).

THRIMBY

Township in Morland parish, West ward, Westmorland.

Acreage: 1,574 acres [637 ha]. Act to enclose commons passed 1830. ***Population:*** in range 50-80 across 19th and earlier 20th centuries (peak of 87 in 1861), then declining to 30 by 2001. ***Landownership:*** held by Thrimby family in 13th century, passing to Harringtons in later 14th century, and, by later 17th century, to Lowther family, who emparked much of township. ***Economy:*** predominantly agricultural; some lime-burning in 19th century; afforestation by Lowther Estates in 20th. ***Places of worship:*** chapel of ease; in state of dilapidation by 1681 when it was rebuilt; demolished 1814, on enlargement of Lowther Park, and replaced by new chapel at Little Strickland (q.v.). ***Schools and other institutions:*** curate endowed to teach in chapel 1684. Separate school room built before 1777; demolished c.1814 and replaced by school at Little Strickland.

THURSBY

Ancient parish in Cumberland ward, Cumberland, containing three townships of High & Low Thursby, Crofton, and Parton & Micklethwaite.

Acreage: 3,142 acres [1,272 ha], divided between constituent townships thus: High and Low Thursby: 1,474 acres [596 ha]; Crofton: 1,013 acres [409 ha]; Parton and Micklethwaite: 654 acres [264 ha]. Thursby and Micklethwaite commons (75 acres [30 ha]) enclosed 1819. ***Population:*** estimated at 535 in 1688; stood at 403 in 1801, rising to 591 in 1851, when bulk of population lived in Thursby; Crofton and Parton & Micklethwaite townships contained around 100 inhabitants each. Parish population remained fairly constant to 1931; then rose to 917 in 1951, continuing to grow to 1,185 by 2001. ***Landownership:*** *manor of Thursby* held by Guy de Boyvill of Kirksanton in mid-13th century and by Ogle family from mid-14th century to later 15th. Later held by Dacres as part of barony of Burgh. *Manor of Crofton* held by de Crofton family in 13th and 14th centuries, passing by marriage of Margaret Crofton to Isold Brisco in late 14th century to Brisco family, in which it descended until sold in 1908 (Crofton Hall demolished c.1958). *Manor of Parton* passed from Parton family though various families until sold by Robert Carliell to William Denton of Cardew late 14th century; descended in Denton family until George Denton sold it to Sir John Lowther 1686, from whom it descended to earls of Lonsdale. ***Economy:*** predominantly agricultural: described by Thomas Denton 1688 as rich arable and meadow lands, enriched by River Wampool. Crofton estate taken over by Land Settlement Association 1935 and divided into smallholdings; scheme continued until 1970s. ***Places of worship:*** medieval parish church of St Andrew; rebuilt 1845-6. Mission room at Micklethwaite recorded 1901. ***Schools and other institutions:*** school built by subscription 1740, endowed by Thomas Thomlinson 1798; rebuilt 1848 and extended 1876. Demolished and replaced by modern building on edge of village; now Thursby Primary School. Reading room and library established in school, 1881. Recreation hall in Thursby village (a wooden hut purchased from Gretna) opened 1922; replaced 2010 by new Parish Hall.

THWAITES *see* Millom Rural

TIRRIL *see* Sockbridge

TORPENHOW AND WHITRIGG

Township in Torpenhow parish, Allerdale below Derwent ward, Cumberland. CP abolished 1934, most becoming part of newly created Blennerhassett & Torpenhow CP; remainder (former moorland surrounding Roman fort at Caermote) added to Bewaldeth & Snittlegarth CP.

Acreage: 2,742 acres [1,110 ha], including part of Bothel and Torpenhow commons, totalling 2,200 acres [890 ha], enclosed 1811. Open field ('townfield') at Whitrigg enclosed 1776. ***Population:*** generally in range 250-300 between 1801 and 1931 (last census year for which separate figure available), but rising slightly in mid-19th century (peak of 349 in 1851). ***Landownership:*** *manor of Torpenhow* passed through several families (including Stutevill, Mulcaster, Tilliol, Moresby and Colville) in medieval centuries. By 17th century it was held by Salkelds of Whitehall, from whom it passed by descent to Charltons of Hesleyside (Northumb.) in 18th century. *Whitrigg* was separate estate, held by Skeltons of Armathwaite by 17th century; Sir Wilfred Lawson of Isel and Brayton was lord of manor 1900. ***Economy:*** predominantly agricultural; limestone quarrying and lime-burning in 19th century. Race course on common land between Caermote and Moota in 17th century. ***Places of worship:*** medieval parish church of St Michael; restored 1880-2. Mission hall used for nonconformist services in late 19th century and later by Cumberland Scripture Readers' Trust; sold 1978 and became part of Tindall House field study centre. ***Schools and other institutions:*** Sunday school, founded 1855, became National school 1895; had closed by 1970 when it became dormitory for field study centre for St Bede's Grammar School, Bradford, and was renamed Tindall House. Reading room and library in village by 1900. Village hall built 1951; replaced by new building 2003.

TORVER

Chapelry and township in Ulverston parish, Lonsdale hundred, Lancashire North of the Sands.

Acreage: 3,815 acres [1,544 ha], including extensive unenclosed fells on Torver High Common and smaller areas of common land at Torver Back Common (287 acres [117 ha]) and Low Common (485 acres [196 ha]). ***Population:*** in range 190-210 across 19th century (peak of 263 in 1821); then falling from 207 in 1901 to 135 by 2001. ***Landownership:*** part of William de Lancaster's moiety of Furness Fells and thus within lordship of Ulverston. Part had come into possession of Flemings of Aldingham by 13th century and Conishead Priory also had land in Torver. Descended with Harrington moiety of lordship of Ulverston, linked to their manor of Muchland (see Aldingham). Purchased with Ulverston 1736 by duke of Montagu, descending to dukes of Buccleuch. ***Economy:*** hill farming; coppice woodland with charcoal burning, medieval period to 20th century; bloomery sites (probably medieval). Sunny Bank Mill: bobbin mill in mid-19th century; later making pick and hammer shafts; closed c.1934. Slate quarrying from 18th century; major quarries at Ash Gill and Eddy Scale in 19th century; closed early 20th century. ***Places of worship:*** chapel of ease consecrated 1538; rebuilt 1849 and again 1884, when it was dedicated to St Luke. Baptist chapel at Sunny Bank established 1678; merged with Hawkshead Hill congregation by early 20th century; finally closed 1940s. ***Schools and other institutions:*** school taught in chapel until c.1777, when John Fleming of Coniston endowed and built grammar school; rebuilt 1873; closed 1927. Old School Room used as village hall since 1970.

TROUGH

Township in Stapleton parish, Eskdale ward, Cumberland. Absorbed into Solport CP 1934.

Acreage: 2,338 acres [946 ha], including commons at Mallshill (225 acres [91 ha]) and Wakey Hill (340 acres [138 ha]), enclosed 1817 and 1845 respectively. ***Population:*** peaked at 169 in 1831; then declined to c.100 by end of 19th century, falling to 74 in 1931 (last census year for which separate figure available). ***Landownership:*** part of Solport manor but divided into small estates by 19th century. ***Economy:*** farming; small-scale coal mining at Hagg Beck recorded 1847; afforestation, mid 20th century. ***Places of worship:*** none. ***Schools and other institutions:*** school at Russelgate with date stone reading 'Trough School 1838'; closed.

TROUTBECK

Chapelry and township in Windermere parish, Kendal ward, Westmorland. CP enlarged by absorbing part of Windermere CP 1935.

Acreage: 5,807 acres [2,350 ha] before boundary change. Extensive fell commons, enclosed 1842, were previously divided into three 'hundreds', each carrying a stint of a 'long hundred' (i.e. 120) cattlegates. Wansfell (270 acres

[109 ha]) remains registered common land. ***Population:*** populous village, described by Thomas Machell 1692 as 'long, rambling disorderly village but with some good houses in it'. Population rose from 310 in 1801 to 523 in 1891; stood at 456 in 1971 (last census year for which separate figure available). ***Landownership:*** part of manor of Windermere (q.v.). ***Economy:*** pastoral farming. Slate quarrying (extensive quarry at Troutbeck Park from 1750s to early 20th century). Paper mill recorded 1692. Tourism increasingly important from 19th century: lakeside villas at Briery Close (built 1839; enlarged 1910-12), Brockhole (built 1897-1902; National Park visitor centre from 1969), and Langdale Chase (built c.1889; now hotel). High Cross Youth Hostel built 1935. Town End (Browne family farmhouse) acquired by The National Trust 1948. Limefitt Holiday Park (across township boundary in Applethwaite) established 1980s. ***Places of worship:*** said to have been medieval chapel, dedicated to St Catherine; replaced by chapel of ease (consecrated as Jesus chapel, 1562, and given parochial rights); rebuilt 1736; now dedicated to St James. Church of St Andrew at Ecclerigg in 20th century. ***Schools and other institutions:*** endowed school at east end of chapel built 1637 or 1639; rebuilt 1853; closed 1975; building converted to hostel accommodation 1977. Troutbeck Institute, with library attached, built by subscription 1873; still in use as village hall.

ULDALE

Ancient parish in Allerdale below Derwent ward, Cumberland. Absorbed into Ireby (now Ireby and Uldale) CP 1934

Acreage: 5,814 acres [2,353 ha], including 3,413 acres [1,381 ha] of unenclosed common land on Uldale Fells. ***Population:*** estimated at 520 in 1688. Rose from 234 in 1801 to 388 in 1851; then fell back to 217 by 1931 (last census year for which separate figure available). ***Landownership:*** *manor of Uldale* said to have been granted by Waldeof, lord of Allerdale, to Adam son of Lyulph, from whose family it passed by marriage to the Bunkle ('Bonekill') family of Berwickshire by 13th century. In Crown hands by 1318; soon after granted to Anthony de Lucy, baron of Allerdale, and descended with barony to Percy family, earls of Northumberland. After being conveyed to Crown, it was granted 1543 to Thomas Dalston, in whose family it remained until sold to Thomas Gaff in later 18th century. Purchased by earl of Egremont 1808, thus reuniting it with Allerdale barony. ***Economy:*** predominantly agricultural: described by Thomas Denton 1688 as rich farming country with corn and cattle on lower grounds and sheep on fells. Small seam of coal; some veins of copper exploited in 1780s. Limestone quarrying and lime-burning on Aughertree Fell in 19th century. ***Places of worship:*** old parish church of St James (a mile north of village), medieval in origin; windows date from 1730; chancel rebuilt 1837. New church of St John the Evangelist built in village 1869; demolished 1963. ***Schools and other institutions:*** grammar school endowed by Matthew Caldbeck 1726; rebuilt 1895; closed c.1990 (converted to tea room and gallery).

ULLOCK, PARDSHAW AND DEANSCALES *see* Dean

ULPHA

Chapelry and township in Millom parish, Allerdale above Derwent ward, Cumberland.

Acreage: 13,092 acres [5,298 ha], including 3,248 acres [1,315 ha] of unenclosed common on Ulpha Fell. ***Population:*** rising from 292 in 1801 to stand in range 350-400 in mid-19th century (peak of 405 in 1831); steady decline to 123 in 1961; then modest increase to 159 in 2001. ***Landownership:*** hunting forest of lords of seigniory of Millom (q.v.), with which it descended. ***Economy:*** hill farming; fair held near Easter and on 24 June in later 17th century. Coppice woodland, exploited for hazel nuts and charcoal-burning. Iron forge in Ulpha Park in later 17th century. Woollen textiles recorded in place-name 'Walk Mill' (i.e. fulling mill). Slate quarrying at Common Wood in later 19th century. Copper mine opened c.1842; disused by 1938. Bobbin-turning at New Mill from mid-19th century to c.1910. Afforestation from mid-20th century: Dunnerdale Forest, planted by Forestry Commission from 1937. ***Places of worship:*** parochial chapel of St John, perhaps of 17th-century date. Wesleyan Methodist chapel built 1842; closed and converted to dwelling. ***Schools and other institutions:*** curate licensed to teach school in chapel 1703. School for Ulpha had been built across River Duddon in Dunnerdale with Seathwaite township by 1840s; rebuilt 1874; modern classroom extension 1967; closed 1999. Baptist school built 1855. Parochial library established 1761 but did not survive by mid-19th century. Gunson Almshouses (now Gunson Cottages) endowed by John Gunson 1912 and built 1914.

ULVERSTON

Borough and market town in Ulverston parish, Lonsdale hundred, Lancashire North of the Sands. Became Ulverston UD 1894.

Acreage. 3,120 acres [1,263 ha], including 1,109 acres [449 ha] of common land enclosed 1813.

Landownership. *Manor or barony of Ulverston* held by Turulf 1066; subsequently formed part of honour of Lancaster. On division of honour after 1246, Ulverston was held in moieties; one passing to Furness Abbey by early 14th century; other granted to Roger de Lancaster by 1260s, from whose descendants it was acquired by Harrington family and descended with Muchland (see Aldingham). Both moieties came to Crown in 16th century (on Dissolution of Furness Abbey 1537 and forfeiture of duke of Suffolk 1554, respectively). Former Furness Abbey moiety came to Kirkbys of Kirkby Ireleth; former Harrington moiety had passed to Thomas Fell of Swarthmoor Hall by 1658. Fell's son-in-law, Daniel Abraham, purchased Kirkby moiety 1718 and whole manor was sold to duke of Montagu 1736, descending thereafter with manor of Plain Furness (see Barrow). *Manor of Neville Hall* originated in grant by William de Lancaster to Lawrence de Cornwall in 13th century. By mid-14th century had come to Sir Edmund de Nevill, in whose family it descended until forfeited to Crown in later 16th century. *Trinkeld* (sometimes styled a manor) was held by branch of Fell family of Pennington from 1595. *Conishead* formed core of endowment of Conishead Priory; passing through several hands after Dissolution before coming to Braddyll family by late 17th century. Col. Thomas Richmond Gale Braddyll commissioned mansion to be built on site 1823.

Origins and growth of the town. Borough privileges granted by charter of Gilbert son of Roger son of Reinfred c.1200. Market charter granted 1280. Ulverston became main market centre for Furness peninsula: in 1671 Sir Daniel Fleming called it 'a good market, especially for corne'. In mid-19th century it was described as 'the capital of Furness ... a neat and well-built market town and port' (Mannex's *Directory*, 1849). Its role as market for wide rural hinterland survived into 20th century, with hiring fairs continuing into 1940s. Cattle auction mart established 1877. Covered market hall opened 1878; destroyed by fire 1935 and replaced by present building. Early industries included tanning, woollen textiles and salt-making (recorded in place-name Salt Cotes).

By 1840s Ulverston's industrial base included tanyards on edge of town and at Dragley beck (recorded by mid-18th century); iron foundry at Dragley Beck and iron and brass foundry in Canal Street; paper mill at Ellers Mill; cotton mills (established from late 18th century; ceased during 'cotton famine' of 1860s) at Ellers High Mill and outside town at Low Mill; and linen weaving factory at Stonecross. Canal connecting town to sea, constructed 1793-6, led to increase in shipping trade; Ulverston became chief port for export of Furness slate and iron and copper ores. Shipbuilding also carried out until 1840s. Population doubled across first half of 19th century from 3,172 in 1801 to 6,742 by 1851.

Canal became redundant after arrival of railway and opening of docks at Barrow and was formally abandoned 1945. Furness Railway's line from Barrow reached Ulverston 1854 and Ulverston & Lancaster Railway to Carnforth, opened 1857, linked town to rest of Lancashire. Ulverston station became interchange for branch line to Lakeside, opened 1869, serving increasing numbers of tourists making excursions to Windermere; tourist accommodation expanded in town's numerous hotels.

Major expansion of industry in second half of 19th century, particularly along canal side, which led to growth of South Ulverston. Low Furness ironworks relocated to canal side 1850; closed by 1863 and replaced by Samuel Pollitt's paper mill, established c.1870. Extensive complex of North Lonsdale Iron and Steel works: furnaces blown in 1876; closed 1938. By 1888 chemical works (established c.1861), wire works (recorded by 1882; closed before 1911) and brick works (from 18th century; expansion in later 19th; closed by early 20th) in Sand Hall area, served by mineral railway, which also linked to major expansion of quarrying at Gascow. Older industrial sites found new uses: by 1880s Low Mill had become tannery and iron foundry site at Dragley Beck had become shutter works. Process continued in 20th century: canal-side paper mill site used for servicing aircraft engines during Second World War and for making metal furniture 1945-1955; Glaxo (now GlaxoSmithKline) built pharmaceutical factory on ironworks site 1948; Low Mill tannery site redeveloped as business park 1995. Population continued to rise across later 19th century, from 7,414 in 1861 to 10,064 by 1901. Across 20th century it stabilised and stood at 11,524 in 2001.

Places of worship. Medieval parish church of St Mary; tower rebuilt after 1540; remainder rebuilt 1804; further rebuilding and extension 1864-6 and 1903-4. Conishead Priory, founded as hospital in 1150s, but an Augustinian priory from c.1180; dissolved 1536. Medieval chapel on Chapel Island, established by prior of Conishead as refuge for travellers crossing the Sands. Holy Trinity Church, New Church Lane, built 1829-32; enlarged 1881 and again 1908; closed 1975. Mission church of St Jude at Sandside (corrugated iron church built 1874; closed 2006) and mission-room at Ratton Row, Quay Street, (1867); now Grace Baptist Church, founded 1989. Town cemetery opened 1878 with three mortuary chapels (for Anglicans, Roman Catholics and Protestant nonconformists).

Quaker meeting house, Swarthmoor, built 1688, replacing Swarthmoor Hall which had been used for Quaker worship since 1652. Quaker meetings held in school in New Church Lane from 1864; discontinued 1970. Congregational church, Soutergate, built 1777; enlarged 1847-8; closed 1968. House licensed for Baptist meetings by 1745; Baptist chapel built in Fountain Street 1871; closed. Wesleyan Methodist chapel, The Ellers, built 1814; replaced by new chapel on adjoining site in Neville Street, built 1899-1901; restored 1992; now Ulverston Methodist Church. Mission chapel, Steel Street, opened 1875; closed 1969. Primitive Methodist chapel, Union Street, built 1906; taken over by Church of Christ 1925; sold 1984 when congregation moved to former Victoria Concert Hall; now Emmanuel Christian Centre. Ulverston Church of Christ (seceded from Union Street congregation late 1920s) moved to meeting house in Mill Street, converted from stable 1977. Other nonconformist places of worship included: Free Church of England, Burlington Street (1880s); Salvation Army barracks in Burlington Street (opened 1895; closed 1920); Spiritualist church, County Square (recorded 1911); Bethany Christian Church (non-denominational), Lightburn Road.

Roman Catholic mission in Dalton relocated to Ulverston 1779; small chapel built 1806, followed by church in Fountain Street (later Oddfellows' Hall), built 1821; replaced by church of St Mary of Furness, Lightburn Road, built 1893-5. Convent in Ulverston from 1913 to 1929, when nuns moved to Barrow.

Conishead Priory was acquired by English Buddhists 1976 and became Manjushri Buddhist Centre; Kadampa Buddhist temple built in grounds 1995-7.

Schools. Grammar school at Townbank founded and endowed by legacy of Judge Thomas Fell 1658; replaced by new higher grade school, Hart Street, opened 1900. New Grammar School built in Springfield Road 1930, Hart Street site becoming Victoria Senior School, later Victoria Secondary Modern School. Grammar School and Victoria School combined to create comprehensive school; Springfield Road premises became upper school; Hart Street premises lower school. Combined on former Grammar School site, Springfield Road, 1997; now Ulverston Victoria High School. Hart Street premises served as refugee centre during Balkan wars; demolished 2001.

National school, The Ellers, built by public subscription 1834; enlarged 1884; closed 1930 (later used as labour exchange and, from 1999 to 2012, as Lanternhouse arts centre). Infant school, Church Walk, built 1854; rebuilt 1896; now Church Walk CE Primary School. National school, Dale Street, built 1876; enlarged 1886; junior girls' department amalgamated with Lightburn junior boys by 1987; remaining infant school absorbed by Sir John Barrow Primary School 2009; Dale Street building demolished. Mixed school built at Sandside 1885; closed 1963 (pupils transferred to Lightburn School); reopened 1964 as special school; demolished 1997 and replaced by new premises nearby; now Sandside Lodge Special School. Lightburn Council mixed school, Argyll Street, opened 1915; site redeveloped as Sir John Barrow Primary School when Lightburn and Dale Street junior schools amalgamated. Croftlands housing estate served by Croftlands Infant and Junior primary schools, Oakwood Drive, built 1960s.

Roman Catholic school opened at Tarnside 1824; replaced by St Mary's RC School, Brogden Street, 1887; again replaced by new school in Springfield Park, 1971, when Brogden Street premises became Catholic Youth Club. Other denominational schools were Wesleyan school attached to Neville Street chapel (closed 1930s) and Quaker school in New Church Lane (recorded 1839).

Other institutions. Neville Hall used as parish workhouse from 1753 until construction of new workhouse in Stanley Street for Ulverston Poor Law Union 1838, which became Stanley Hospital, part of Ulverston Hospital, 1948; demolished 2004. Cottage hospital opened 1873; enlarged 1904; closed and demolished 1971. Community Health Centre built on Stanley Street site c.2004. Children's tuberculosis sanatorium at Oubas Hill opened 1928; became maternity home 1950; closed 1971. Conishead Priory (a hydropathic hotel, 'The Paradise of Furness', from 1878) purchased as convalescent home for Durham Miners 1928-9; used as military hospital in Second World War.

Public entertainments held at Assembly Rooms and theatre, Theatre Street, built 1796 (now auction room); Victoria Concert Hall, Mill Street (built 1850; sold 1984 to become place of worship); Temperance Hall; Drill Hall (built 1873; and by 2013 used as Red Rose Social and Recreational Centre); Masonic Hall (built 1905-6) and Coronation Hall (opened 1920; still in use). Palladium Cinema, Victoria Road, opened 1922; closed 1957; demolished 1965 during road improvements. Roxy Cinema, Brogden Street, opened 1937; still in use. Library established 1797; modern library building dates from 1961. Laurel & Hardy Museum opened 1983; moved to larger premises in Roxy Cinema 2009.

UNDERBARROW AND BRADLEYFIELD

Chapelry and township in Kendal parish, Kendal ward, Westmorland.

Acreage: 5,122 acres (2,073 ha) before boundary changes (79 acres [32 ha] were lost to Kendal 1935 and 67 acres [27 ha] to Crook 1951). Common land (2,145 acres [868 ha]) enclosed 1828. ***Population:*** rising from 376 in 1801 to peak at 526 in 1831; then falling back to 405 by 1901 and to 387 in 1931 (last census year before boundary changes). ***Landownership:*** *manor of Cunswick* granted to Henry de Alneto late 12th century, passing by marriage to Leyburne family in 14th century and descending in that family until forfeited to Crown on attainder of John Leyburne 1715. Subsequently purchased by Thomas Crowle; then sold to Sir James Lowther in 1750s, from whom it descended to earls of Lonsdale. Other lands in township held of barony of Kendal. ***Economy:*** predominantly agricultural. Limestone quarrying and lime-burning. Bobbin mill at Grigghall; tannery at Tullythwaite in 19th century. ***Places of worship:*** chapel of ease, recorded 1547; rebuilt 1708; replaced by new church of All Saints, built 1869 on old foundations. Methodist chapel recorded 1860. ***Schools and other institutions:*** school built 1859; enlarged 1875 and 1880; closed c.1985 and converted to dwelling. Institute built c.1902; still in use as village hall.

UNDERMILLBECK *see* Windermere and Bowness

UNDERSKIDDAW

Township in Crosthwaite parish, Allerdale below Derwent ward, Cumberland. Part of CP (Great Crosthwaite) transferred to Keswick UD 1894; this entry treats areas remaining in Underskiddaw CP after that transfer. CP enlarged by absorbing Skiddaw CP 1934.

Acreage: 6,026 acres [2,439 ha] before 1894; 5,747 acres [2,326 ha] after loss of Great Crosthwaite. Commons of Brundholme manor (3,488 acres [1,412 ha]) enclosed 1815. ***Population:*** rose from 338 in 1801, to stand in range 460-520 in mid- and later 19th century (peak of 549 in 1841); reduced to 262 in 1901 after loss of Great Crosthwaite. ***Landownership:*** *manor of Brundholme*, part of barony of Allerdale, given 1531 by 6th earl of Northumberland to Henry VIII, who granted it to Thomas Dalston 1543. Held by Tolsons of Bridekirk by 1688, later passing to Hasells of Dalemain; then purchased by

Robert Watson, bishop of Llandaff, before 1810. ***Economy:*** farming; forestry. Woollen textiles: fulling mills for 'Skiddaw grey' cloth by 17th century; woollen mill complex at Millbeck, owned by Dover family, developed from c.1760; closed 1894. Woollen mill at Applethwaite in mid-19th century. ***Places of worship:*** church room built by Daniel Dover of Millbeck 1829; chancel added 1837; still in use. For Crosthwaite parish church and other places of worship, see Keswick. ***Schools and other institutions:*** school built at Scalebeck 1875; closed c.1920. Reading room and library built at Millbeck 1896; now village hall. For other schools and institutions, see Keswick.

UNTHANK *see* Skelton

UPPERBY *see* St Cuthbert Without

UPPER DENTON *see* Denton, Upper

URSWICK

Ancient parish in North Lonsdale hundred, Lancashire North of the Sands, comprising townships of Bardsea, Great Urswick and Little Urswick (itself divided into three subdivisions).

Acreage: 3,905 acres [1,561 ha], divided between townships thus: Bardsea: 828 acres [331 ha]; Great Urswick: 1,232 acres [493 ha]; Little Urswick: 1,845 acres [738 ha]. Subdivisions of Little Urswick were: Little Urswick (621 acres [248 ha]); Bolton with Adgarley (688 acres [275 ha]) and Stainton (536 acres [214 ha]). Common land at Skelding Moor (90 acres [36-ha]) in Little Urswick and Scales enclosed by act of 1820. Urswick parish also had rights, in common with neighbouring parish of Aldingham, on Birkrigg Common (276 acres [110 ha]), which was transferred to Urswick CP 1906. ***Population*** rising from 633 in 1801 to peak at 1,287 in 1881; then remaining fairly stable, standing at 1,128 in 1961 (last census year for which separate figure available). ***Landownership:*** part of manor of Muchland (see Aldingham). *Manor of Bolton with Adgarley*, originally part of manor of Muchland, passed by marriage to Coupland family; forfeited by Sir Thomas Broughton 1487; then granted to earl of Derby and descended with earls of Derby to 20th century. ***Economy:*** largely agricultural; fishing and coastal trade at Bardsea. Short-lived iron ore mining at Stainton in 19th century. Limestone quarrying and lime-burning; extensive quarries at Stainton, opened 1868 to supply ironworks at Barrow; continue in production. ***Places of worship:*** medieval parish church of St Mary and St Michael; surviving pre-Conquest sculpture indicates early origin. Site of medieval chapel at Bolton. Holy Trinity church (Anglican), Bardsea, built 1843-53. Congregational chapel (iron building) at Stainton opened 1873; replaced by new building in Long Lane 1902; closed 1951. Mission room at Stainton built 1904 (also used as school, reading room and working men's club); closed and converted to dwelling. Mission room at Great Urswick in early 20th century. Tarn Close Church of Christ, Great Urswick, a 'tin tabernacle' built 1914; became United Reformed Church 1981; still in use. ***Schools and other institutions:*** Urswick Grammar School founded by bequest of William Marshall 1585; replaced by Low Furness CE Primary School on new site, opened 1994. Endowed school at Bardsea founded 1781; replaced by new school, built 1851; enlarged 1897; closed c.1927. Infants' school at Stainton opened 1904; closed c.1927. Urswick Recreation Hall, midway between Great and Little Urswick, built 1929. Village hall at Bardsea (former malt kiln) established as memorial hall 1946; refurbished 2003-4. Stainton Recreation Hall (former Congregational chapel) opened 1953.

VICKERSTOWN *see* Walney Island

WABERTHWAITE

Ancient parish in Allerdale above Derwent ward, Cumberland. CP enlarged by absorbing Corney CP 1934.

Acreage: 1,771 acres [717 ha], including 638 acres [258 ha] of unenclosed common on Waberthwaite Fell. Waberthwaite Wood stinted pasture (267 acres [108 ha]) enclosed 1806. ***Population:*** estimated at 375 in 1688; in range 100-150 across earlier 19th century, rising to around 200 later in century; peak of 279 in 1921. Modern CP (which includes Corney) had 338 inhabitants in 1951, falling to 274 in 2001. ***Landownership:*** *manor of Waberthwaite* held by Wybergh family in medieval centuries; acquired before 1688 by Penningtons of Muncaster, who remained lords of manor to 20th century. ***Economy:*** predominantly agricultural; bacon curing and sausage making since late 19th century. ***Places of worship:*** medieval parish church of St John, of pre-Conquest origin, to judge by surviving Viking age cross. ***Schools and other institutions:*** school recorded 1818; old schoolhouse replaced by new building on different site 1938, later extended; now Waberthwaite CE Primary School. Institute and reading room built 1911; now Waberthwaite, Corney & District village hall.

WAITBY

Township in Kirkby Stephen parish, East ward, Westmorland. CP enlarged by absorbing Smardale CP 1894.

Acreage: 1,183 acres [479 ha]; Waitby intake (220 acres [89 ha]) enclosed 1852 and Waitby common (500 acres [202 ha]) in 1854. ***Population:*** in range 40-60 across 19th century, rising as result of railway construction to peak of 101 in 1871. Population of enlarged CP declined from 96 in 1901 to 60 by 2001. ***Landownership:*** *manor of Waitby* held by de Waitby family c.1200; divided into moieties before 1310, held by Parving and Dacre families in 14th century. Both moieties passed through several families across 15th to 17th centuries. Manor (or part of it) sold by James Lowther to Richard Monkhouse of Winton 1713; held by Wakefield family in later 19th century. ***Economy:*** largely agricultural; limestone quarrying and lime-burning in 19th century. ***Places of worship:*** traces of medieval chapel reported in mid-19th century. ***Schools and other institutions:*** school built 1680 and endowed by James Highmoor to educate poor children of Waitby and Smardale; rebuilt by subscription 1868; closed c.1910. School building then used as village hall until abandoned in 1950s; later converted to holiday accommodation.

WALNEY ISLAND

Island off Barrow-in-Furness, historically part of Hawcoat township in Dalton-in-Furness parish, Lonsdale hundred, Lancashire North of the Sands.

Acreage: 3,065 acres [1,240 ha], divided between hamlets of North Scale (1,598 acres [647 ha]) and Biggar (1,467 acres [594 ha]). ***Population:*** 339 in 1871 rising to 474 in 1891 but expanding rapidly as result of building of Vickerstown in early 20th century: had doubled to 961 by 1901 and rose to estimated 4,800 in 1911 and 7,203 by 1931. Continued to rise, reaching 9,811 by 1961 and 11,388 by 2001. ***Landownership:*** part of 'bierley' of Hawcoat in Furness Abbey's manor of Plain Furness with which it descended (see Barrow-in-Furness). ***Economy:*** farming; fishing; gathering seaweed ('tangle') as manure; lime-burning. Rabbit warrens at north and south ends, probably from medieval times. From late 19th century closely bound to economy of Barrow-in-Furness. Salt extraction at South Walney 1896-1909;

site subsequently became gravel pit which closed 1963. Plan to build airships proposed in First World War and sheds built near North Scale. After Second World War two factories employing female labour in clothing trade on Church Lane and on Promenade near North Scale. Both had closed by 1970s. Tourism and recreation from later 20th century (chalet and caravan parks). ***Places of worship:*** chapel of ease of St Mary founded c.1568; rebuilt 1852 but proved inadequate after growth of Vickerstown; new building on adjacent site consecrated 1908 and fully completed 1928-31. Mission room at Biggar built 1897; sold 1947 and transported to Millom for use as social hall on housing estate. Baptists met in sod hut at Southend Farm c.1853. Methodist New Connexion chapel at North Scale opened 1881; closed 1907 following opening of Wesleyan Methodist chapel on The Promenade at Vickerstown 1905; replaced by modern building in Warren Street 1998. Presbyterian chapel (St Andrew's) at Vickerstown opened 1908; closed 1977. St Columba's Roman Catholic chapel opened 1916; replaced by new church 1958. Former Presbyterian chapel purchased by Jehovah's Witnesses 1979; sold into private ownership 1999. ***Schools:*** school recorded 1849. School close to St Mary's church built in memory of Thomas Y. P. Michaelson 1856; closed early 20th century. Schoolroom at Biggar by 1900. Vickerstown School opened 1902; moved to new building on new site 2011; now Vickerstown Primary School. Ocean Road School opened 1905 in temporary structure before permanent building opened 1917; demolished 1993. New secondary modern school at Sandy Gap opened 1958; amalgamated with Ocean Road School 1979 to form Walney Comprehensive School. Growing population catered for by South Walney Primary School, opened 1920s, and North Walney Primary School, opened 1960s. St Columba's RC School built 1916. Richard Brunskill Special School, Moor Tarn Lane, opened 1963; closed 1991; replaced by George Hastwell Special School. ***Other institutions:*** lighthouse at Haws Point built 1790. Biggar Bank acquired by Barrow Corporation 1877 and became town's main outdoor recreation ground. James Dunn Park on Promenade opened 1903. Walney Island airfield built 1940; air gunnery school opened 1941; closed as military airfield 1955; civilian use, including gliding, since. Vickerstown Institute, Gateacre Street, opened 1902; moved to new premises on Church Lane 1917; closed 2009. Walney Cinema, Dominion Street, opened as theatre and picture house 1915; closed 1959; demolished. Branch library on Promenade, opened 1940; replaced by new building in Central Drive.

WALTON

Ancient parish in Eskdale ward, Cumberland, comprising townships of High Walton and Low Walton.

Acreage: 3,811 acres [1,542 ha], divided between townships thus: High Walton (including detached portion): 1,680 acres [680 ha]; Low Walton: 1,591 acres [644 ha]. Walton Moss (common to both townships) covered 540 acres [219 ha]. Walton commons (1,200 acres [486 ha]) enclosed 1742. ***Population:*** estimated at 345 in 1688; rising from 421 in 1801 to peak of 481 in 1831; then declining across later 19th and 20th century to stand at 278 in 2001. ***Landownership:*** *manor of Walton* granted to Lanercost Priory, passing with priory estate after Dissolution to Sir Thomas Dacre 1543, descending in Dacre family until sold by William Dacre 1789 to John Johnson, who built Castlesteads (or Walton Hall) 1789-94. Descended in Johnson family to 20th century. ***Economy:*** farming; bacon curing in mid-19th century. Brick and tile works (1840s to late 1860s). ***Places of worship:*** parish church first recorded 1687; rebuilt 1811; rebuilt again (as St Mary's church) 1869-70. Primitive Methodist chapel built 1858; now closed. ***Schools and other institutions:*** two schools recorded 1818. Parish school built 1859; closed 1970 and became village hall. Reading room opened 1893; used as meeting room 1970s; now café.

WARCOP

Ancient parish in East ward, Westmorland, embracing townships of Warcop, Sandford, Burton and Bleatarn.

Acreage: 11,489 acres [4,650 ha] divided between constituent townships thus: Warcop: 4,708 acres [1,905 ha]; Sandford: 2,113 acres [855 ha]; Burton: 2,666 [1,079 ha]; Bleatarn: 2,002 acres [810 ha]. Sandford 'mire' enclosed 1773; Bleatarn Common (720 acres [291 ha]) enclosed 1791; commons in Warcop township (720 acres [291 ha]) enclosed 1831. ***Population:*** around 700 in early 19th century, surging to 906 in 1861, then falling back to 652 by 1891; downward trend across 20th century to stand at 491 in 2001. ***Landownership:*** *manor of Warcop* held by Warcop family from early 13th century but had passed to Ralph Neville, earl of Westmorland, by 1422. Said to have been regained by Warcop family 1554; then sold to Thomas Braithwaite of Ambleside 1590, with whose descendants it remained until 1740; then held by Preston and Wild families until 1957. *Manor of Sandford* held by Sandford family from 12th century to early 15th when passed by co-heirs to Bardsey and Warcop families; tenants enfranchised 1592 and purchased manorial rights 1690. *Manor of Burton* held by Hilton family by 13th century, descending in that family until marriage of Mary Hilton to Thomas Wybergh of Clifton 1713; descended through Wybergh family to 20th century; manorial rights bought by Ministry of Defence. *Bleatarn* was grange of Byland Abbey (Yorks.), passing after Dissolution successively to families of Bellasis, Salvin, Fothergill and Musgrave. ***Economy:*** farming. Brough Hill livestock fair (in Warcop township), held annually on 30 September and 1 October; noted as cattle fair in later 17th century and as largest horse fair in northern England in later 19th. Lead mining from 17th century; lime-burning in 19th century. Woollen textile manufacture recorded in place-name Walk Mill (a saw mill in 1859). Warcop army training camp established 1942; continues as military training base. Tourism from later 20th century, including Eden Valley Railway since 2006, on restored section of railway (closed 1962). ***Places of worship:*** medieval parish church of St Columba; chancel rebuilt 1854-5; bellcote rebuilt 2006. Chapel of ease at Bleatarn built 1865. Inghamite chapel at Birks flourished c.1754 to 1813, after which used by other nonconformist groups; disused by 1860. Wesleyan Methodist chapel at Warcop built 1821; rebuilt 1872; still in use. Methodist New Connexion chapel in Warcop built 1845; closed and converted to dwelling. Methodist Association chapel at Sandford built 1848; closed and converted to cottage c.2010. ***Schools and other institutions:*** two schools recorded 1818. National school built 1863; now Warcop CE Primary School. Reading room and library established 1859; new premises built 1877; extended 1920; closed 2009. Temperance hall, also used as market, built 1865; became village hall; sold 2013 and replaced by new village hall, opened 2015. Village hall at Sandford.

WARWICK

Ancient parish in Cumberland ward, Cumberland, containing townships of Warwick-on-Eden and Aglionby. Absorbed into Wetheral CP 1934.

Acreage: 1,858 acres [765 ha], divided between townships thus: Warwick: 1,295 acres [524 ha]; Aglionby: 564 acres [228 ha]. Common land at Warwick Moor (535 acres [220 ha]) enclosed 1801. ***Population:*** estimated at 225 in 1688; rising from 287 in 1811 to 373 in 1831, then falling to 269 in 1931 (last census year for which separate figure available). ***Landownership:*** *manor of Warwick* held in 12th century by Odard of Corby, whose descendants took surname Warwick. Descended in that family until death of Francis Warwick

1772; then passed by descent to Maddison and Bonner families, successively, until bankruptcy of Robert Bonner (who took surname Warwick) in 1822, when sold to Thomas Parker, who built Warwick Hall 1828 (destroyed by fire 1930); remained in Parker family until c.1900. Bought by Charles Liddell, whose son-in-law R. G. G. J. Elwes rebuilt Warwick Hall 1934. *Manor of Aglionby* held by Aglionby family, descendants of Agillun (from whom township took its name) from 12th century to c.1900. ***Economy:*** predominantly agricultural; salmon fishery in River Eden recorded late 17th century. Carlisle golf course, near Aglionby, opened 1940. ***Places of worship:*** medieval parish church of St Leonard. Sunday School (built before 1823) converted to Wesleyan Methodist chapel before 1847; now closed. ***Schools and other institutions:*** school taught in church 1703; schoolroom built on waste before 1814. Two day schools built by Thomas Parker early 19th century. Warwick-on-Eden Memorial Hall built c.1920; still in use as village hall.

WARWICK BRIDGE *see* Wetheral

WASDALE *see* Eskdale and Wasdale; Netherwasdale

WATERHEAD

Township in Lanercost parish, Eskdale ward, Cumberland, enlarged by addition of part of Banks township in 19th century.

Acreage: 4,617 acres [1,868 ha]. Commons enclosed under Lanercost enclosure award 1807. ***Population:*** rising from 234 in 1801 to 473 in 1831; then declining to 234 by 1901 and 124 by 2001. Inflated to 637 in 1921 by presence of staff and patients at convalescent home. ***Landownership:*** part of estate of Lanercost Priory (see Burtholme). ***Economy:*** pastoral farming; afforestation in 20th century. Gilsland Spa, noted for its curative qualities by later 17th century, became popular resort by later 18th century: hotel (The Shaws) established by 1740s; burnt down 1859; replaced by Spa Hotel, built 1859-60; bought by Co-operative Society 1901 and used as convalescent home until 1972; now Gilsland Hall Hotel. Archaeological tourism to Hadrian's Wall (particularly Birdoswald Roman fort) since 19th century. ***Places of worship:*** chapel at Tryermaine in 12th century. Church of St Mary Magdalene, serving new ecclesiastical parish of Gilsland, built 1852-4. ***Schools and other institutions:*** none.

WATERMILLOCK

Chapelry and township in Greystoke parish, Leath ward, Cumberland. Absorbed into Matterdale CP 1934.

Acreage: 9,735 acres [3,940 ha], including detached part of Dacre parish (69 acres [28 ha]) at Wreay. Watermillock commons and stinted pasture (4,591 acres [1,858 ha]) enclosed 1835. ***Population:*** rising from 333 in 1801 to 598 in 1851; then drifting down to 448 by 1931 (last census year for which separate figure available). ***Landownership:*** part of barony of Greystoke (q.v.), with which it descended. Lyulph's Tower built 1781 by duke of Norfolk as lakeside villa. ***Economy:*** farming; fishing with nets (recorded 1472). Pencil manufacture mid-19th century (recorded in name Pencil Mill Beck). Tourism significant from late 18th century: 19th-century villas on lake shore of Ullswater. ***Places of worship:*** medieval chapel of ease (chaplain recorded early 13th century), known as Old Church, beside Ullswater. Parochial chapel, 'New Kirk', built 1558 on higher ground near Priest's Craggs; rebuilt 1749 and again 1884 (as All Saints church). Wesleyan Methodist chapels at Beauthorn (recorded 1861) and Ulcat Row (founded 1830s; closed c.1940). ***Schools and other institutions:*** grammar school recorded from 1550s; rebuilt on new site as boys' elementary school 1860; closed 1954. Girls' school built 1832; closed 1913. Rev. John Tinkler ran boarding academy 1829. Village hall built 1926.

WAVERTON

Township in Wigton parish, Cumberland ward, Cumberland.

Acreage: 4,163 acres [1,685 ha], including common moorland in eastern part of township enclosed 1817. Lessonhall Meadow and Horse Park enclosed 1874. ***Population:*** rising from 375 in 1801 to peak of 574 in 1861; then declining gently to 326 in 2001. ***Landownership:*** part of barony of Wigton. *Manor of Waverton* had been divided into eight parts before 1316; Wyndham family retained manorial rights, as overlords of Wigton barony, in 19th century. *Manor of Lessonhall or Little Waverton* held by Dalston family by 16th century. Sir George Dalston enfranchised tenants 1747. ***Economy:*** predominantly agricultural; brick- and tile-making at Parkgate in mid-19th century. ***Places of worship:*** Christ Church (Anglican) built 1865. ***Schools and other institutions:*** schoolmaster recorded 1847. Waverton Board School built 1874; closed and converted to dwelling. Lecture room built in village 1864.

WEDDICAR

Township in St Bees parish, Allerdale above Derwent ward, Cumberland. CP enlarged by absorbing part of Hensingham CP 1934.

Acreage: 1,150 acres [456 ha] before boundary change; enlarged CP contained c.1,760 acres [712 ha] after 1934. Common land on Weddicar Moor (510 acres [206 ha]) enclosed 1809. ***Population:*** in range 50-70 from 1801 to 1934, living in just eight scattered farmsteads. Transfer of Keekle and Padstow area of Hensingham increased population by c.400; it stood at 442 in 2001. ***Landownership:*** held by Fleming family in early 14th century and by Patrickson family by 1578; sold by William Patrickson 1652; acquired by Sir James Lowther of Whitehaven in 18th century, descending to earls of Lonsdale. ***Economy:*** farming; small-scale coal mining (recorded 1688) and quarrying. Oil and paint mill at Weddicar Hall in mid-19th century, run by Randelson and Forster, Whitehaven chemists. Over half surface of original township subject to opencast coal mining in 1980s-90s (subsequently reinstated and planted with trees). ***Places of worship:*** none (for Keekle Mission, see Hensingham). ***Schools and other institutions:*** none.

WELTON *see* Sebergham

WESTLINTON

Township in Kirklinton parish, Eskdale ward, Cumberland.

Acreage: 3,420 acres [1,384 ha]. Commons enclosed 1810. ***Population:*** rising from 519 in 1801 to peak of 623 in 1821; then downward drift to 402 in 1901 and 359 in 2001. ***Landownership:*** held by Boyvills of Kirklinton barony from 12th century to 15th, passing by marriage of Boyvill heiress to Alexander Highmore of Harby Brow; sold, probably by Francis Highmore (d. 1610), to Blencowe family; later held by earls of Lonsdale. ***Economy:*** largely agricultural; brick- and tile-making in 19th century. Ministry of Agriculture Food and Fisheries field research station at Blackford mid-20th century. ***Places of worship:*** Anglican church of St John the Baptist at Blackford, built 1870. ***Schools and other institutions:*** endowed parochial school at Blackford, built 1831; enlarged 1882; now Blackford CE Primary School.

WESTMORLAND (barony) *see* Appleby

WESTNEWTON
(otherwise 'Longnewton' or 'Newton in Allerdale')

Township in Bromfield parish, Allerdale below Derwent ward, Cumberland. United with Allonby as West Newton and Allonby CP, 1866 to 1894.

Acreage: 2,411 acres [976 ha]. ***Population***: in range 300-350 for most of 19th and early 20th centuries (peak of 389 in 1871); then falling to 199 in 1971 before recovering to stand at 247 in 2001. ***Landownership:*** *manor of Newton* held by de Newton family from 12th century, passing by descent to Roger Martindale, who married daughter of Thomas de Newton in 14th century. Descended through Martindales to Isabel, daughter and heiress of James Martindale, who married William Musgrave (d. 1596) of Hayton; then through Musgrave family to Eleanor, daughter of Sir Richard Musgrave-Hylton, who married William Jolliffe 1769; remained in hands of Jolliffe family until early 20th century. ***Economy:*** largely agricultural; small-scale quarrying; brewing (Yates Brewery established 1986). Wind farm near Warwick Hall established 2014. ***Places of worship:*** St Matthew's church (Anglican) built 1857, funded by John Todd, Manchester merchant and native of Westnewton. ***Schools and other institutions:*** National school established 1848; enlarged 1858; now St Matthew's CE Primary School. Almshouses (St Matthew's Cottages) built c.1858. Village hall (formerly reading room) built c.1920.

WESTWARD

Parish in Allerdale below Derwent ward, Cumberland, embracing townships of Brocklebanks & Stoneraise and Rosley & Woodside.

Acreage: 13,230 acres [5,354 ha], around half of which (7,246 acres [2,932 ha]) was formerly common land, enclosed 1822. Township of Brocklebanks and Stoneraise contained 7,425 acres [3,005 ha]; Rosley and Woodside contained 5,805 acres [2,349 ha]. ***Population:*** estimated at 610 in 1688; rising from 918 in 1801 to peak of 1,311 in 1841; then steady decline to 718 in 1981; rising again to stand at 814 in 2001. ***Landownership:*** in Crown hands as part of Inglewood Forest (it was 'west ward' of forest) from 12th century until granted 1344 to Anthony de Lucy, uniting it with barony of Allerdale, with which it descended (see Cockermouth). ***Economy:*** largely agricultural. Fair at Rosley, held fortnightly from Whitsuntide to Michaelmas for cattle and horses, as well as linen cloth, described as 'best fair' in north of England for Irish and Scottish horses and for cattle and sheep 1688; still held mid-19th century but had declined almost to extinction by 1901. Fairs for sheep and wool at Red Dial recorded 1860. Race courses on Harethwaite Common and Woodcock Hill in late 17th century. Several coalmines near Chalkhead recorded 1688; mining had ceased by 1847. Quarrying for red and white freestone at Shawk and Howrigg quarries, said in 19th century to be 'esteemed the best in Cumberland'. ***Places of worship:*** early 13th-century hermitage at Isle Kirk, recorded as 'Hildkirk' chapel 1374. Chapel called 'New Kirk' built in Stoneraise township between 1557 and 1563; had acquired parochial status by 1563; rebuilt as parish church of St Hilda 1785-6. Anglican church of Holy Trinity, Rosley, built 1839-41. ***Schools and other institutions:*** two schools by 1815, one by churchyard, taught by curate; other at Craggs in Rosley. School adjacent to church rebuilt 1828; closed 1920s and became parish room (village hall). Craggs school closed 1895, having been replaced by Rosley and Woodside School, built 1873 on new site at Carwath near Holy Trinity church; this in turn replaced by modern building opposite church 1961; now Rosley CE Primary School. Rosley Recreation Hall (ex-army hut from Longtown) opened 1926; replaced by new village hall built 1986-8.

WETHERAL

Ancient parish in Cumberland ward, Cumberland, comprising four townships: Wetheral, Scotby, Cumwhinton & Cotehill, and Warwick Bridge & Great Corby (each treated separately below). CP enlarged by absorbing Warwick CP 1934. Part of CP transferred to Carlisle 1951.

Acreage: parish covered 11,489 acres [4,731 ha]. ***Population:*** estimated at 515 in 1688; rising from 1,697 in 1801 to 3,165 by 1851, then fairly stable to stand at 3,324 in 1931 (last census year before boundary changes). For population of individual townships in earlier 19th century, see below.

WETHERAL: 4,478 acres [1,812 ha], including detached portion of 2,313 acres [936 ha], largely consisting of former moorland at southern extremity of parish. Commons enclosed under Inglewood Forest enclosure award 1819. ***Population:*** rising from 376 in 1801 to 666 in 1861. ***Landownership:*** core of endowment of Wetheral Priory, founded c.1100, passing after Dissolution to Dean and Chapter of Carlisle. ***Economy:*** farming; market gardening; shops and services in village. ***Places of worship:*** medieval parish church of Holy Trinity; tower added 1790 (restored 1882); chancel rebuilt 1872. Wesleyan Methodist chapel built 1873; still in use. ***Schools and other institutions:***

school endowed 1760; rebuilt 1854; had closed by 1978 when buildings taken over for play school. Ladies' boarding school established at Lime House c.1856; became boys' preparatory school 1899; moved to Hawkesdale (see Dalston) 1946. Working men's reading and recreation room built 1891; became village hall; sold 2013 and replaced by new Wetheral Village and Community Hall.

SCOTBY: 1,681 acres [680 ha]. Common fields at Scotby Holme (144 acres [59 ha]) and Holme Gate enclosed 1777 and 1788. ***Population:*** rising from 275 in 1801 to 520 in 1861. ***Landownership:*** part of honour of Penrith (q.v.), with which it descended. ***Economy:*** farming; market gardening; tanneries established by Elihu Sutton 1781; closed 1930. ***Places of worship:*** Quaker meeting house built 1718; closed 1913. All Saints' church (Anglican) built 1854. ***Schools and other institutions:*** school recorded 1814; endowed parochial school built 1853-4; replaced by modern building 1963; now Scotby CE Primary School. Parish workhouse for Warwick and Wetheral parishes at Wheelbarrow Hall, Holme Lane, late 18th century. Newsroom and library built by mid-19th century; said to have been one of earliest working men's reading rooms in northern England. Welcome Institute converted from inn 1917; rebuilt as village hall 1926; extended 1953.

CUMWHINTON & COTEHILL: 2,570 acres [1,040 ha]. Common field at Cotehill (115 acres [47 ha]) enclosed 1840. Commons in Cumwhinton enclosed under Inglewood Forest enclosure award 1819. ***Population:*** rising from 365 in 1801 to 633 in 1861. ***Landownership:*** lordship of manor divided at an early date: by late 19th century half belonged to Armathwaite Castle Estate and half to Aglionby family. ***Economy:*** Cumwhinton noted for rich corn grounds in later 17th century. Brick and tile works in mid-19th century. Gypsum quarrying recorded from 1695; major expansion with alabaster works at Knott Hill and Cocklakes from 1873 to 1920s. ***Places of worship:*** school used as chapel before church of St John the Evangelist, Cotehill, built 1868. Wesleyan Association chapel at Cumwhinton built 1816; replaced by adjacent Free Methodist chapel 1904; closed 1965. Wesleyan Methodist chapel at Cotehill by 1900; demolished for road widening 1930s; new chapel opened 1937; still in use. Mission hall at Cumwhinton built 1908; now village hall. ***Schools and other institutions:*** endowed school at Cumwhinton built 1839; rebuilt on new site 1895; now Cumwhinton Primary School. National school at Cotehill built 1852; open in late 20th century; now closed. Private boarding school at Englethwaite Hall in 1901. Village halls at Cotehill (1930) and Cumwhinton (former mission hall).

WARWICK BRIDGE & GREAT CORBY: 2,760 acres [1,117 ha]. ***Population:*** rising from 561 in 1801 to 1,180 in 1851. ***Landownership:*** *manor of Corby* granted to Richard Salkeld 1337 after attainder of Sir Andrew de Harcla, earl of Carlisle; descended through Salkeld family until divided between two heiresses in later 15th century. Sir William Howard of Naworth acquired both halves 1605 and 1625 and gave Corby to his second son, Sir Francis Howard, in whose family it remained until later 20th century. ***Economy:*** salmon fishery in River Eden since medieval period (with coops at Corby Castle). Ferry across River Eden at Corby, recorded 1688. Textile industry suggested by name Walk Mill (later Glencairn Mill); site was cotton mill in 1861. Cotton spinning mill and dye works at Warwick Bridge, established 1790-3; closed c.1977. Paper manufacture at Allenwood mill from mid-19th century to early 20th. Woollen mill at Broadwath from later 19th century; disused by 1962. Broadwath estate divided into Land Settlement Association smallholdings 1930s. ***Places of worship:*** Roman Catholic church of Our Lady and St Wilfred at Warwick Bridge built 1840-1. Anglican church of St Paul, near Warwick Bridge, built 1845-6. Primitive Methodist chapels at Great Corby (built 1889; closed 1980s) and Warwick Bridge (built c.1907; closed). ***Schools and other institutions:*** school at Corby endowed 1720; rebuilt 1845; extended 1882; now Great Corby Primary School. School at Warwick Bridge by early 20th century; now Warwick Bridge Primary School. School at Warwick Bridge cotton mill in mid-19th century. Reading room at Great Corby established 1850; rebuilt 1877; now village hall. Down-a-gate community centre, Warwick Bridge, built 1997.

WHARTON

Township in Kirkby Stephen parish, East ward, Westmorland.

Acreage: 1,501 acres [607 ha], of which almost half remains unenclosed common land on Wharton Fell (715 acres [289 ha]). ***Population:*** village cleared when Wharton park enclosed c.1547 (tenants resettled at Wharton Dikes). Population in range 70-80 in early 19th century; then downward drift to stand at 31 by 2001 (apart from temporary surge to 100 in 1871 during railway construction). ***Landownership:*** held by Wharton family from late 13th century; sold to Robert Lowther 1728 after attainder of Philip, 1st duke of Wharton, subsequently descending to earls of Lonsdale. ***Economy:*** predominantly agricultural; small-scale quarrying and lime-burning in 19th century. ***Places of worship:*** none. ***Schools and other institutions:*** none.

WHICHAM

Ancient parish in Allerdale above Derwent ward, Cumberland, including village of Silecroft. CP enlarged by absorbing Whitbeck CP 1934.

Acreage: 4,883 acres [1976 ha], including part of 4,303 acres [1,742 ha] of unenclosed common fell on Black Combe and White Combe in Whicham and Whitbeck parishes. ***Population:*** estimated at 445 in 1688; in range 230-330 for much of 19th century, rising to peak of 455 in 1891, from which it fell to 319 in 1931 (last census year before boundary change). ***Landownership:*** *manor of Whicham* held by Bethom family in 13th century and Corbet family by early 14th; by 19th century lordship had passed to earls of Lonsdale. 'Satherton' (north-eastern part of parish) retained under direct lordship of lords of Millom. ***Economy:*** predominantly agricultural; brick- and tile-making at Silecroft in later 19th century; lead mining (recorded in place-name 'Lead Mine Breast'). Pencils of Black Combe slate manufactured in parish in early 19th century. ***Places of worship:*** medieval parish church of St Mary; restored and enlarged 1858. ***Schools and other institutions:*** endowed National school beside church, formerly Whicham and Millom Grammar School, apparently founded before 1540; rebuilt 1862; closed 1970s. Parish room built at Silecroft by 1922; now Silecroft village hall.

WHINFELL (Cumberland)

Township in Brigham parish, Allerdale above Derwent ward, Cumberland. Absorbed into Blindbothel CP 1934.

Acreage: 1,747 acres [699 ha], including 817 acres [330 ha] of common land on Whin Fell, enclosed 1826. ***Population:*** rising from 84 in 1801 to peak of 132 by 1841; then declined to stand at 71 in 1931 (last census year for which separate figure available). ***Landownership:*** part of honour of Cockermouth. Ownership divided from c.1260 when Alice de Lucy gave moiety of Whinfell to her son John de Lucy; that share was held by Allan Pearson of Bridekirk in early 19th century. Other moiety descended with honour of Cockermouth (q.v.), being one of group of manors in hands of Wharton family 1530 to 1739. ***Economy:*** predominantly agricultural. Iron smelting in later 12th century (Holm Cultram abbey granted leave to erect bloomery at Whinfell). Woollen textiles: fulling mill recorded c.1282. ***Places of worship:*** none. ***Schools and other institutions:*** none.

WHINFELL (Westmorland)

Township in Kendal parish, Kendal ward, Westmorland. CP enlarged by absorbing Patton CP 1986.

Acreage: 4,346 acres [1,759 ha]. Commons enclosed 1842. ***Population:*** rising from 184 in 1801 to peak of 214 in 1831, then declining to 167 in 1891 and falling to low of 96 in 1951. ***Landownership:*** divided from early period: land granted to Gervase Deincourt late 12th century descended to Stricklands of Sizergh who continued to hold manor (or part of it) in later 16th century. Matthew de Redman held half manor of Whinfell 1283, which descended through Redman family to early 17th century; in 1627 it was held of heirs of Richard Redman by William Thornburgh. ***Economy:*** predominantly agricultural; fishery at Whinfell tarn, with eel trap ('eel ark') at outflow. ***Places of worship:*** Primitive Methodist chapel, adjacent to former school, built 1903; closed 1970s and converted to dwelling. ***Schools and other institutions:*** school near Whinfell Tarn in mid-19th century.

WHITBECK

Ancient parish in Allerdale above Derwent ward, Cumberland. Absorbed into Whicham CP 1934.

Acreage: 4,482 acres [1,814 ha], including surviving unenclosed common fell on slopes of Black Combe (see Whicham). ***Population:*** estimated at 435 in 1688, but below 200 for most of 19th century. Rising from 180 in 1801 to peak of 234 in 1831; then falling to 150 in 1931 (last census year for which separate figure available). ***Landownership:*** *manor of Whitbeck* held by Morthing family in 13th century; granted by Sir William Morthing to Conishead Priory, from which it passed to Crown at Dissolution. Purchased by Lawrence Parke 1687; sold by his descendant Charles Parke to earl of Lonsdale 1807. *Manor of Monkfoss* originated in grant to Furness Abbey 12th century; granted on Dissolution to Hudlestons of Millom who sold it to Edmund Gibson 1738. ***Economy:*** predominantly agricultural; medicinal spa well at Gutterby. ***Places of worship:*** medieval parish church of St Mary; partly rebuilt 1883. ***Schools and other institutions:*** hospital for six poor people built before 1722 under charitable endowment of Henry Park in 1631; building said to be 'insufficient' 1779 but still in use in mid-19th century.

WHITEHAVEN

Port and industrial town in St Bees parish, Allerdale above Derwent ward, Cumberland. This entry treats area of MB as established 1894, including Bransty, Harras and Corcickle areas and northern section of Preston Quarter. See also Hensingham.

Acreage. Whitehaven township covered 176 acres [71 ha] before 1878, when boundary extended to include parts of Hensingham and Bransty. Extended further when became MB 1894, covering 1,813 acres [734 ha], including northern part of Preston Quarter CP, Bransty and Harras areas of Moresby CP and Corcickle and northern section of Hensingham CP. Further enlarged 1934 by absorbing parts of Moresby, Sandwith, Hensingham and Preston Quarter CPs.

Landownership. Part of manor of St Bees, passing to Crown on Dissolution 1539; granted to Sir Thomas Chaloner (d. 1565) 1553; most of manor bought by Thomas Wybergh 1599, who conveyed part to Lowthers; Wyberghs continued to hold an interest until bought out by Lowthers 1670 and 1684. On death of Sir John Lowther 1637, Whitehaven passed to his second son Christopher Lowther and descended to his son Sir John (d. 1706) (who rebuilt The Flatt as his seat) and grandson Sir James (d.1755). Passed to Sir James Lowther of Lowther 1756, who enlarged The Flatt, converting it into Whitehaven Castle; ownership descended thereafter through earls of Lonsdale.

Origins and growth of the town. Whitehaven had been small harbour and fishing village from 13th century or earlier. Expansion began in mid-17th century with building of piers by Lowthers 1632-4 and 1679-81 and granting of market charter 1660. By 1680s it had grown rapidly, expanding from village of c.30 households in early 17th century to new town of over 1,000 inhabitants by 1685, which more than doubled to 2,281 by 1696. Sir John Lowther had laid out grid of streets by 1680s, making Whitehaven earliest planned new town in post-medieval Britain. Thomas Denton, writing 1688, recorded the 'great encrease of trade within this twenty years; the houses being build of pure white stones ... the town running along the shore in a figure like an halfe moon, between the cliffe & Branstee-brow'. Trade was based on exports of coal from pits at Hensingham and Moresby, salt from saltpans at Bransty and Workington, grindstones from local quarries, and corn (oats and barley). Imports by 1680s included French wine and brandy; softwood, pitch and tar from Baltic; tobacco and sugar from West Indies.

Peak of wealth came in 18th century, when, as well as supplying coal to Dublin, it was major port in Atlantic trade and leading tobacco-importing centre. By 1762 town contained 9,062 inhabitants. Though losing out to Liverpool in Atlantic trade from later 18th century, Whitehaven remained major coal port across 19th century. Several deep collieries were sunk: Saltom pit (first to mine coal under sea floor; sunk 1729; closed 1848); William pit (sunk 1804-12; closed 1965; site of major disaster 1947, when 104 killed); Wellington pit (sunk 1838; closed 1933; site of major disaster 1910 when 136 killed) and Haig pit (sunk 1914-18; closed 1986). Other important industries in 19th century included shipbuilding (shipyards at Bransty); sail-making (until 1931); pottery manufacture (until 1915; three potteries in Ginns area in 1840s); iron making (Newtown Foundry and Phoenix Foundry); fire brick manufacture; copperas/vitriol manufacture (established at Ginns 1718); flax spinning and linen manufacture, and bone manure works. Town's role as regional centre reflected in publication of two major newspapers: *Cumberland Pacquet*, established 1774, and *Whitehaven Herald,* established 1830. Population continued to grow, rising from 10,628 in 1801 to 12,438 by 1821 and 14,190 by 1851. Long terraces of workers' housing at New Houses built 1788. Population of area which became MB in 1894 remained fairly steady until after First World War, standing at 18,916 in 1851, 19,003 in 1891 and 19,535 in 1921. Town expanded onto higher ground to north (Bransty) and south (Kells) in later 19th/early 20th century, with new housing estates (associated with slum clearance) from 1930s.

After decline of coal mining, Whitehaven acquired new industries. Marchon Chemicals, established 1941 (factory at Kells from 1943), initially manufacturing firelighters, soon specialised in soap powders, using local anhydrite (see Sandwith) and importing rock phosphate from Morocco from 1960s to 1990s; employed 2,250 in late 1970s; factory closed 2005 and demolished. H. Edgard, military uniform manufacturer, moved from London 1940; major employer, employing 580 in late 1970s. Quaker Oats company manufactured cereal at Beacon Mills from 1949 to 1972. Harbour closed for commercial purposes in 1990s, but developed as thriving marina. Tourist industry of growing importance to town's economy from late 20th century.

Places of worship. Chapel of ease by 1642; replaced by St Nicholas' church, Lowther Street, built 1693; enlarged 1745; rebuilt 1883; nave and chancel destroyed by fire 1971, tower survives as chapel and community centre. Holy Trinity church built 1714; demolished 1949. St James' church, Queen Street, built 1752-3; now town's parish church. Christ Church: 1847-1977, demolished 2004 after a fire. St Peter's, Kells, built 1938-9, replacing mission established 1908 and relocated 1920. St Mary's, Mount Pleasant, built by

Mr Hogarth 1789 but not consecrated; became Primitive Methodist chapels (see below, Hogarth's Mission).

First Roman Catholic church built 1761 in Chapel Lane; replaced by St Mary's church, Duke Street, opened 1780; enlarged 1824; in turn replaced by St Bega's church, Coach Road, built 1865-8. Older church of St Gregory, Coach Road, built 1835, became school when new mission church of St Gregory and St Patrick, Quay Street, opened 1890. Modern Roman Catholic churches at Kells (St Mary, built 1927; rebuilt 1960-1) and Mirehouse (St Benedict, built 1973-6).

Protestant nonconformist presence from foundation of new town: Presbyterian worship from 1672; chapel ('Low Meeting') built 1695 for all Protestant Dissenters; enlarged 1749; renovated 1856; and rebuilt 1905; now United Reformed Church. Scotch Presbyterian 'High Meeting', Howgill Street, from 1755; then High Street from 1760 to 1895 (reopened later as Kirk Mission Hall – see below). Quaker meeting house, Sandhills Lane, built 1725; closed 1926; re-established in new meeting house (former Methodist Sunday school), Scotch Street; opened 1984; closed 2013. Congregational chapel (Providence Chapel), Duke Street, built 1793; enlarged 1838; moved to Scotch Street 1874; closed and demolished 1969 when URC Church formed, using former Presbyterian building. English Baptist chapel, Charles Street, built 1758; enlarged 1842; closed after 1851.

Wesleyan Methodist chapel at Michael Street from 1761; closed because of subsidence 1791; used Hogarth's Mission until chapel rebuilt 1795; rebuilt again 1818; closed 1877 and replaced by new church, Lowther Street, 1877; closed 1996. Primitive Methodists met in Hogarth's Mission from 1828 and also had chapel in Fox Lane until 1859, when Howgill Street chapel built; closed 1940. Wesleyan Association (from 1858 United Methodist) chapel, Catherine Street, opened 1836; closed 1934 after Methodist reunification.

Other places of worship established in 19th century included Christian Brethren (from 1854; moved to former Primitive Methodist chapel, Fox Lane, 1859; then to former Quaker meeting house, Sandhills Lane, 1931; still open) and Salvation Army (from 1878; moved to former Providence chapel, Duke Street, 1880; moved again to former United Methodist chapel, Catherine Street, 1937; rebuilt on same site 1971; still open). Non-denominational missions: Town Mission, Rosemary Lane, established 1862; rebuilt 1919; closed c.2015. Colliery Mission opened 1865; moved 1876 to former Glasshouse School, Ginns; apparently re-founded 1898; relocated 1960; still open. Holy Trinity Church ran mission at West Strand from 1885 to 1899, run by Wesleyans in closing years; re-opened 1899 in long-abandoned St Mary's church (Hogarth's Mission): although staffed by Wesleyans was technically independent until 1910; closed 1954 and demolished in slum clearance programme. Kirk Mission Hall (former Presbyterian chapel) opened 1905; united with Hogarth's Mission 1908; became Wesleyan mission 1910; closed 1965. Ashkenazi Orthodox synagogue opened 1874; date of closure unknown. New places of worship in 20th-century include Primitive Methodist chapel, Kells (opened 1920; closed 1982) and Elim Pentecostalist church (founded 1957; church at George Street built 1961; closed 2012).

Schools. Plot for school for new town laid out on north side of Lowther Street 1690s. Private schools included mathematics and surveying school taught by Andrew Pellin from 1690s and several private academies recorded across 18th century. Sunday schools important in providing schooling for working classes in earlier 19th century: Wesleyan Methodist Sunday school opened 1811; Sunday school at Ginns opened 1817.

Provision of day schools expanded in early/mid-19th century: National school founded 1814; school built on Wellington Row 1824; became Elementary School 1874; rebuilt 1965; now St James' CE Junior School. Marine School founded by Matthew Piper 1817; schoolroom built on High Street 1822; closed 1908. School at Ginns from 1818 (Infants) and 1845 (British School). Trinity School, Newtown, built 1837; replaced by Holy Trinity National School, Howgill Street, built 1852. St Nicholas School established 1846; enlarged 1874; closed 1911; replaced by Irish St School which later became Richmond Secondary Modern School and moved to former grammar school site, Catherine Street, 1969; closed 1980s.

Colliers' School converted from former glasshouse at Ginns 1853; replaced 1876 by Earl of Lonsdale's School; enlarged 1922; now Monkwray Junior School. Market Place Refuge School, James Street, built 1852; enlarged 1859; replaced by Crosthwaite Memorial School 1901; closed 1985. National Infant school, High Street, built 1875; enlarged 1900; modified 1928; rebuilt 1965; now St James' CE Infant School. County Secondary School, Catherine Street, opened 1908 on site bought with proceeds of sale of Marine School, becoming Whitehaven Grammar School 1948; moved to Hensingham (q.v.) 1968 where all secondary education now concentrated. In 20th century new schools built to serve housing estates at Bransty (1930), Kells Infants (1938), Kells Secondary Modern (1951; closed 1978) and Jericho (1969).

Roman Catholic school at Ginns by 1847. School established 1868 in former St Gregory's RC church, Coach Road; now St Begh's Catholic Junior School.

School at mission church of St Patrick & St Gregory, Quay Street, from 1889; rebuilt on new site 1960; now Ss Gregory & Patrick Catholic Community School. St Mary's RC Primary School, Kells, built 1936.

Other institutions. Customs house in 17th century on Custom House Street (later King Street); customs house, West Strand, built 1811. Market halls (one for fish; one for poultry, butter and eggs) built 1813; replaced by new market hall 1881. Public Office, Lowther Street (with court house on lower floor and newsroom on upper floor in 1840s) built 1807. Town Hall, Duke Street, converted from former mansion of Gale family ('The Cupola') 1851; replaced by Civic Centre, Lowther Street, built 1953; extended 1969. Copeland Borough Council offices, Catherine Street, built 2003-5.

Town workhouse, Scotch Street, opened 1743; enlarged 1795 when it became female workhouse; male house built at Ginns in Preston Quarter. New Union workhouse south of town built 1854-5; closed progressively from mid 1930s, becoming Meadow View Geriatric Home; closed 1966 when services moved to new West Cumberland Hospital (see Hensingham). Dispensary, Queen Street, opened 1783; replaced by Infirmary, Howgill Street, opened 1830 and later extended; moved to Whitehaven Castle 1924 (former seat of Lowther family) until opening of West Cumberland Hospital at Hensingham (q.v.) 1966. Fever hospital at The Ginns in early 19th century. Smallpox hospital at Bransty, rebuilt 1895 for infectious diseases, open until mid-20th century. Almshouses (Aged Miners homes), totalling 31, opened progressively at Arrowthwaite, Kells, from 1926.

Assembly rooms, Howgill Street, built 1736. Other 19th-century public halls included Temperance Hall (rebuilt 1859) and Oddfellows' Hall, Lowther Street. Empress Ballroom, Duke Street, built 1933. Subscription library opened 1797 in Catherine Street, moving to Lowther Street 1808. Mechanics' & Apprentices' Library & Institution established 1825 but short-lived; reopened as Mechanics' Institute, Queen Street, 1844, containing library, reading room and newsroom. Temperance newsroom, Tangier Street, established 1847. Free public library 1888; replaced by Carnegie Library, Catherine Street, built 1905-6. Modern library opened 1970. Library at Kells Welfare from 1933 to 1963; replaced by new library 1965. Kells Community and Miners' Welfare Centre built 1939; demolished 1970s.

Museums at Mechanics' Institute, Duke Street, and in King Street by 1847. Museum at Whitehaven Natural History Society, Queen Street, 1867; moved to old Assembly Rooms, Howgill Street, 1874; then to former subscription library 1921; closed 1957 (building demolished 1961); museum reopened in Market Hall 1974, relocated to Civic Hall 1986, then to 'The Beacon' 1996.

Theatre, Roper Street, built 1769; rebuilt 1909; screened films until 1931; demolished 1976. Films screened in Market Hall by 1908; renamed Palace Cinema during First World War; replaced by Queens Cinema, opened 1928; closed c.1976. Empire Cinema opened 1914 (in former Chapel Street theatre); closed 1963. Gaiety Cinema, Tangier Street, opened 1922; closed 2003; reopened 2015. Race course on Harras Moor by mid-19th century.

WIGGONBY *see* Aikton

WIGTON

Market town in Wigton parish, Cumberland ward, Cumberland. Township of Wigton combined with Woodside Quarter to form Wigton-cum-Woodside CP 1887, which was then re-divided into separate administrative units 1894, Wigton UD (from 1934 CP) forming an island ringed by Woodside CP.

Acreage. Township of Wigton, parts of which lay intermixed with lands in Woodside township, contained 3,141 acres [1,271 ha] before boundary changes. After 1894 Wigton UD covered 1,650 acres [668 ha]. Common moorland around Wigton, mostly in what was to become Woodside CP, enclosed 1817.

Landownership. Wigton was caput of *barony of Wigton*, said to have been granted by lord of Allerdale in 12th century to Odard de Logis, from whom it descended to John de Wigton. After death of his daughter Margaret de Wigton 1349, barony passed to Thomas de Lucy, lord of Allerdale, and descended thereafter with barony of Allerdale and honour of Cockermouth (q.v.).

Origins and growth of the town. Wigton was market town but not a borough. Market probably originated at early date with trading at entrance to medieval parish church: original market place probably represented by Corn Market, in front of church. Grant of market charter 1262 probably merely confirmed status quo. By early modern period, its role as market centre for rich Solway lowlands was well established. Thomas Denton's vivid description of town in 1680s shows that specialisation in linen trade had developed as an extension of its agricultural market function: 'The principall traffick vended in this mercate is corn & yarne, for it is a great corne countrey all round this mercate, the soyl being naturally fertill. ... Great quantities of linnen yarn are likewise sold here by the neighbouring women, who make it their sole employment to spin all the year long (harvest time onely excepted). And there are a great number of linnen weavers in the adjacent villages, as well as in this town, who buy the same & work it into webbs, and have it bleached by their wives & fitted for Roseley & Carlisle fairs'. Weekly markets held on Tuesdays with two fairs annually, on St Thomas' day and Good Friday. By mid-19th century fairs were held 20 February (then said to be one of largest horse fairs in northern England) and 5 April (for cattle and merchandise). Post-medieval market was located in Market Place, at junction of major routes, where wooden pole served as market cross until burnt down during celebrations after battle of Trafalgar.

Wigton's rich agricultural hinterland was reflected in presence of several corn mills, of which there were four (two water mills and two wind mills) by c.1800. Early textile activity suggested by name Tenters (referring to frames on which woollen cloth was stretched). Yet Wigton does not appear to have grown wealthy by late 17th century: Denton reported that most houses were 'walled with clay & covered with straw'. Clay Dubs, water-filled pits which survived in 1860s to north-east of town, probably represent clay digging for building purposes.

Linen industry grew in 18th century to include wide range of linen products: 'osnaburghs', tow-cloths, coarse linen, striped checks and ginghams. Linen check manufactories established from 1748 with expansion in 1790s. Calico printing began 1790 and fustian manufacture 1795. In mid-19th century there were shirt factory, old print works (recorded by 1832; burnt down 1845), Brookside Works ('cotton, linen and woollen') and three dyeworks. Other industries included two tanneries, four breweries and three nail factories. Linen trade had declined by 1900, when principal industry was tanning. Shirt factory was taken over by Samuel Redmayne 1868, who established Wigton Clothing Factory, which relocated to Station Road 1875 and remained leading employer until late 20th century (factory demolished 1987). Construction of Maryport & Carlisle Railway, with station just outside town from 1845, drew Wigton's industrial focus towards land between medieval town and railway. As well as Redmayne's factory, Burnfoot Preserves Works ('jam factory'), had opened by 1901. Population, which had risen from 2,450 in 1801 to peak of 4,885 in 1831, fluctuated at around 4,000 for most of later 19th and early 20th century.

From 1934, Wigton's industrial mainstay became manufacture of cellophane wrapping materials. Production of cellulose film began 1934 when British New Wrap Co. established factory on site of jam factory. New factory built 1963 and further major expansion took place in 1990s; now global head office of Innovia Films Ltd. From low point of 3,521 in 1931, Wigton's population grew again in later 20th century, to stand at 5,360 in 2001.

Places of worship. Medieval parish church of St Mary; rebuilt 1788. St Mary's parish room, built 1899, later became Kingdom Hall of the Jehovah's Witnesses. Quaker meeting house built c.1705; replaced by larger building to accommodate pupils at Brookfield School 1830; closed 2010. Congregational chapel, Market Hill, built 1819; replaced by new Independent chapel, Water Street, built 1834; became United Reformed Church; closed 2009. Presbyterian chapel, Meeting House Lane, built c.1790; congregation moved to former Congregational chapel on Market Hill 1834, which became United Presbyterian church. Methodist presence in town from 1759; Wesleyan Methodist chapel, George Street, built 1828, replaced by new chapel on High Street 1883; now Wigton Methodist Church. Primitive Methodist chapel (Potter's chapel), New Street, built 1864; amalgamated with High Street congregation 1957. Roman Catholic church of St Cuthbert, begun 1836; extended 1857, with adjacent convent of the Sisters of Mercy. Lowmoor Evangelical Church, built c.1991. Town cemetery on Station Hill, with mortuary chapels for Episcopalians and Dissenters, laid out 1856.

Schools. School, probably in parish church, recorded from 1660s. Free grammar school, Market Hill, founded 1730; replaced by Nelson School (for boys), at Floshfield, established 1896. Thomlinson Girls' Grammar School founded 1899; combined with Nelson School to become Nelson Thomlinson Grammar School 1952. Secondary modern school built c.1952; merged with Nelson Thomlinson to become Nelson Thomlinson Comprehensive School 1969 and combined on Floshfield site 1995. National School, High Street, opened 1820; moved to former Thomlinson girls' school site 1968; now Thomlinsion Junior School. British school (in room under Independent chapel, Water Street) by 1847. Modern infant school built on Longthwaite Road. Non-denominational school attached to Catholic mission may have existed before 1847; Roman Catholic school at St Cuthbert's church, opened 1855; rebuilt on new site nearby 1972-4.

Institutions. Place-name Spital probably records site of medieval leper hospital of St Nicholas. Widows' Hospital, a row of almshouses by parish church, founded by John Thomlinson 1724 (date stone reads 1723); sold 1884. Parish workhouse, Old Lane, recorded 1829. Workhouse for Wigton Poor Law Union built outside town on Station Hill 1838-42; later became cottage hospital; now Wigton Community Hospital. Mechanics' Institute reading room built 1850s (now West Green Social Club). Market Hall, used as venue for entertainments, built 1882. Swimming baths, built for town by Edwin Banks of Highmoor, 1901.

WINDERMERE AND BOWNESS

Resort towns in Windermere parish, Kendal Ward, Westmorland. Windermere and Bowness became separate UDs 1894, merging to become Windermere UD 1905. This entry covers former townships of Applethwaite and Undermillbeck, out of which UDs were carved.

Acreage. Applethwaite township covered 9,122 acres [3,692 ha], of which c.3,000 acres [c.1,200 ha] comprised lake bed of Windermere; Undermillbeck township covered 4,183 acres [1,693 ha] and included chapelry of Winster (in Kendal parish). Most of Applethwaite became Windermere UD (4,345 acres [1,758 ha]) 1894; Bowness-on-Windermere UD (983 acres [398 ha]) was carved out of Undermillbeck; remaining parts of Undermillbeck CP were absorbed into Bowness and Crook CPs 1935. Common land in Undermillbeck (2,000 acres [810 ha]) enclosed 1822; Applethwaite commons enclosed 1842.

Landownership. *Manor of Windermere*, which included both townships, together with Troutbeck and Ambleside, was head of Richmond Fee of barony of Kendal (q.v.); manor house was on Long Holme (now Belle Isle). *Winster* was held by Philipson family of Calgarth until sold to tenants 1717.

Origins and growth of the towns. Origins of towns were quite distinct: Bowness was long-established village, clustering around late-medieval parish church; Windermere a new settlement which grew up around rail terminus from 1847. Since 19th century, economy of both towns has been based on tourism.

Before arrival of railway, main sources of wealth lay in cattle and sheep farming but stone quarrying, woodland industries and fishing had developed by early modern period. Informal market grew up around church at Bowness in 17th century, at first under a yew tree in churchyard on Sundays. Thomas Machell described Bowness 1692 as 'a small country village, with a schoolhouse and with nine or ten other houses about it, where there is a sort of market or meeting place lately begun'. Public ferry across lake has existed for over 500 years. Early water-powered industry included fulling and paper mills at Troutbeck Bridge. By end of 18th century, 'discovery' of Lake District led to growth of tourism.

Arrival of railway brought significant change. Terminus of branch line from Oxenholme and Kendal opened 1847 at Birthwaite (in Applethwaite township), leading to development of resort town at rail head: Windermere

Hotel built by 1849; newsroom, library and picture gallery by 1855. Windermere quickly became magnet for day trippers from industrial towns of Lancashire: by 1885 it had two hotels and 45 lodging and boarding houses. Bowness also grew rapidly into town and resort, with pleasure boats and other leisure attractions, to become one of major tourist resorts in Lake District by later 19th century. Villas overlooking lake proliferated around both towns. Population grew rapidly: Applethwaite township contained 343 inhabitants in 1801, rising to 436 by 1841; growth of Windermere saw its population rise to 1,285 by 1861 and 1,926 by 1891. Undermillbeck's population stood at 500 in 1801 and had doubled to 1,033 by 1841, growing to 2,361 by 1891.

Industrial growth also took place: former paper mill at Troutbeck Bridge converted to bobbin manufacture by 1820s, with another bobbin mill at Thickholme by mid-19th century. Hydro-electric power generation at Troutbeck Bridge bobbin mill, supplying Windermere from 1893; still in operation 1948. Bridge iron works at Troutbeck Bridge in late 19th and early 20th century. Slate quarrying on Applethwaite Common expanded in later 19th century. Short-lived lead mine at Winster, opened 1848. During Second World War flying boat factory, building Sunderland flying boats, established at Calgarth, opening 1941 with extensive temporary housing estate for workforce; site became White Cross Bay caravan park 1954.

Population continued to grow across 20th century, though census figures often inflated by presence of visitors. Bowness grew from 2,682 in 1901 to 4,099 by 1971. By 2001 combined population of Windermere and Bowness stood at 8,245.

Places of worship. Parish church of St Martin in Bowness, originally chapel of ease in Kendal parish, acquired parochial status 1348; present building dates largely from c.1483; restored 1870. Ancient chapel of St Catherine, near Browhead in Applethwaite, had been converted to house by mid-18th century. Additional Anglican churches built as towns grew: St Mary's in Applethwaite (i.e. Windermere), built as private chapel 1847-8; bought by town 1855; enlarged in several stages 1857-82. St John's, Lake Road, built 1886; closed 1995 and converted to flats. Mission room beside Windermere Hotel built by 1897. Quaker meeting house at Mislet in Applethwaite township, built 1701; closed 1822. Wesleyan Methodist chapel in Bowness built 1859; rebuilt 1882; closed after 1991; now Lakes Christian Centre (independent church). Wesleyan Methodist chapel, College Street, Windermere, built 1867; still in use. Christian Brethren chapel at Bowness built c.1839. Congregational chapel at Troutbeck Bridge opened 1858; closed late 20th century. Carver Memorial chapel (now United Reformed Church) built 1879-80 in memory of William Carver (d. 1875) of The Priory, a Mancunian. Roman Catholic church, Lake Road, opened 1884; replaced by church of Our Lady of Windermere and St Herbert on adjacent site, built 1962-3. Christian Science church, Birthwaite Road, opened 1978. **Winster:** chapel, possibly medieval in origin, granted burial rights 1722; rebuilt as church of Holy Trinity, 1874-5.

Schools. Free grammar school in Bowness established c.1637; rebuilt on new site at Brantfell 1836; replaced by new building on Princes Road 1885; became one of first comprehensive boys' schools 1945; closed after The Lakes School, Troutbeck Bridge (purpose-built comprehensive school) opened 1965. College of St Mary, Windermere, opened 1850s for sons of clergymen, survived as independent boys' school until 1938; buildings taken over as Old College Secondary Modern School for girls 1950; closed 1965 on opening of The Lakes School. National school, Windermere, endowed 1856; replaced by new primary school on new site at Princes Road 1960s; now St Martin's & St Mary's CE Primary School. St Cuthbert's Catholic Primary school (opened 1954). Goodly Dale Community Primary School opened 1975. Windermere School (independent boarding school) originated as St Anne's school, which moved to the town from St Annes, Lancashire, 1924. **Winster:** schoolmaster licensed 1691; school by chapel built 1849; closed c.1958.

Other institutions. 'Circulating library and gentlemen's newsroom' established at Bowness c.1823. Institute 'for the working classes' built at Windermere c.1873; had closed by 1885. Working Men's Institute, Bowness, opened 1875. Hydropathic establishment opened at Thornburrow House by 1858; Hydro Hotel opened 1881. Community centres at Ladyholme Centre, Lakes Road, and Marchesi Centre, Holly Road (built as Windermere Old People's Community Centre 1969-71). Steamboat Museum opened at Windermere 1977. Private villas converted to other uses: Holehird became Leonard Cheshire Foundation home 1961; Blackwell opened as Arts & Crafts House 2001.

WINSCALES

Township in Workington parish, in Allerdale above Derwent ward, Cumberland. Winscales CP gained parts of Stainburn and Workington Rural CPs 1934.

Acreage: 974 acres [394 ha] before boundary changes. Workington and Winscales Common (c.1,000 acres [c.400 ha]), straddling boundary between townships, enclosed 1815. ***Population:*** 165 in 1811, declining to c.100 by 1881 and remaining at that level until 1931 (last census year before boundary change). ***Landownership:*** part of manor of Workington (q.v.). ***Economy:*** largely agricultural until 19th century. Distington ironworks (opened 1879; closed 2007) located at Lillyhall in southern corner of Winscales township; site and surroundings subsequently became Lillyhall Industrial Estate and business park. Opencast coal mining in 1970s; wind farms from 1999. ***Places of worship:*** none. ***Schools and other institutions:*** Board school, built 1891, near Lillyhall (now demolished). Lakes College, West Cumbria, moved from Workington to new purpose-built campus at Lillyhall 2001.

WINSTER *see* Windermere and Bowness

WINTON

Township in Kirkby Stephen parish, East ward, Westmorland.

Acreage: 5,110 acres [2,068 ha]. Moorland boundary between Winton, Kaber and Muker (Yorks. North Riding) finally determined 1911 and moorland in Winton and Kaber regulated 1915. ***Population:*** around 250 across 19th century, inflated to 358 in 1841 by 136 residents of boarding academies, and to 301 in 1861 as result of railway construction. Declined to low point of 166 in 1971, rising again to stand at 213 in 2001. ***Landownership:*** part of manor of Brough (q.v.). ***Economy:*** predominantly agricultural; limestone quarrying and lime-burning in 19th century. Coal pits on Green Fell, disused by 1850s. Iron mining in 1870s (short-lived). ***Places of worship:*** Baptist chapel built 1837 or 1842; rebuilt 1863; now Winton United Baptist Church; ecumenical union with Methodists since 1989. Wesleyan Methodist mission room opened 1895. ***Schools and other institutions:*** school built by subscription 1659 and endowed 1681; rebuilt 1862; closed 1977 and converted to holiday accommodation. Boarding academies at Winton Hall (ladies' school) and Manor House, mid-19th century. Village hall.

WITHERSLACK

Chapelry and township in Beetham parish, Kendal ward, Westmorland.

Acreage: 4,791 acres [1,939 ha], including c.1,800 acres [c.730 ha] of common land enclosed 1829. ***Population:*** rising from 306 in 1801 to peak of 541 in

1881; then fluctuating decline to 330 in 1971, rising again to 482 in 2001. ***Landownership:*** *manor of Witherslack* held by Harrington family until their attainder 1485; briefly held by Sir Thomas Broughton of Broughton (d. 1495); then granted to Stanley family, earls of Derby, in 1495, who lost it to Leyburns during Interregnum but regained possession 1747. ***Economy:*** farming; quarrying and lime-burning in 19th century. Woodland industries: tanning suggested by place-name Bark House. ***Places of worship:*** chapel recorded 1542, when dedicated to St Mary; taken down and rebuilt on new site c.1669 (consecrated 1671); now church of St Paul. ***Schools and other institutions:*** school built by trustees of will of Rev. John Barwick (d. 1664), dean of St Paul's, 1678; room for girls added 1824; boys' school rebuilt 1874 and girls' school rebuilt on new site 1876; latter now Dean Barwick Primary School. Witherslack Hall housed Sandford School for Girls from 1946 to 1968; in 1973 it became residential and day school for boys with special educational needs. Institute at Town End (now Witherslack village hall) built before 1897.

WOODSIDE (formerly Woodside Quarter)

Township in Wigton parish, Cumberland ward, Cumberland. Combined with Wigton to form Wigton-cum-Woodside CP 1887, which was then re-divided 1894, the new Woodside CP covering rural areas surrounding town of Wigton. CP enlarged by absorbing Oulton CP 1934.

Acreage: township contained 1,839 acres [744 ha] before boundary changes: reconfiguration of Wigton and Woodside CPs in 1894 increased Woodside's extent to 3,703 acres [1,499 ha] before absorption of Oulton CP. Common moorland totalling 3,205 acres [1,297 ha] in Wigton, Woodside and Waverton (bulk of it in Woodside Quarter) enclosed 1817. ***Population:*** rapid increase from 238 in 1801 to peak of 750 in 1831, before declining to 589 in 1881 (last census year before boundary changes). Population of largely rural post-1894 CP stood at 359 in 1901 and remained little changed in 1931 (last census year before enlargement). ***Landownership:*** part of Wigton barony and manor. *Dockray* said by Thomas Denton to have been separate subsidiary manor, held by Dalston family from c.1540 to 1672. ***Economy:*** predominantly agricultural: said in later 17th century to produce abundant corn and hay. ***Places of worship:*** Quaker meeting house at Tiffinthwaite from 1674; short-lived. ***Schools and other institutions:*** National school at Rosewain, converted from barn, established 1872; closed. Quaker school established at Highmoor 1815 by Cumberland Quarterly Meeting; moved 1826 to new premises at Brookfield; closed 1984.

WOODLAND *see* Kirkby Ireleth

WORKINGTON

Industrial town in Workington parish, Allerdale above Derwent ward, Cumberland. This entry treats township of Workington and extra-parochial place of Cloffocks, which formed core of Workington MB, established 1883. For adjacent areas, see Harrington; Seaton; Stainburn.

Acreage. Township of Workington covered 3,355 acres [1,358 ha] and Cloffocks (extra-parochial meadow or common land beside River Derwent) 108 acres [44 ha]. In 1894 those parts of Workington township not included in MB became Workington Rural CP, which was, in turn, abolished 1934, when part was reunited with Workington, and remainder transferred to Winscales CP. On expansion of its boundaries in 1934 Workington MB also gained parts of Harrington and Stainburn.

Landownership. By early 12th century Workington seems to have been part of extensive estates of Ketel son of Eldred; what became fee of Workington also included Harrington, Winscales and Seaton, on coast, and Lamplugh and Kelton inland. Ketel's descendants took surname Curwen (from Colvend in Galloway) and Workington Hall became seat of Curwens from 13th to 20th century. Male line ended on death of Henry Curwen 1778, when estate passed to his daughter Isabella, who married John Christian (1756-1828) of Ewanrigg (q.v.), who added surname Curwen to his name; Workington descended through their heirs to 20th century. Workington Hall, a fortified house from late 14th century, remodelled in 16th and 17th centuries; Curwen family left 1929; Hall bequeathed to town 1946; now derelict.

Origins and growth of the town. Workington appears to have been centre of some significance from early date, as suggested by wealth and early foundation of its church (attested by finds of Northumbrian age sculpture) and manor's position as focal point of fee of Workington. By 16th century Workington was port and small fishing community – 'where shyppes cum to wher ys a lytle prety fyssher town' as John Leland described it c.1540 – particularly noted for its salmon fishery at mouth of River Derwent. New dock, between existing 'harborye for botes' at mouth of Derwent and 'fyssh houses' on shore, was planned 1569. Market charter granted 1573, reflecting growth of trade; salt-making (salt pan on coast) and coal shipments to Ireland recorded 1604-5. Sir Daniel Fleming, writing 1671, described Workington as 'good haven for ships which carry forth coals, salt and other commodities'. Thomas Denton (writing 1688) valued salmon fishery at £300 per year and implied that sea fish and wildfowl were also exploited. He valued Curwen's coal mines at £200 per year and wrote of Workington Hall and its extensive demesnes: 'I do not know of any one seat in all Brittain so commodiously scituate for beauty, pleasure & plenty as this place is'.

By early 18th century Workington was firmly established as coal port, shipping coal from Curwens' collieries in Workington, Harrington, Seaton and Clifton. Coal mining in Workington township expanded in later 18th century, main pits being in Moorbanks, Banklands and Chapel Bank areas. However, by 1820s flooding and catastrophic inflows of sea water into under-sea workings had become serious problems and collieries began to decline; a disaster at Chapel Bank colliery in 1837, when 27 men and boys and 28 horses were drowned, put paid to that colliery. Jane Pit was sunk inland in 1843 but by 1850 coal seams in township were becoming exhausted.

Shipbuilding, recorded by 1738 with shipyards at mouth of Derwent, expanded across later 18th and early 19th centuries. Ancillary maritime industries, such as rope walks and sail-making, grew. By mid-19th century other industries included paper-making (at Derwent Paper Mills), brewing (established 1792), iron founding, brick-making and quarrying (for freestone and limestone). Town grew up hill to south-east of early core beside Derwent: planned grid of new streets, centred on Portland Square, laid out 1780s. Workingon's population stood at 5,716 in 1801 but rose only modestly to 6,280 in 1851. Outside town, John Christian Curwen established his renowned experimental model farm at Schoose c.1800.

By mid-19th century coal mining was in decline but Workington experienced major expansion as centre of iron and steel making. By 1860 only Jane Pit was in operation and Workington Colliery ceased production 1875. (Costly attempt was made to re-start coal production by sinking new pit, Solway Pit, 1937; it closed 1973.) Opening of ironworks in Seaton (q.v.) by Workington Haematite Co. 1856 heralded expansion of iron and steel industry and major growth in population, which leapt from 8,413 inhabitants in 1871 to 14,361 in 1881 and 23,749 by 1891. Arrival of railway 1847 linked Workington to other industrial centres in west Cumberland and port's capacity expanded with construction of Lonsdale Dock (on Seaton side of river) 1865 and further harbour works 1917. Lonsdale Dock deepened and made wider to accommodate larger vessels (and renamed Prince of Wales Dock) 1927. However, coastal trade declined in late 19th century: 80-90 ships had belonged to port c.1870, but only a mere 'seven to nine' by 1900. Moss Bay Haematite and Steel Co., established 1871, erected two Bessemer converters at Moss Bay 1872, adding rolling mills 1877. This massive steelworks continued until later 20th century: last converter blown out 1974 and production ceased 1982, though railway rails continued to be made using steel manufactured elsewhere until 2006. Other ironworks c.1900 were Derwent Haematite Iron Co., between Moss Bay and New Yard, which was 'new' in 1883; and Kirk Brothers ironworks at New Yard, Marsh side and Derwent Rolling Mills. Shipbuilding continued in second half of 19th century, making merchant vessels. Population growth levelled off in early 20th century, to stand at 24,751 in 1931 (last census year before boundary changes of 1934). On eve of Second World War, *Kelly's Directory* summed up Workington's economy thus: 'The chief industries here are concerned with the manufacture of iron, steel rails and tinplates; there are also works for the manufacture of iron bridges, fences, gates, rivets, boilers, and railway spikes ... There is also a brewery and a ship-building yard'. Port remained crucial to town's economy, importing iron ore and exporting iron and steel, as well as coal and lime. Decline of heavy industry after Second World War resulted in downward drift in population across later 20th century, from a peak of 29,552 in 1961, to 24,295 by 2001.

Places of worship. Medieval parish church of St Michael (dedication recorded as St Mary in 12th century), an early foundation to judge by discovery of pre-Conquest sculpture; rebuilt 1770-2 and again 1887-90; reconstructed again after fire 1994. Chapel, called the 'watch chapell' in 1569, on coast at Chapel Hill (otherwise 'How Michael' or 'St Michael's Mount'); roofless in 1883, when known as 'Old Chapel'. With growth of population, second Anglican church, St John's, Washington Street, built at opposite end of town from parish church 1821-3; tower rebuilt 1847. St Mary's church, Westfield, built 1889, as chapel of ease to St John's. Protestant cemetery on Harrington Road consecrated 1879.

Roman Catholic mission established at Banklands 1811 and Benedictine priory opened 1813, with chapel dedicated to St Michael. Increasing immigration from Ireland led to need for larger building; church of Our Lady Star of the Sea built adjacent to priory 1873-6. Catholics were buried at Bank Cemetery. Presbyterian congregation formed in town 1742; chapel built on Thompson Street 1750; enlarged 1858 but replaced by larger building on same site 1888; closed 1972, then used by other denominations; now demolished. Independent (Congregational) chapel, William Street, built 1780; enlarged 1855 and rebuilt 1884; became United Reformed Church after merger with Presbyterians; congregation merged with Methodists 2008; now Workington United Church. Methodist chapel, Tiffin Lane, built 1791 and replaced by larger building in William Street 1840, which burnt down 1889 and was replaced by Trinity Methodist Church on same site, built 1890; closed after merger with United Reformed congregation 2008; sold 2016. Primitive Methodist chapel built 1827; replaced by larger chapel in Gothic style 1882; closed 1965 and demolished. United Free Methodist church, Victoria Road, built 1890-1; closed and sold 1939; now Bolton Street Community Hall. Wesleyan Methodist chapel, Queen Street, an iron church, built 1894; closed 1950s (later used by amateur operatic society). Baptist chapel, Harrington Road, built 1886; demolished 1980s and replaced by modern building on same site on corner of Gray Street; now Grace Evangelical Baptist Church. Seamen's Christian Friends' Society built Bethel chapel on quayside 1899; now used by Christian Spiritualists. Other nonconformist places of worship by end of 19th century included Salvation Army barracks, Edkin Street; several mission halls and Christian Brethren meetings in Dent Hall. By 1938 there were also places of worship for Plymouth Brethren and Church of Christ and church on Vulcan's Lane, now Emmanuel church of Evangelical Connexion of Free Church of England.

Schools. School endowed 1664 by will of Sir Patricius Curwen (d. 1664), probably forerunner of National school, Portland Square, established by John Christian Curwen 1808. School of Industry, Guard Street, founded 1816, had object of encouraging industriousness in poor girls and making them 'notable housekeepers and good Christians'. Infant school (which became Wilson's Infant School) added at Guard Street site 1831. By 1847 there was a second infant school, Bell Street. Spacious new National schools built c.1860 close to Anglican churches of St Michael and St John, later becoming Board schools. Further Board schools on Marsh and at Westfield (built c.1883). School run by Roman Catholic church by 1860s; St Joseph's RC High School opened 1929. Guard Street premises of Wilson's Infant School used for new Workington Higher Grade School, established 1894, later known as Central Boys' and Girls' Schools, which moved to new premises on High Street 1932. These amalgamated 1947 to become mixed secondary modern school (Newlands School), which became girls-only 1967 and closed on reorganisation 1984. County Secondary School and Cumberland Technical College, Park Lane, built 1912 and enlarged 1933-4. After reorganisation of secondary schooling 1984, site became Workington Technical College until 2001, when it was replaced by new Lakes College campus at Lillyhall, Winscales. Park Lane site demolished 2005 and Workington Community

Hospital built on it. New schools, including a secondary school, Southfield Technology College, built on housing estates at Moorclose and Salterbeck as town expanded in later 20th century.

Other institutions. Parish workhouse opened just outside town 1793; continued to house Workington's poor until 1853 (by which time Workington had become part of Cockermouth Union). Building converted to fever hospital (Ellerbeck Hospital for Infectious Diseases) 1888 and later became part of Workington Infirmary; closed 1965. Workington Infirmary, opened 1886, closed 2005; replaced by Workington Community Hospital, on site of former Technical College.

Public halls included Assembly Room (known alternatively by 1900 as Drill Hall), Portland Square, built before 1847; Albert Hall, Fisher Street, built 1883 (it and Public Hall, Hag Hill, were said to be chief venues for auctions and concerts, 1901); and Jubilee Hall (or Queen's Opera House), built 1887. By 1883 there were also halls in Savings Bank (built 1844), Good Templar's Hall, Free Gardeners' Hall and Victoria Hall. Parochial library in infant school, Guard Street; lending library at Methodist chapel and news room in Portland Square by 1847. Mechanics' Institute opened in Savings Bank 1847; new library and reading room opened there 1891; moved to new premises when Carnegie Library and Lecture Hall opened 1904; now closed. Helena Thompson Museum opened 1949 in Park End, a house bequeathed to town 1940. Theatre in Christian Street said in 1847 to be 'seldom visited by a company of comedians'. Lyceum Theatre, Washington Street, opened 1866; renamed Theatre Royal 1874; still open. Cinemas included Opera House (which replaced Jubilee Hall after fire 1927) and Ritz Cinema (opened 1938; closed 1988).

WRAY *see* Claife

WREAY

Chapelry and township in St Mary's parish (Carlisle), Cumberland ward, Cumberland. Absorbed into St Cuthbert Without CP 1934.

Acreage: 1,110 acres [449 ha], including c. 490 acres [c.200 ha] of common land, enclosed 1778. ***Population:*** rising from 118 in 1801 to peak of 181 in 1881, then falling back to 131 by 1931 (last census year for which separate figure available). ***Landownership:*** manorial rights held by Carlisle Priory, passing to Dean and Chapter of Carlisle at Reformation. Losh family of

Woodside, Brisco, held much of land in chapelry from later 18th century until 1904. ***Economy:*** predominantly agricultural; tile works (to west of Ravenside) mid-19th century. ***Places of worship:*** chapel of ease recorded 1319; it or successor chapel consecrated 1739. Replaced by St Mary's church, a highly original creation built by Sarah Losh in memory of her sister 1841–42. Cemetery chapel built c.1835. ***Schools and other institutions:*** reader in chapel served as schoolmaster, maintained by chapel endowment, in later 17th century. School built 1751 and endowed by John Brown of Woodside 1763; replaced by new school nearby, built and further endowed by Losh family 1830; now Wreay CE Primary School. Girls' school built c.1840 by Losh family. Working men's reading room established 1856. Parochial library adjoining school, erected 1874. Village hall, near Chapel Hill, built c.1956.

WYTHOP

Township in Brigham parish, Allerdale above Derwent ward, Cumberland.

Acreage: 3,353 acres [1,357 ha], including 1,292 acres [523 ha] of common fell, enclosed 1830. ***Population:*** falling from 137 in 1801 to 87 in 1861, then rising to peak of 131 in 1921 before falling across later 20th century to 51 in 2001. ***Landownership:*** *manor of Wythop,* in forest of Derwentfells and honour of Cockermouth; granted c.1260 to John de Lucy, younger son of lord of Cockermouth. Had passed to Robert Lowther by 1429 and descended in Lowther family until sold 1606 by Sir Richard Lowther to Richard Fletcher, from whom it descended to Fletcher-Vane family of Hutton-in-the-Forest (q.v.). ***Economy:*** hill farming; forestry (Wythop woods noted for valuable oak timber in later 17th century). Silica mine and brick manufacturing plant south of Wythop Hall in operation 1935-9. ***Places of worship:*** chapel of ease recorded 1552; rebuilt (or replaced by) chapel near Kelswick, built 1673; in turn replaced by church of St Margaret of Antioch on new site, built 1865-6. ***Schools and other institutions:*** for Embleton and Wythop school, see Embleton.

YANWATH AND EAMONT BRIDGE

Township in Barton parish, West ward, Westmorland.

Acreage: 1,299 acres [526 ha]. Commons enclosed 1818. ***Population:*** rising from 198 in 1801 to 327 in 1831; remained in range 250-350 across rest of 19th and first half of 20th century. Temporary surge to 569 in 1951, then decline to 184 by 1981 before rising again to stand at 457 in 2001. ***Landownership:*** *manor of Yanwath* held under barony of Westmorland by barons of Greystoke, and under them by Threlkeld family, who probably built Yanwath Hall in 15th century. On death of Sir Lancelot Threlkeld (d. 1506-10), passed by marriage of his daughter Grace to Thomas Dudley; Sir Christopher Dudley sold it 1654 to Sir John Lowther, from whom it descended to earls of Lonsdale. ***Economy:*** farming; quarrying and lime-burning in 19th century. Eamont Bridge river crossing (bridge recorded 1425) with inns on main north-south route (A6). Textile manufacture recorded by place-name Walk Mill (i.e. fulling mill), recording woollens; its later name, Bleach Green, suggesting linen or cotton (cotton spinner/candlewick manufacturer recorded 1829). Snuff mill at Eamont Bridge (on Cumberland side of river) from 1835; closed 1937. Bobbin mill on River Lowther in mid-19th century. Tourism and leisure businesses increasingly important from later 20th century: chalet park at Southwaite Green Mill, opened 1994; archaeological tourism to Neolithic henges of Mayburgh and King Arthur's Round Table. ***Places of worship:*** St John's mission room (an 'iron church'), Eamont Bridge, built 1871; in use until 1932 when Brougham and Eamont Bridge parishes merged. ***Schools and other institutions:*** for Yanwath school, see Sockbridge. Workhouse at Eamont Bridge established 1837 in building known as 'The Mansion House' (built 1685); closed 1874. Jubilee House rehabilitation and therapy centre for injured firefighters, opened 1995. Village hall at Eamont Bridge built 1933 on site of mission room.

HOW THE GAZETTEER WAS COMPILED

Scope and Sources

The information presented in this volume was gathered from a prescribed selection of readily accessible sources; it was not based on full archival research, nor does it claim to be comprehensive. The entries aim to cover all periods from the earliest written record to the present (information being updated to 2015 wherever possible). Details of the scope of the entries and the core sources used to compile each section are listed below. Further information on local sources specific to particular places will be found on the Cumbria County History Trust web pages for each parish or township (at **http://www.cumbriacountyhistory.org.uk/places**).

1. Administrative history

Based on Frederic A. Youngs, *Guide to the Local Administrative Units of England: Volume II: Northern England* (London: Royal Historical Society, 1991). The major boundary changes made to CPs in Cumberland in 1934 are detailed in *Cumberland Review Order, 1934* [copy in Cumbria Archive Centre (Carlisle), CA/C/b/150].

2. Acreage and enclosure history

Acreage of the administrative unit is generally that given on the first edition of the Ordnance Survey Six-Inch (1:10,560) county series maps, surveyed for Cumberland 1859-65; for Westmorland 1856-60 and for Lancashire and the West Riding of Yorkshire in the 1840s. Acreages and boundaries of the hamlets or quarters into which some CPs were subdivided are given on the first edition maps of Lancashire and Yorkshire, surveyed in the 1840s, but not on those of Cumberland and Westmorland, surveyed later: where such detail is available, the acreages of subdivisions have been included. It should be noted that many CPs included detached portions until implementation of the Divided Parishes Act 1883, which transferred detached areas to the administrative unit by which they were surrounded. As a result, the acreages of many CPs had changed by the end of the 19th century. For current purposes, only transfers of major blocks of territory have been noted. Details of the enclosure of common land are from W. E. Tate, *A Domesday of English Enclosure Acts and Awards*, ed. M. E. Turner (Reading: University of Reading, 1978). Acreages of surviving common land are from the 'Common Land in England' website: **www.commonland.com**

3. Population

A one-line summary of population trends across the past two centuries has been drawn from the published reports of the decennial censuses, 1801-2001. For urban areas, population data is included in the section headed 'Origins and growth of the town'. Estimates of population in 1688 are from Thomas Denton's *Perambulation of Cumberland* (see below).

4. Landownership

This section is restricted to a very brief summary of the descent of the lordship of manors; it does not attempt a full survey of non-manorial landed estates. The starting point for places in Cumberland and Westmorland has been William Whellan, *The History and Topography of the Counties of Cumberland and Westmorland* (Pontefract, 1860); for the Furness and Cartmel areas manorial descents are available in *VCH Lancashire*, Volume VIII. Additional information has been drawn from C. Roy Hudleston and R. S. Boumphrey, *Cumberland Families and Heraldry* (CWAAS Extra Series XXIII; Kendal, 1978) and R. S. Boumphrey, C. Roy Hudleston and J. Hughes, *An Armorial for Westmorland and Lonsdale* (CWAAS Extra Series XXI; Kendal, 1975). Earlier antiquarian sources have also been used, specifically *John Denton's History of Cumberland*, ed. A. J. L. Winchester (Surtees Society 213 and CWAAS Record Series XX; Woodbridge, 2010); *Thomas Denton: a Perambulation of Cumberland 1687-1688, including descriptions of Westmorland, The Isle of Man and Ireland*, ed. A. J. L. Winchester with M. Wane (Surtees Society 207 and CWAAS Record Series XVI; Woodbridge, 2003); *Antiquary on Horseback*, ed. Jane M. Ewbank (CWAAS Extra Series XIX; Kendal, 1963); Joseph Nicolson and Richard Burn, *The History and Antiquities of the Counties of Westmorland and Cumberland* (London, 1777; facsimile reprint with introduction by B. C. Jones, Wakefield, 1976); Daniel & Samuel Lysons, *Magna Britannia, Vol. IV:*

Cumberland (London, 1816); Samuel Jefferson, *The History and Antiquities of Leath Ward* (Carlisle, 1840) and *The History and Antiquities of Allerdale Ward Above Derwent* (Carlisle, 1842); William Farrer and John F. Curwen, *Records Relating to the Barony of Kendale* (CWAAS Record Series IV, V and VI; Kendal, 1923-6); and John F. Curwen, *The Later Records Relating to North Westmorland or the Barony of Appleby* (Kendal, 1932).

5. Economy (for urban areas this section is entitled 'Origins and growth of the town')

The aim of this section is to give a brief indication of the main areas of economic activity in the place in question from the earliest record to the time of writing. Some areas of economic activity found in most rural communities (village shops and tradesmen, inns and public houses, for example) have not been noted, nor have water corn mills, unless they were converted to other uses (as a woollen mill, for example). Sources used as a starting point have been successive editions of the Ordnance Survey 1:10,560 (Six-Inch) maps and 19th-century trade directories. These have been supplemented, where possible, by earlier sources, notably Thomas Denton's *Perambulation of Cumberland* from 1687-8 and Sir Daniel Fleming's accounts of Cumberland, Westmorland and Lancashire North of the Sands in 1671 (printed in Edward Hughes (ed.), *Fleming-Senhouse Papers* (Carlisle, 1961), pp. 3-64).

The key trade directories used were Parson & White's *History, Gazetteer and Directory of Cumberland and Westmorland* (1829); Mannix & Whellan's *Directory of Cumberland* (1847); Mannex & Co.'s *Directory of Westmorland with Lonsdale and Amounderness* (1851); Bulmer's *History and Directory of Cumberland* (1901) and *Directory of Westmorland* (1885 and 1906); and *Kelly's Directory of Cumberland & Westmorland* (1938). Information dated 'mid-19th century' is usually drawn from first edition Ordnance Survey maps and the directories of Mannix and Whellan; information dated 'late 19th century' from second edition Ordnance Survey maps and Bulmer's directories.

Secondary sources yielding data on individual industrial sites include J. D. Marshall, and M. Davies-Shiel, *Industrial Archaeology of the Lake Counties* (2nd edn, Whitehaven, 1977) and Oliver Wood, *West Cumberland Coal 1600-1982/3* (CWAAS Extra Series XXIV; Kendal, 1988).

The weight of evidence in these sections of the Gazetteer inevitably comes from the 19th and 20th centuries. Again, full research for the VCH will not only amplify but also amend these entries.

6. Places of worship, schools and other institutions

The aim here has been to record places of worship of all denominations and the schools and main public institutions which had premises in the place in question. Places of worship are arranged as follows: the parish church or chapel of ease is followed by other Anglican churches and then by nonconformist places of worship, arranged by denomination. Where no denomination is specified, the church in question was Anglican. The heading 'Other institutions' embraces legal and administrative buildings (court houses, moot halls etc); social support institutions (hospitals, workhouses, almshouses etc) and cultural bodies (libraries, reading rooms, village halls, theatres, cinemas etc).

Again, the starting points have been successive editions of the Ordnance Survey Six-Inch (1:10.560) maps, directories (see above) and M. Hyde and N. Pevsner, *Cumbria* (Buildings of England series, Yale University Press, 2010). Specific sources for places of worship and earlier schools include: C. M. L. Bouch, *Prelates and People of the Lake Counties* (Kendal, 1948) (which contains lists detailing evidence for the dates of foundation of churches and chapels); William Nicolson, *Miscellany Accounts of the Diocese of Carlile: with the Terriers Delivered to Me at My Primary Visitation*, (CWAAS, 1877); *The Cumbria Parishes 1714-1725 from Bishop Gastrell's Notitia, with additions by Bishop Porteous 1778-1779*, ed. L. A. S. Butler (CWAAS Record Series XII; Kendal, 1998) and *The Diocese of Carlisle, 1814-1855: Chancellor Walter Fletcher's 'Diocesan Book' with additional material from Bishop Percy's parish notebooks*, ed. Jane Platt (Surtees Society 219 and CWAAS Record Series XXII; Woodbridge, 2015). Sources for nonconformist places of worship include David M. Butler, *The Quaker Meeting Houses of Britain* (London, 1999) and the growing body of material on individual Methodist chapels online at **http://www.myprimitivemethodists.org.uk/** and **http://www.mywesleyanmethodists.org.uk/**

Finally (but by no means of least importance), the riches of the internet have been tapped, particularly to pin down information about the recent past. The sheer volume of material available online can be overwhelming and discernment is required to identify reliable sources of information, but the following sites have proved particularly useful. Googling – not least taking a virtual walk through the contemporary landscape using Google maps and 'street view' – has yielded many answers. Use has also been made of the burgeoning array of community websites, including those of local history societies, many of which contain a wealth of well-researched material. Images of buildings, notably the superb photographic record compiled by

Martin and Jean Norgate on their 'Old Cumbria Gazetteer' website (**http://www.geog.port.ac.uk/webmap/thelakes/**), have been invaluable for the 'Places of worship', 'Schools' and 'Other institutions' sections.